COMPUTATIONAL SEISMOLOGY

A Practical Introduction

HEINER IGEL

OXFORD

Elastic wave equation (displacement)

$$\rho \partial_t^2 u_i = \partial_j \left(\sigma_{ij} + M_{ij} \right) + f_i$$

Stress-strain relation (anisotropic)

$$\sigma_{ij} = c_{ijkl} \varepsilon_{kl}$$

Stress-strain relation (isotropic)

$$\sigma_{ij} = \lambda \delta_{ij} \varepsilon_{kk} + 2\mu \varepsilon_{ij}$$

Strain tensor

$$\varepsilon_{ij} = \frac{1}{2} \left(\partial_i u_j + \partial_j u_i \right)$$

Elastic wave equation 1D

$$\rho \partial_t^2 u_y = \partial_x \left(\mu \partial_x u_y \right) + f_y$$

Acoustic wave equation

$$\partial_t^2 p = c^2 \Delta p + s$$

Elastic wave equation (velocity-stress)

$$\rho \partial_t v_i = \partial_j \left(\sigma_{ij} + M_{ij} \right) + f_i$$

$$\partial_t \sigma_{ij} = \lambda \delta_{ij} \partial_t \varepsilon_{kk} + 2\mu \partial_t \varepsilon_{kk}$$

$$\partial_t \varepsilon_{ij} = \frac{1}{2} \left(\partial_i v_j + \partial_j v_i \right)$$

Planar free surface

$$t_i = \sigma_{ij} n_j = 0 \quad \Rightarrow \quad \sigma_{xz} = \sigma_{yz} = \sigma_{zz} = 0$$

Internal boundary condition (interface)

$$\sigma_{ij} n_j^{(1)} = \sigma_{ij} n_j^{(2)}$$

$$u_i^{(1)} = u_i^{(2)}$$

Seismic velocities

$$v_p = \sqrt{\frac{\lambda + 2\mu}{\rho}} \qquad v_s = \sqrt{\frac{\mu}{\rho}}$$

Phase velocity, dispersion

$$c = \frac{\omega}{k} = \frac{\lambda}{T}$$

Volumetric change (divergence)

$$\Theta = \nabla \bullet \mathbf{u} = \partial_x u_x + \partial_y u_y + \partial_z u_z = \varepsilon_{ii}$$

Rotation (curl)

$$\Omega = \frac{1}{2} \nabla \times \mathbf{u} = \begin{pmatrix} \partial_y u_z - \partial_z u_y \\ \partial_z u_x - \partial_x u_z \\ \partial_x u_y - \partial_y u_x \end{pmatrix}$$

Moment tensor

$$M_{ij} = \begin{pmatrix} M_{xx} & M_{xy} & M_{xz} \\ M_{yx} & M_{yy} & M_{yz} \\ M_{zx} & M_{zy} & M_{zz} \end{pmatrix}$$

Scalar moment

$$M_0 = \mu A d$$

$$M_0 = \frac{1}{\sqrt{2}} \left(\sum_{ij} M_{ij}^2 \right)^{1/2}$$

Moment magnitude

$$M_w = \frac{2}{3} \left(\log_{10} M_0 - 9.1 \right)$$

Reciprocity

$$G_{ij}(\mathbf{x}, t; \mathbf{x}_0, t_0) = G_{ji}(\mathbf{x}_0, -t_0; \mathbf{x}, -t)$$

Green's function, convolution

$$u_i(\mathbf{x}, t) = G_{ij}(\mathbf{x}, t; \mathbf{x}_0) \otimes s(t)$$

$$u_i(\mathbf{x}, t) = IFT\{G_{ij}(\omega) S(\omega)\}$$

COMPUTATIONAL SEISMOLOGY

Computational Seismology

A Practical Introduction

Heiner Igel

Department of Earth and Environmental Sciences,
Ludwig-Maximilians-University of Munich

OXFORD

UNIVERSITY PRESS

Great Clarendon Street, Oxford, OX2 6DP,
United Kingdom

Oxford University Press is a department of the University of Oxford.
It furthers the University's objective of excellence in research, scholarship,
and education by publishing worldwide. Oxford is a registered trade mark of
Oxford University Press in the UK and in certain other countries

First Edition published in 2017

Impression: 1

Published in the United States of America by Oxford University Press
198 Madison Avenue, New York, NY 10016, United States of America

British Library Cataloguing in Publication Data

Data available

Library of Congress Control Number: 2016940973

ISBN 978–0–19–871740–9 (hbk.)
ISBN 978–0–19–871741–6 (pbk.)

Printed and bound by
CPI Group (UK) Ltd, Croydon, CR0 4YY

To Maria,
and our wonderful kids
Anna, Jonas, Lukas, and Clara.
Thanks for always bringing me
back to Earth.

Preface

When I was an undergraduate student in Scotland in the late eighties working on seismic anisotropy I was lucky to have been able to attend an international meeting that took place at the University of Berkeley, California. I saw a presentation of a Stanford PhD student showing 2D simulations of anisotropic wave propagation using the finite-difference method. I was totally struck by the beauty of the graphics, the ease with which one could develop intuition about wave phenomena, and the elegant and simple maths underlying the simulation method. This became what I wanted to learn, master, and apply! Luckily, later I was offered a PhD position in Paris, where the development of finite-difference-based simulation methods and their application to inverse problems became my topic.

It is important to note that at the time (parallel) computer codes needed for our research were basically written from scratch. The primary goal was to ensure that the codes were correct (rather than making them readable by others through heavy commenting). As computers grew larger and architectures became more complex, this *heroic* (unprofessional) coding style no longer worked. Today, parallel codes must have a different quality. To be able to obtain computational resources on large supercomputer facilities, one has to demonstrate proper parallel scaling properties and in many cases this is not possible without interaction with computational scientists.

This implies a paradigm shift in the approach to simulation technology. Today, the shortest time to results for students involves the use of community software provided by projects like CIG (<http://www.geodynamics.org>), individual researchers or groups (see appendix), or community platforms as developed within the VERCE project (<http://www.verce.eu>). This creates problems. Numerical methods are not necessarily featured with much detail in Earth science courses. On the other hand, it is usually straightforward to use community simulation tools and obtain synthetic seismograms. However, without experience, quality control is difficult. Not seldom am I presented with simulation results where there are obvious problems with the set-up.

How can we fix this? Students and researchers should have at least a basic understanding of what is under the bonnet of current simulation technologies used to solve interesting research problems. They should understand what problems to look out for, and how to properly design simulation tasks, and to ensure the results are correct. In addition, today there is a zoo of different methods, and it is difficult to choose the right method for a particular problem. In this volume I try to provide some guidelines.

The strategy is to keep the maths as simple as possible, extensively using graphics to illustrate concepts, while at the same time presenting the link between theory and computer program (using the Python language and Jupyter notebooks). This concept should be beneficial to both students and lecturers. While it is advisable to write own codes as much as possible (and compare with the solutions presented here), lecturers can start right away using the supplementary electronic material and the online platform provided.

This volume should be considered a starting point. There are many excellent books for each of the numerical methods presented in this text. These references should be consulted when more detail is required. The focus here is to present the fundamental concepts of the various methods, their inter-relations, and pros and cons for specific applications in seismology (and other fields).

My hope is that you become equally excited about this fascinating field of Earth science, and use your knowledge to further our understanding of this amazing planet!

Heiner Igel
Munich, July 2016

Acknowledgements

Thanks to Sonke Adlung from Oxford University Press for suggesting this project, to Ania Wronski and the production staff at Integra (Marie Felina Francois) for their help and support, and to Henry MacKeith for copy-editing.

I would like to express my gratitude to those who helped me to get off the ground in science. Stuart Crampin of the University of Edinburgh, taught me to 'think science', and became a lifelong friend. This volume would never have been possible without the vision of Albert Tarantola and Peter Mora of the Institut de Physique du Globe Paris, who—decades ago—foresaw the impact of parallel computing in the Earth (and other) sciences. I consider myself very lucky having had these scientists as supervisors.

The work presented in this volume benefitted from research projects funded by the German Research Foundation, the European Union, the European Research Council, the Bavarian Government, the German Ministry of Research, the Volkswagen Foundation, and the European Science Foundation. I gratefully acknowledge the strong support from the Leibniz Supercomputing Centre Munich.

The concepts for this volume were born out of workshops organized during the SPICE and QUEST training networks funded by the European Union between 2003 and 2013. The people who invented the training network funding instruments should be awarded! We learned so much through these projects on both seismic forward and inverse modelling, which pushed the limits in seismology substantially, and at the same time shaped careers for dozens of young scientists who today are scattered around the world, many in senior positions.

Infinite thanks to the principal investigators and associates of the SPICE and QUEST projects—they were so much fun: Chris Bean, Lapo Boschi, Johana Brokesova, Michel Campillo, Torsten Dahm, Ana Ferreira, Domenico Giardini, Alex Goertz, Matthias Holschneider, Raul Madariaga, Martin Mai, Valerie Maupin, Peter Moczo, Jean-Paul Montagner, Andrea Morelli, Tarje Nissen-Meyer, Guust Nolet, Johan Robertsson, Barbara Romanowicz, Malcolm Sambridge, Geza Seriani, Karin Sigloch, Eleonore Stutzmann, Jeannot Trampert, Colin Thomson, Jeroen Tromp, Jean-Pierre Vilotte, Jean Virieux, John Woodhouse, Aldo Zollo, and many others; the enthusiastic administrators Erika Vye and Greta Küppers; and all doctoral and postdoctoral researchers involved.

Part of the material benefitted enormously from other people's work; for example, Bernhard Schuberth's diploma thesis, course material by Martin Käser, and Andreas Fichtner's book on modelling and inversion. Thanks to Peter Shearer and Cambridge University Press for giving permission to use some of the graphics from Peter's excellent introductory work in seismology. I also want to thank Wolfgang Bangerth who introduced me to the finite-element method many years ago using just the blackboard and a pen.

This volume got started during the phenomenal RHUM-RUM cruise in the Indian Ocean in autumn 2013, coordinated by Karin Sigloch and Guilhem Barruol. Thanks to Yann Capdeville with whom I shared many day and night shifts. He helped me a lot with the spectral-element method. According to him I was his worst student ever.

Thanks to Florian Wölfl, who helped tremendously getting the Latex project started. Sebastian Anger, Bryant Chow, Jonas Igel, Lion Krischer, David Vargas, and Moritz Goll helped with the graphics, slide material, and the notebooks. Special thanks to Stephanie Wollherr: her mathematical skills and good humour helped substantially in getting the finite-volume and discontinuous Galerkin chapters (and the codes) in shape. Sujana Talavera provided great comments on these parts, too. I am also grateful to Matthias and Thomas Meschede for creating the graphics for the title page.

Thanks to all the participants of the Munich Earth Skience School 2015 (<http://www.geophysik.lmu.de/MESS>) in Sudelfeld, Christine Thomas for collecting all the comments and the staff Renate and Winfried Löffler and Michael Sponi for always creating a wonderful atmosphere over the years. Thanks to the participants of the 2015 seminar on computational seismology and all their comments and suggestions on the draft: Michael Bader, Esteban Bedoya, Christoph Heidelmann, Eduard Kharitonov, Jiunn Lin, Martin Mai, Sneha Singh, Taufiq Taufiqurrahman, Tushar Upadhyay, Vasco Varduhn, and Donata Wardani.

Thanks to Toshiro Tanimoto, Alain Cochard, and Nicolas Brantut who went through the manuscript and provided many useful comments and corrections. Thanks to Josef Kristek and Peter Moczo for reviewing the chapter on the finite-difference method, and making many excellent suggestions.

Special thanks to Lion Krischer and Tobias Megies for pushing me towards Python and the Jupyter Notebooks, opening new ways for the training of numerical methods in seismology, and bringing <http://seismo-live.org> to live.

Many have helped with the project, in various forms (comments, graphics, ideas, codes): Robert Barsch, Moritz Bernauer, Jacobo Bielak, Alex Breuer, Emanuelle Casarotti, Josep de la Puente, Michael Dumbser, Kenneth Duru, Michael Ewald, Ana Ferreira, Bob Geller, Sarah Hable, Celine Hadziioannou, Kasra Hosseini, Alice Gabriel, Verena Herrmann, Gunnar Jahnke, Brian Kennett, Lane Johnson, Fabian Lindner, Dave May, Christian Pelties, Michael Reinwald, Johannes Salvermoser, Stefanie Schwarz, Robert Seidl, Geza Seriani, my wife Maria Stange, Simon Stähler, Marco Stupazzini, Ulrich Thomas, Jeroen Tromp, Maria Tsekhmistrenko, Martin van Driel, Jean-Pierre Vilotte, Haijiang Wang, Joachim Wassermann, Moritz Wehler, Stefan Wenk, and Djamel Ziane. Thanks to the Geophysics IT and HPC team, Jens Öser, Gerald Schroll, and Marcus Mohr for their support.

Thanks to Tamiko Thiel for sharing stories with me about the early days of Thinking Machines and the design of the parallel Connection Machine.

I greatly appreciate the help of my Edinburgh flatmate and friend Sean Matthews, who proofread the entire manuscript for language problems.

Thanks to Elisabeth and Ernst Ullmann for providing me with the right Colnago frames for my cycling workout, and for their friendship. Finally, I would like to thank my family: my father Hans, Inge, my sister Bärbel, Jürgen, Moritz, Felix, and my parents-in-law Ortrud and Karl, for their support and love. In loving memory of my mother.

Thanks also to all those whom I have forgotten to acknowledge.

Contents

About Computational Seismology

1.1 What is computational seismology?

For seismologists the calculation of *synthetic* (or *theoretical*) *seismograms* is a key activity on the path to a better understanding of the structure of the Earth's interior, or the sources of seismic energy. There are many ways of doing this, depending in particular on the assumptions made in the geophysical model. In the most general case—an Earth in which the properties vary in three dimensions—analytical solutions do not exist.

The complete solution of the governing 3D partial differential equations descriptive of elastic wave propagation requires the adaptation of *numerical methods* developed in the field of applied mathematics. For the purpose of this volume I define *computational seismology* such that it involves the complete *numerical* solution of the seismic wave-propagation problem for arbitrary 3D models (a pictorial example is shown in Fig. 1.1). A further restriction is that we focus on so-called *time-domain* solutions rather than *frequency-domain* approaches. In time-domain approaches the space-dependent seismic wavefield is extrapolated time step after time step into the future. In frequency-domain approaches the wave equations are transformed into the spectral domain and solved for each frequency. Seismograms can then be obtained by inversely transforming the spectra into the time domain. Another numerical approach that is not discussed here is the boundary-integral-equation method that is being used in fracture mechanics. Time domain solutions are the most commonly used computational tools today to calculate seismograms and to solve seismic inverse problems in 3D.

The definition of the field of *computational seismology* would be incomplete without making a reference to the mind-boggling evolution of computational hardware. Your smartphone today is almost as powerful as the supercomputers that were used when I started my PhD in 1990 at the Institut de Physique du Globe in Paris. At that time we were lucky to be able to lay our hands on one of the first (then called *massively*) parallel computers, built by the company *Thinking Machines Corp.*[1]

The calculation of seismograms for any even vaguely realistic problem in 3D is computationally expensive. This implies that the actual *implementation* of a numerical algorithm that solves the seismic wave-propagation problem is involved, requires so-called *parallelization*, and is potentially strongly hardware-dependent.

Fig. 1.1 *Snapshot of global seismic wave propagation through a 3D mantle convection model. Inside the Earth iso-velocity surfaces indicate flow patterns. The seismic wavefield—simulated with the spectral-element method—is indicated with green and pink colours. Figure courtesy of B. Schuberth.*

---

[1] In his later life legendary physicist and Nobel Prize winner Richard Feynman designed the communication scheme for this machine.

Computational Seismology. First Edition. Heiner Igel.
© Heiner Igel 2017. Published in 2017 by Oxford University Press.

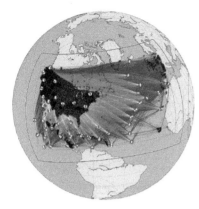

Fig. 1.2 *Source–receiver raypaths for earthquake-receiver configuration involving part of the US-Array portable network (stations shown as small grey triangles). The figure illustrates a projection of the ray density to the Earth's surface, giving a qualitative indication of how well structure could be recovered with tomographic inversion. Figure courtesy L. Krischer.*

It is beyond the scope of this volume to deal with implementation issues. However, the necessity to parallelize numerical solutions has had an important impact on the design, evolution, and survival of specific approaches.

There are several other—let us call them *classic*—ways to calculate synthetic seismograms, and there are excellent textbooks on many of them (see end of this chapter). Each of these methods is making specific assumptions that—in general—are not necessary when applying numerical methods to the wave-propagation problem. For example, almost everything we know today about the Earth's interior, much of the dynamics of the mantle, and the use of our planets' (once) gigantic energy resources (e.g. hydrocarbons) is based on the observations of arrival (travel) times of seismic phases that are analysed with *ray theory*.

Ray theory is based on the assumption that high (even infinite) frequencies are travelling through the Earth and along their way only see long-wavelength structures (for an illustration see Fig. 1.2). The advantage is that these calculations are very fast and thus allow an efficient solution of the inverse problem (ray tomography). In recent years, the inclusion of *finite-frequency effects* into the calculation of arrival times and ray-based synthetic seismograms has led to an exciting novel approach to the waveform inversion problem at the high-frequency end of the seismic spectrum.

Other so-called *modal solutions* are quasi-analytical[2] solutions to the governing partial differential equations mostly based on the assumption of one-dimensional variation of the parameters (e.g. seismic velocities only vary with depth and are laterally homogeneous). On a global scale the normal-mode solutions based on spherical harmonics fall into this category. In Cartesian coordinates the reflectivity method for layered Earth models is another example. The two aforementioned methodologies are still workhorses for many applications.

At this point I would like to make a bold statement: Even though this is a volume on *computational seismology* (in the way I have defined it), it is important to stress that the *classic* methods just described will continue to play an extremely important role. To understand certain parts of the seismogram it is sometimes advantageous (and computationally cheaper) to use classic methods. Also, when verifying whether your computer program does the right thing the only way to find out is to compare it with well-established (quasi-) analytical solutions, where this is possible.

In the light of this—and this is part of the rationale for this volume—*use 3D simulation tools with care*, and make sure you also gain experience with other classic techniques. This will help you to develop an in-depth understanding of seismic wave propagation, and to efficiently solve your research problem.

1.2   What is computational seismology good for?

So, you want to calculate a seismogram? Whether you are an exploration geophysicist, global seismologist, volcanologist, rock physicist, or a seismologist

---

[2] You might be able to write down quasi-analytical solutions with pen and paper but eventually when you do the computations some approximations are required (e.g. when calculating series). Therefore the term *quasi*.

working in an insurance company trying to quantify hazard, you are faced with the question of which numerical approach to choose for your particular problem. There is no simple answer to this question, and that is a key reason for writing this volume. It is not likely that one approach will develop into the one-and-only solution for all wave-propagation problems. Therefore it is important to understand some of the properties of the various schemes in order to make the right choice.

In the following I want to briefly highlight the role of 3D simulations in the various application domains and will point to some of the issues involved. A detailed discussion of applications can be found in Part III of this volume.

In the absence of any serious hope of predicting earthquakes in the near future, the calculation of the ground shaking for potential earthquake scenarios is one of the most important strategies in *ground-shaking hazard* studies. In order for these calculations to be useful for earthquake engineers, they have to reach frequencies that are relevant for building responses. This is hard to do given the uncertainties of the near-surface small-scale structure, which, however, might have important effects on the shaking characteristics. Already, this implies that we have strong material variations in the model, indicating that we might have to use a methodology that allows varying the grid density across the model.

Equally relevant for hazard studies is the field of *earthquake physics* in which we seek an understanding of the processes governing seismic rupture. While in general seismology is a data-rich science, this is less the case for earthquake source studies. Even though we now have some very well-recorded large seismic events (like the M9.0 Tohoku-Oki earthquake in Japan in 2011), we are lacking observations close to the source that would allow us to put tighter constraints on the rupture properties. In this field it is important to be able to properly implement the specific frictional boundary conditions that act on a fault. Finding accurate and efficient solutions to this problem is still a very active research field.

Active source seismology in *exploration geophysics* usually aims at generating body waves and avoiding surface waves if possible. The focus on body waves has some consequences on the choice of an appropriate solver. For some methods the implementation of accurate free-surface boundary conditions is more difficult. In reservoir situations it might also be necessary to involve more complicated rheologies such as anisotropy and poroelasticity, or to generate meshes for highly complex models (see Fig. 1.3).

For seismic wave propagation beyond a certain scale ($\approx 1,000$ km) the Earth's curvature can no longer be neglected. This concerns *regional or continental wave propagation* as well as *global or planetary seismology*. Spherical coordinates would be one method, but there are certain restrictions concerning the model domain. An alternative is the use of the *cubed-sphere* approach, which is applicable for both regional and global wave propagation and has been very successfully employed using spectral-element or discontinuous Galerkin methods.

Other domains of application include *volcanology*. On volcanoes a tremendously wide spectrum of seismic signals is observed, and modelling them is

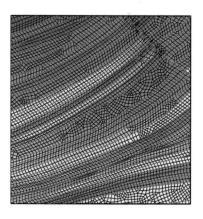

Fig. 1.3 *Exploration problems. Rectangular mesh for the Marmousi benchmark model for reservoir simulations. The mesh honours fine layering and internal faults. Figure from Capdeville et al. (2015).*

a challenging task. Depending on the frequency range involved, the often substantial topography needs to be incorporated. Topography combined with a free-surface boundary condition is something that is basically impossible to implement with some methods, but easy with others.

Finally, it is important to mention the field of *seismic tomography* and in particular *full waveform inversion*. We are currently going far beyond extracting only a few bytes of information from observed seismograms like travel times and explaining them using ray theory. The future is in matching complete waveforms, that is, calculating synthetic seismograms through 3D models and directly comparing them with observations. An example comparing theory with observations is given in Fig. 1.4. Full waveform inversion is an iterative procedure that requires the calculation of a great many forward problems, progressively improving the fit between synthetic and observed seismograms. I predict that in a few years from now we will routinely do full waveform inversion on all scales with tremendously improved images of the Earth's interior. Eventually it might lead to new theories on how our planet Earth works. But there is a long way to go!

This volume aims at providing you with a very basic introduction to the various numerical methods that are currently used for seismic wave-propagation problems. The focus is on the fundamental principles of the methods that lead to the properties that are relevant to the question which solver works best for a certain problem. To present a complete 3D *implementation* goes beyond the scope of this volume. Relevant references are provided at the end of each chapter.

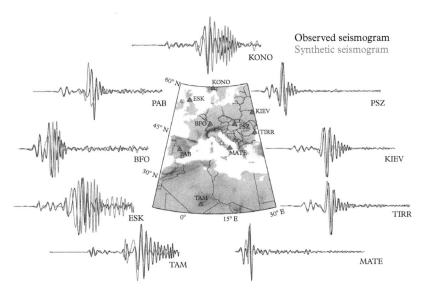

Fig. 1.4 *One of our ultimate goals is to reduce the misfit between observations (black traces) and theoretical seismograms (red traces). In this case, seismograms are compared for the damaging M6.3 l'Aquila, Italy earthquake in 2009 (yellow star) at several European stations (red triangles) with calculations using the discontinuous Galerkin method. From Wenk et al. (2013).*

1.3   Target audience and level

This volume is based on lectures given by the author and his colleagues at the Ludwig-Maximilians University, Munich, as well as in short courses at foreign institutions and during the workshops of two EU-funded training networks, SPICE (2003–7) and QUEST (2009–13) in the field of computational seismology. The volume can be used as a basis for a one-semester or two-semester course for senior undergraduates, or junior postgraduates of physics, Earth sciences, or engineering with sufficient background in mechanics and analysis. It also addresses experienced researchers who intend to use some of the community codes that are on offer today (see the Appendix), and seek a quick overview of numerical methods applied to the elastic wave equation.

Experience with teaching numerical methods and guiding students who are starting to using community codes for 3D seismic wave propagation for many years led me to conclude that *an excellent preparation for a research project in computational seismology with any method is to start by coding the wave equation from scratch in 1D and to explore its capabilities and traps.* With some of the methods this is done in almost no time, while with others this simple problem might already be quite involved. Yet, it is worth the effort!

To help achieve this goal substantial supplementary electronic material is provided (see Appendix). This involves elementary ingredients of numerical methods such as finite-differencing, numerical integration, and function approximation, as well as complete 1D (some 2D) solutions for each of the numerical methods introduced. I strongly recommend that you first try out coding a solution yourself, and then consult the solutions! Alternatively, the codes can be used as a starting point to solve the many computer exercises given at the end of each chapter.

To enable easy, fast, and direct access to the practical material, computer codes are provided as Python-based *Jupyter notebooks* through a dedicated server[3] (i.e. no download necessary; codes can be run anywhere with internet access, even on your smart phone). This should enable lecturers to easily and immediately use the volume plus the supplementary material for teaching. Many of the codes are also available in Matlab®.

This is neither a *seismology* volume nor a *maths* volume. It is somewhat in between. The term *practical* in the title refers to the priority given to presenting the numerical algorithms for each method in combination with implementation in a computer code. Sometimes the path from a mathematical algorithm to a code can be painful, in particular if one has to rely on algorithms presented in research papers where details are often omitted.

The mathematical background required to understand the numerical tools presented here strongly depends on the methods themselves. We use elements of calculus, linear algebra, functional analysis, and partial differential equations that should be covered in undergraduate or postgraduate lectures like *Mathematical Methods for Earth Scientists (Physicists, Engineers)*. These elements include

[3] http://www.seismo-live.org

- Taylor series, Fourier series
- Fourier transforms, convolution theorem
- Exponential functions, complex numbers
- Function interpolation
- Polynomial functions (e.g. Chebyshev, Lagrange, Legendre)
- Numerical integration
- Vector, matrix calculations
- Eigenvector analysis
- Vector field operations (curl, div, grad)

It might, therefore, be useful to have your favourite maths handbook at hand.

The specific sequence of numerical methods presented—from simple to more complicated—somehow also reflects the chronology of the evolution of numerical methods applied to wave-propagation problems, and is intentional.

1.4 How to read this volume

The volume is divided into three parts.

Part I serves as an introduction to fundamental aspects of seismic wave propagation, the discrete world, and computations. These chapters are written with a clear focus on what is relevant when you use numerical solutions to wave-propagation problems. What are the governing equations? What are analytical solutions to simple problems? How do we describe seismic sources? What boundary conditions apply? In addition, we introduce some basic concepts of describing wavefields in a discrete way and the consequences for large-scale computations.

Part II is the heart of the volume, with six chapters on specific numerical methods. The *finite-difference method* is covered in Chapter 4 in quite some detail, including an analytical way of treating the numerical approximations (von Neumann analysis), leading to some fundamental results (e.g. stability criterion, numerical anisotropy) which are also relevant for the other numerical methods. All methods rely on a finite-difference type approximations for the time-dependent (extrapolation) part. Therefore, this chapter is essential.

In the chapter on the *pseudospectral method*, the concepts of exact interpolation and cardinal functions are introduced. These will play an important role later in high-order Galerkin methods. Both Fourier and Chebyshev methods are presented.

Despite their intimate relation, there are separate chapters on the *finite-element* and the *spectral-element method*. The finite-element method is introduced in the simplest possible form using the static elastic case with linear basis functions. The spectral-element method is developed using Lagrange polynomials and Gauss integration leading to the attractive explicit scheme that is so popular today.

The chapter on the *finite-volume method* takes the scalar advection equation as a starting point, and shows that it is formally equivalent to the problem of wave propagation. The concept of numerical fluxes is introduced, based on analytical solutions of the Riemann problem. The finite-volume method is also discussed as a direct application of Gauss's theorem, allowing numerical solutions for finite volumes of arbitrary shape.

Finally, the most recent numerical approach is the *discontinuous Galerkin method*, which joins the best part of the spectral-element method with the flux scheme developed for finite-volume method. We focus on the nodal form of the method using the same Lagrange polynomial basis functions as for the spectral-element method.

One of the fascinating aspects of this zoo of numerical methods, which sometimes have entirely different starting points, is the fact that, in their simplest (linear) form, most of them are basically identical. Whenever possible this is highlighted in the text. Nevertheless, their pros and cons become apparent as soon as more realistic simulation scenarios are envisaged.

The many domains of applications of the methods presented in this volume are discussed in Part III. For each application, domain-specific requirements are presented with some indications as to which methodologies might work best. This is complemented by a discussion of some open issues and current hot topics in computational seismology.

In the Appendix, a list of links to currently accessible community numerical solvers for seismic wave-propagation problems, data analysis, visualization, and data access is provided. Furthermore, access to the supplementary electronic material and some information on content is detailed.

The technical chapters on the various methods in Part II each have the same structure. A brief historical overview is followed by presenting the method *in a nutshell*, highlighting its most important aspects. Each chapter then explores the details of the method in question. Towards the end, a *Road to 3D* section provides some information on where to find readable complete 3D algorithms in the literature. A *Summary* is then given, followed by a *Further reading* section. The chapter then concludes with *Exercises,* divided into (1) comprehension questions, (2) theoretical problems, and (3) programming exercises. This structure makes the technical chapters quite self-contained, with the drawback that there is some repetition.

1.5 Code snippets

The volume contains fragments of Python codes taken from the supplementary material that can be downloaded or run as interactive Jupyter notebooks online (<http://www.seismo-live.org>). It is important to note that they are presented in a style that optimizes readability in the sense that someone not so familiar with Python (and maybe more familiar with Matlab®) can understand what's going on. In any case it is advisable to spend some time with the Python introductory material available online at the site given above.

Nevertheless, a few important points are given here. Even though it is considered bad practice, we import (sub-) libraries such as *numpy* using the commands

```
from numpy import *
from numpy.fft import *
```

This allows us to use Matlab® style calls to intrinsic routines such as

```
# Fast Fourier transform of vector f
F = fft(f)
```

or to initialize vectors with

```
time = linspace(0, nt * dt, num=nt)
```

A feature that enables quite dense coding and mimics the implicit matrix-vector calculations of Matlab® is the @ sign used in Python 3.5 versions (or higher). For example,

```
from numpy.linalg import *
# [...] Initialize matrices
 A = R @ L @ inv(R)
```

would correspond to matrix operations for an eigenvalue problem

$$\mathbf{A} = \mathbf{R}\,\mathbf{L}\,\mathbf{R}^{-1},$$

where $\mathbf{A}$ is a square matrix, $\mathbf{R}$ contains its eigenvectors, and $\mathbf{L}$ is a diagonal matrix with eigenvalues.

The material presented here is about waves that propagate. It is so much fun bringing life to the figures in this volume: seeing waves propagate and scatter after writing your own code or using the available supporting material. By doing so you will be able to better understand the underlying numerical mathematics and learn wave-propagation phenomena. Have fun!

....

FURTHER READING

There follows a list of books covering general seismology, basic maths, and *classic* approaches such as the reflectivity method, ray theory, and normal mode solutions.

- Kennett (1983) presents the theory for the calculation of synthetic seismograms in stratified media (e.g. reflectivity method).

- Cerveny (2001) is the classic textbook on seismic ray theory.

- Nolet (2008) provides a comprehensive introduction to seismic tomography, including the basic theory of wave propagation using ray and Born approximations.

- Dahlen and Tromp (1998) is an exhaustive textbook on global wave propagation using normal mode techniques.

- Chapman (2004) presents a comprehensive introduction to the propagation of high-frequency body waves in elastodynamics.

- Snieder (2015) provides an entertaining tour of the mathematical knowledge and techniques that are needed by students across the physical sciences, with very illustrative examples.

- The Encyclopedia of solid Earth geophysics (ed. Gupta, 2011) covers many fundamental research problems in seismology and also contains several review papers on methods discussed in this volume.

- The Treatise on Geophysics (Schubert, 2015) is maybe the most complete collection of review articles in solid Earth geophysics. Volume 1 (Deep Earth Seismology) contains a collection of articles on various approaches to the seismic forward problem.

- The International Handbook of Earthquake and Engineering Seismology (Lee et al., 2002) contains about 50 review-style articles on many fundamental aspects of seismology.

- The New Manual of Seismological Observatory Practice (ed. Bormann, 2012) is an open-access collection of articles on many aspects of seismic wave propagation, seismic sources, instrumentation, networks, and processing.

Part I

Elastic Waves
in the Earth

Seismic Waves and Sources

So, you want to calculate synthetic seismograms, or dive into a seismological research problem with a complicated Earth model or earthquake source mechanism? Most likely you will be making use of a community code to get going, or a program that is handed to you by other researchers or your supervisor. There will be a parameter file that allows you to change the set-up, some of the entries being obvious, some less so. It is usually quite easy to obtain the first results. But how do you check that the results are correct? Is every wiggle really accurate in travel time and amplitude? Is your Earth model correctly implemented? Are the sources and receivers correctly positioned?

These questions are difficult to answer for complex models and we will later devise some strategies on how to overcome this. When you start using any solver for seismic wave propagation (well-tested or not) it is wise to begin with a simple earthquake source and Earth model (e.g., a homogeneous half-space) and to check whether the seismograms make sense. This chapter aims at providing some hints as to what you should expect for simple media, and what fundamental strategies are available to help you obtain correct results. This will require a basic understanding of seismic sources and elastic wave propagation.

This introduction cannot replace a textbook on seismology (see recommendations at the end of this chapter). The topics are introduced with a view to their relevance for seismic simulation problems. Key issues are: (1) What are the governing equations for elastic wave propagation? (2) What phenomena do we expect in simple Earth models? (3) What boundary conditions apply? (4) How are seismic sources described? and (5) What are the consequences of linearity and reciprocity of the elastic wave equation for simulations?

Let us introduce some fundamental concepts relevant for wave simulations by looking at real global seismic wavefield observations. The simple principles discussed in what follows can easily be adapted to other scales (e.g. reservoirs, sedimentary basins, volcanoes, rock samples). In Fig. 2.1 the vertical component of a broadband velocity seismogram is shown representing the ground motion at station WET in southern Germany following the devastating Sumatra-Andaman earthquake with moment magnitude M9.1 that occurred on 26 December 2004. Before we start discussing these observations, let us introduce one of the most important relations that you need in order to plan, check, and understand

Computational Seismology. First Edition. Heiner Igel.
© Heiner Igel 2017. Published in 2017 by Oxford University Press.

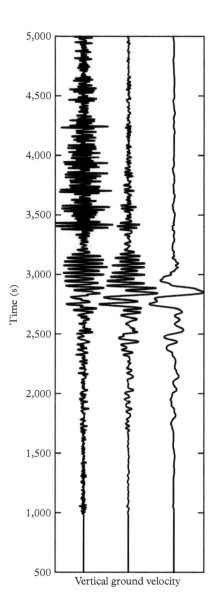

Fig. 2.1 *Vertical component ground velocity seismogram of the M9.1 Sumatra-Andaman earthquake of 26 December 2004, recorded in southern Germany (WET) with an STS2 seismometer.* **Left:** *Original broadband seismogram.* **Middle:** *Low-pass filtered seismogram with corner period 40 s.* **Right:** *Low-pass filtered with corner period 100 s. Amplitudes are normalized.*

seismic simulation results: the connection between wavenumber k and angular frequency ω (or frequency f), or wavelength λ and period T:

$$c = \frac{\omega}{k} = \frac{2\pi f}{2\pi/\lambda} = \frac{\lambda}{T}, \tag{2.1}$$

where c is phase velocity. Let us play around with this simple relation given the seismograms in Fig. 2.1.

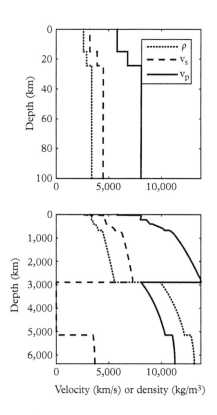

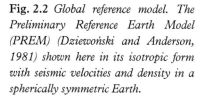

Fig. 2.2 *Global reference model. The Preliminary Reference Earth Model (PREM) (Dziewoński and Anderson, 1981) shown here in its isotropic form with seismic velocities and density in a spherically symmetric Earth.*

Velocity (km/s) or density (kg/m³)

First, by visual inspection of the original broadband record we can appreciate that a wide range of frequencies make up the entire seismogram. The initial high-frequency part consists of body waves, mainly compressional P- and shear SV-wave energy (shear-wave polarized in the vertical plane) with a maximum frequency of around 1 Hz. What does this imply for the spatial wavelengths inside the Earth? The Preliminary Reference Earth Model (PREM) (Dziewoński and Anderson, 1981) shown in Fig. 2.2 tells us that near the Earth's surface P-velocities are around 6 km/s and shear velocities close to 3 km/s (ignoring the oceans). That means the *shortest* wavelength λ_{min} is likely to be around 3 km. As velocity increases with depth (at least down to the core–mantle boundary), wavelengths will almost always be longer. This is important as in all the numerical solvers discussed later we discretize the Earth's interior with grid points or elemental cells, and need to make sure that the *smallest* wavelength is accurately sampled.

Sampling a wave-like function is characterized by a concept called the *number of grid points per wavelength*, which will play a central role in all numerical methods (see Fig. 2.3). It is instructive to calculate how many cells (or points) you would need for the entire Earth if you were to discretize her regularly with cubic

Fig. 2.3 *Sampling waves. Sinusoidal wave function sampled with 10 grid points per wavelength (more precisely 10 segments or elements per wavelength, and 11 grid points, including the boundaries).*

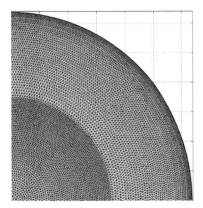

Fig. 2.4 *Triangular-element mesh for global seismic velocity structure. The mesh is designed such that the number of points per wavelength for a given frequency is approximately constant everywhere (Käser and Dumbser, 2006).*

elements, and—for example—require that the shortest wavelength is sampled with at least 10 points. Let's use the numbers above:

- The shortest wavelength is 3 km and we require it to be sampled with at least 10 points. That means the maximum grid spacing (or cubic element side) should be h = 300 m.
- As we are in 3D the volume of our cube element $V_e = h^3$.
- The volume V_E of the entire Earth with radius r_E = 6,371 km can be calculated as $V_E = 4/3\pi r_E^3$.
- The number of required elements is $V_E/V_e = 4 \times 10^{13}$.
- One space-dependent field (e.g. density, displacement, Lamé parameters) at double precision (8 bytes per number) would thus require 320 TBytes.

Wow! Clearly, even on today's supercomputers this is a tremendous challenge, and besides, we would oversample large parts of the Earth's interior, where velocities reach 13 km/s (how many points per wavelength would that correspond to?). This suggests that for problems with strongly varying velocity models regular grids do not make sense. An example of a spherical mesh (reduced to 2D) with point- or (element-) density adapted to the velocity model is shown in Fig. 2.4. In principle, meshes for heterogeneous media can be designed such that the number of grid points per wavelength is approximately constant throughout the model. More on this later.

Let us have a look at the spectral content of our observations (see Fig. 2.5). The largest amplitudes appear between the frequencies 0.02–0.06 Hz, characteristic of very large earthquakes. The low-pass filtered spectra (and seismograms) reveal that substantial energy with high signal-to-noise ratio is present at low frequencies.

These dominant amplitudes are Rayleigh-type surface waves that start around t = 2500 s, lasting for several hundreds of seconds. We note strong *physical dispersion*.[1] Long-period surface waves arrive much earlier than short-period surface waves. The period range of Rayleigh waves spans an interval from about tens to hundreds of seconds.

Performing a similar calculation as presented above for surface waves with appropriate phase velocities (e.g. for cut-off period T = 40 s) we obtain model sizes that fit on today's common institutional computer clusters. Note that—depending on epicentral distance—we would only have to discretize the upper mantle and crust. An important point is that surface waves are a consequence of the stress-free boundary condition. This implies that when using numerical methods and when surface waves are the target, this boundary condition should be implemented with high accuracy.

This illustration with observed seismograms serves to show that (1) target frequencies, (2) seismic phases to be modelled (body waves or surface waves), and (3) seismic velocity heterogeneity are dominant factors on how Earth models need to be discretized. Seismograms are affected by both source and structure.

[1] In the wave-propagation context the term *dispersion* denotes the frequency (or wavenumber) dependence of seismic velocities.

In the following sections we will present the relevant equations and theoretical concepts used in computational seismology, some of which we will discretize and approximate with numerical methods later on. This is kept at a very fundamental level. For further details please refer to the general seismology textbooks listed at the end of this chapter.

2.1 Elastic wave equations

In the following we will present several forms of the elastic wave equation, starting with the complete set of equations in 3D, ending with the specific form in 1D, which we will use as the central equation to be solved employing the various numerical techniques in the subsequent chapters. Throughout the volume we alternate between assuming space–time dependencies implicitly or stating them explicitly. Often this is a matter of space and/or clarity. The reader is encouraged to return to this chapter in case the dependencies are not clear.

Let us introduce the key players in our problem: our unknown field that we would like to determine is either the displacement field $u_i(\mathbf{x}, t)$ or its time derivative the velocity field $v_i(\mathbf{x}, t) = \partial_t u_i(\mathbf{x}, t)$. The displacement field determines the strain field $\epsilon_{ij}(\mathbf{x}, t)$, that in turn is proportional to the stress field $\sigma_{ij}(\mathbf{x}, t)$ with the general fourth-order tensor of elastic constants $c_{ijkl}(\mathbf{x})$ as proportionality factors. Elastic constants and space-dependent density $\rho(\mathbf{x})$ constitute the geophysical properties of an elastic Earth model. The seismic sources are characterized either by the seismic moment tensor $M_{ij}(\mathbf{x}, t)$ or by the volumetric forces $f_i(\mathbf{x}, t)$. In summary, the dependencies are

$$
\begin{aligned}
u_i &\rightarrow u_i(\mathbf{x}, t) & i &= 1, 2, 3 \\
v_i &\rightarrow v_i(\mathbf{x}, t) & i &= 1, 2, 3 \\
\sigma_{ij} &\rightarrow \sigma_{ij}(\mathbf{x}, t) & i, j &= 1, 2, 3 \\
\epsilon_{ij} &\rightarrow \epsilon_{ij}(\mathbf{x}, t) & i, j &= 1, 2, 3 \\
\rho &\rightarrow \rho(\mathbf{x}) & & \\
c_{ijkl} &\rightarrow c_{ijkl}(\mathbf{x}) & i, j, k, l &= 1, 2, 3 \\
f_i &\rightarrow f_i(\mathbf{x}, t) & i &= 1, 2, 3 \\
M_{ij} &\rightarrow M_{ij}(\mathbf{x}, t) & i, j &= 1, 2, 3,
\end{aligned}
$$

where i, j, k, l are determined by the dimensionality of the problem (1D, 2D, or 3D).

These players make up the elastic wave equation which we first show in the *displacement* form for isotropic media:

$$
\begin{aligned}
\rho \partial_t^2 u_i &= \partial_j(\sigma_{ij} + M_{ij}) + f_i \\
\sigma_{ij} &= \lambda \epsilon_{kk} \delta_{ij} + 2\mu \epsilon_{ij} \\
\epsilon_{kl} &= \frac{1}{2}(\partial_k u_l + \partial_l u_k),
\end{aligned}
\tag{2.2}
$$

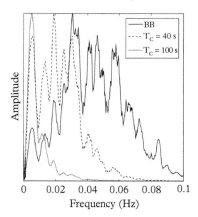

Fig. 2.5 *Normalized amplitude spectrum of the original broadband (BB) vertical-component velocity seismogram of Fig. 2.1 (black line) and the spectra after filtering with cut-off period $T_c = 40$ s (dashed line), and $T_c = 100$ s (dotted line). The spectra are shown for the low-frequency part below 0.1 Hz.*

where the Einstein summation convention[2] applies. $\lambda(\mathbf{x})$ and $\mu(\mathbf{x})$ are the Lamé parameters, the latter being the shear modulus; δ_{ij} is the Kronecker delta. We will discuss anisotropic elastic parameters in the section on *rheologies*.

In principle, these nested relations can be merged into one equation. As an example, we show this for one component (i = 2, i.e. the y-component)

$$
\begin{aligned}
\rho \partial_t^2 u_y \;=\; & \partial_x \left[\mu(\partial_x u_y + \partial_y u_x) + M_{yx} \right] \\
& + \partial_y \left[(\lambda + 2\mu)\partial_y u_y + \lambda(\partial_x u_x + \partial_z u_z) + M_{yy} \right] \\
& + \partial_z \left[\mu(\partial_z u_y + \partial_y u_z) + M_{yz} \right] \\
& + f_y,
\end{aligned}
\tag{2.3}
$$

with equivalent expressions for the other two motion components (see exercises). We anticipate that the temporal and spatial derivatives in these equations cannot be solved analytically in the general case, which is the reason why we have to employ numerical methods. As the focus of this volume is to introduce the concepts of a variety of numerical methods for this equation, we keep the specific form of the elastic wave equation as simple as possible. Therefore, we reduce this equation to a 1D problem by assuming (1) propagation in x-direction, (2) displacement perpendicular to the propagation direction (transverse motion in y-direction), and (3) initialization of wave propagation either by external forcing f_y or by an appropriate initial condition. With these assumptions applied to Eq. 2.3, and an external force term f_y, we obtain

$$
\rho \partial_t^2 u_y \;=\; \partial_x(\mu \partial_x u_y) + f_y.
\tag{2.4}
$$

This equation is descriptive of transversely polarized elastic waves propagating in x-direction (e.g. motion of a string, see Fig. 2.6). Analytical solutions to this equation that are extremely useful for validating numerical results are presented in the following sections.

There is another form of wave equation that can be derived from Eq. 2.2, assuming constant density and vanishing shear modulus μ. This so-called *acoustic wave equation* describes the propagation of compressional waves in media like fluids and gases and is still widely used today to describe P-wave fields, for example in exploration problems. The acoustic wave equation reads

$$
\partial_t^2 p \;=\; c^2 \Delta p + s,
\tag{2.5}
$$

where $p(\mathbf{x}, t)$ is the unknown pressure field, $c(\mathbf{x})$ is acoustic velocity, $s(\mathbf{x}, t)$ is a pressure source field, and $\Delta = \nabla^2 = [\partial_x^2 + \partial_y^2 + \partial_z^2]$ is the Laplace-operator.[3]

Yet another form of the elastic wave equation that plays an important role in computational seismology is the so-called *velocity–stress* formulation. We replace the displacement field $u(\mathbf{x}, t)$ by its time derivative, the velocity field $v(\mathbf{x}, t) = \partial_t u(\mathbf{x}, t)$, and obtain a set of coupled equations

Fig. 2.6 *The 1D elastic wave equation for transverse motion is descriptive of vibration problems of strings.*

[2] Summation over the index or indices that appear twice in a term, e.g. $\epsilon_{kk} = \epsilon_{11} + \epsilon_{22} + \epsilon_{33}$. Indeed, Einstein introduced this compact notation in his treatise on general relativity in 1916.

[3] You might wonder why we do not use the acoustic equation to introduce the various numerical methods. The point is that the elastic version (Eq. 2.4) contains derivatives of the elastic parameters leading to specific features (e.g. grid staggering) common in today's numerical solvers.

$$\rho \partial_t v_i = \partial_j (\sigma_{ij} + M_{ij}) + f_i$$
$$\partial_t \sigma_{ij} = \lambda \partial_t \epsilon_{kk} \delta_{ij} + 2 \mu \partial_t \epsilon_{ij} \qquad (2.6)$$
$$\partial_t \epsilon_{ij} = \frac{1}{2} (\partial_i v_j + \partial_j v_i).$$

It is important to note that this is a first-order partial-differential equation of hyperbolic form (more on this later) and both velocities and stresses are considered unknown. Another point is that we are not directly taking derivatives of the elastic parameters. Most current finite-difference approaches are based on solutions to these coupled equations. With the mapping $v = v_y$, $f = f_y$, and $\sigma = \sigma_{yx}$ (and an external force only) this reduces to the following system of equations in 1D, which we will use extensively in later chapters

$$\rho \partial_t v = \partial_x \sigma + f$$
$$\partial_t \sigma = \mu \partial_x v. \qquad (2.7)$$

For the sake of simplicity we will restrict ourselves to elastic isotropic media when discussing various numerical solutions to the above equations. In the following we will take a purely mathematical approach and discuss the few analytical solutions that are available for the most basic wave equations. These solutions are extremely useful for checking the numerical solutions to be developed later for the case of homogeneous media.

2.2 Analytical solutions: scalar wave equation

In this section we present analytical solutions for the scalar (acoustic) wave equation

$$\partial_t^2 p(x, t) - c^2 \Delta p(x, t) = s(x, t) \qquad (2.8)$$

assuming constant velocity c and infinite space. Note that in 1D and 2D this equation is mathematically equivalent to the problem of *SH* wave propagation (i.e. shear waves polarized perpendicular to the plane through source and receiver). In 3D it is (only) descriptive of pressure (sound) waves. There are several ways to initiate propagating waves. The most simple case is when there is no external source, thus $s(x, t) = 0$, but an initial pressure (or displacement) field exists at $t = 0$ without time variation

$$p(x, t = 0) = p_0(x)$$
$$\partial_t p(x, t = 0) = 0. \qquad (2.9)$$

The solution to this problem is

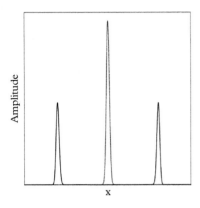

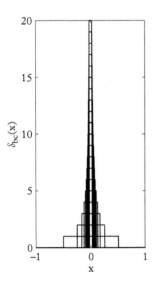

Fig. 2.7 *Illustration of the analytical solution for a pressure (displacement) initial condition $p_0(x)$ (dashed line). The solution $p(x, t)$ corresponds to the initial waveform propagating in both directions undisturbed with velocity c (solid line).*

Fig. 2.8 *Boxcar functions for $dx \in [1 \ldots 20]$ as an example of a function that converges to a δ-function. As their width goes to zero their amplitude goes to infinity. See text for details.*

$$p(x, t) = \frac{1}{2} p_0(ct - x) + \frac{1}{2} p_0(ct + x), \qquad (2.10)$$

corresponding to the waveform $p_0(x)$ being advected undisturbed (with half the initial amplitude) in positive and negative x-direction. This is illustrated in Fig. 2.7 with a waveform of Gaussian shape.

Analytical solutions for inhomogeneous partial differential equations (i.e. with non-zero source terms) are usually developed using the concept of Green's functions $G(\mathbf{x}, t; \mathbf{x}_0, t_0)$. Green's functions are the solutions to the specific partial differential equations for δ-functions as source terms evaluated at $(\mathbf{x}, t)$ and activated at $(\mathbf{x}_0, t_0)$. Thus we seek solutions to

$$\partial_t^2 G(\mathbf{x}, t; \mathbf{x}_0, t_0) - c^2 \Delta G(\mathbf{x}, t; \mathbf{x}_0, t_0) = \delta(\mathbf{x} - \mathbf{x}_0)\delta(t - t_0), \qquad (2.11)$$

where Δ is the Laplace operator. We recall the definition of the δ-function as a *generalized function* with

$$\delta(x) = \begin{cases} \infty & x = 0 \\ 0 & x \neq 0 \end{cases} \qquad (2.12)$$

and

$$\int_{-\infty}^{\infty} \delta(x)dx = 1, \quad \int_{-\infty}^{\infty} f(x)\delta(x)dx = f(0). \qquad (2.13)$$

When comparing numerical with analytical solutions the functions that, in the limit, lead to the δ-function will become very important. An example is the boxcar function

$$\delta_{bc}(x) = \begin{cases} 1/dx & |x| \leq dx/2 \\ 0 & \text{elsewhere}, \end{cases} \qquad (2.14)$$

fulfilling these properties as $dx \to 0$. These functions are used to properly scale the source terms to obtain correct absolute amplitudes. The convergence of the boxcar function thus defined is illustrated in Fig. 2.8.

To describe analytical solutions for the acoustic wave equation we also make use of the unit step function, also known as the Heaviside function, defined as

$$H(x) = \begin{cases} 0 & x < 0 \\ 1 & x \geq 0 \end{cases}. \qquad (2.15)$$

The Heaviside function is the integral of the δ-function (and vice versa the δ-function is defined as the derivative of the Heaviside function). Omitting their derivation (see references at the end of this chapter) the Green's functions for Eq. 2.11 are presented in Table 2.1 for any number of spatial dimensions. These analytical solutions are illustrated in Fig. 2.9. It is worth spending some time

Table 2.1 *Green's functions for the inhomogeneous acoustic wave equation after Rienstra and Hirschberg (2016).*

1D	2D	3D
$\frac{1}{2c}H(t - \frac{\|r\|}{c})$	$\frac{1}{2\pi c^2}\frac{H(t - \frac{\|r\|}{c})}{\sqrt{t^2 - \frac{r^2}{c^2}}}$	$\frac{1}{4\pi c^2 r}\delta(t - r/c)$
$r = x$	$r = \sqrt{x^2 + y^2}$	$r = \sqrt{x^2 + y^2 + z^2}$

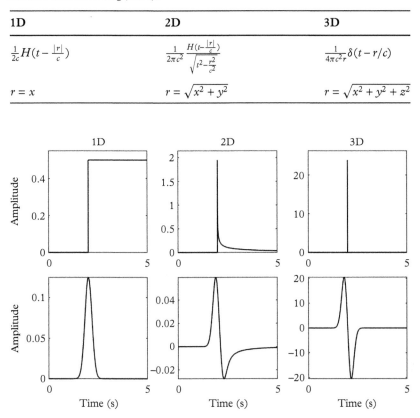

Fig. 2.9 *Analytical solutions to the scalar wave equation.* **Top:** *Green's functions in 1D, 2D, and 3D.* **Bottom:** *Green's functions obtained after convolution with the first derivative of a Gaussian with 1 Hz dominant frequency (see text for details). Note that the source time function is centred around t = 0.*

discussing these extremely important results. In the 1D case the Green's function is proportional to a Heaviside function. As the response to an arbitrary source time function can be obtained by convolution this implies that the propagating waveform is the integral of the source time function. This is illustrated in Fig. 2.9 (bottom left) where the response is shown for a source time function with a first derivative of a Gaussian.

A special situation occurs in 2D. An impulsive source leads to a waveform with a coda that decreases with time. This is a consequence of the fact that the source actually is a line source (see discussion in the next chapter). From a computational point of view this is extremely important. Numerical solutions in 2D Cartesian coordinates cannot directly be compared to observations in which we usually have point sources.

Finally, in 3D the Green's function is proportional to a δ-function scaled by the distance from the source (geometrical spreading). Beware of the practical significance of this result for sound wave propagation! Ideally, whatever your lecturer tells you should physically arrive undisturbed at your ears (unfortunately the same does not generally apply to the meaning of the message carried by the sound waves).[4]

To recover these results with numerical simulations for arbitrary spatio-temporal sources requires some care as to the scaling of point sources in the discrete world. This is discussed in more detail in connection with the specific numerical methods.

2.3　Rheologies

Non-isotropic rheologies are important for realistic applications and are usually covered in current solvers. Therefore we briefly discuss them in this section. The term *rheology* actually originates from the description of flowing material. In the context of solid material it describes how deformations (here: strains) are related to forces (here: stresses). In the following we briefly discuss (1) viscoelastic material, (2) anisotropic material, and (3) poro-elasticity. It is difficult to prioritize these effects in terms of their relevance to modelling real observations. However, the first two are almost equally important on all spatial scales. As we will not use these rheologies in our numerical approximations to be developed later, they are mentioned for completion. The adaptation of rheologies to the specific numerical solutions is sometimes easy (anisotropy), and sometimes not so easy (attenuation or poroelasticity). The interested reader is referred to the references given for each numerical method.

2.3.1　Viscoelasticity and attenuation

There is no such thing as perfect elasticity in nature. Thus, seismic wave fields are permanently losing energy, for example due to micro-damage created during the passage of waves or friction-induced conversion into heat. Such phenomena lead to *intrinsic* attenuation,[5] an energy loss that is described using the letter Q (from *quality factor*).

Q is a dimensionless quantity and is defined by the fractional energy loss per cycle

$$\frac{1}{Q(\omega)} = -\frac{\Delta E}{2\pi E}, \tag{2.16}$$

where $-\Delta E$ is the energy loss per cycle and E is the peak strain energy.[6] Q is usually considered independent of frequency, low values (e.g. $Q = 10$ in a reservoir) corresponding to strong attenuation, high values (e.g. $Q = 600$ in the mantle) to small attenuation.[7]

The decay $A(x)$ for initial amplitude A_0 as a function of propagation distance x for a monochromatic plane wave of frequency ω, velocity c, and quality factor Q is given by

[4] Furthermore, imagine a different kind of Green's function. I don't want to know what the integral of Beethoven's Fifth Symphony would sound like!

[5] Here the term attenuation is clear and related to the loss of energy due to intrinsic processes. Sometimes (e.g. in earthquake engineering) the term attenuation is also used to describe the decay of amplitudes with distance due to geometrical spreading.

[6] The strain energy function is defined by $W = 1/2\,\sigma_{ij}\epsilon_{ij}$ with summation over both indices.

[7] Note that we owe the high Q inside our Earth the fact that we know so much about its interior from seismic tomography.

$$A(x) = A_0 e^{-\frac{\omega x}{2cQ}}. \tag{2.17}$$

This equation is illustrated in Fig. 2.10 for frequency $\omega = 1$ Hz, a variety of Q factors, and propagation distances up to 100 km.

Note that constant, frequency-independent Q has important consequences. As Q describes the energy loss per cycle, this implies that for a given propagation distance the high-frequency (short wavelength) part of the wavefield is substantially more attenuated than the low-frequency part, progressively altering the waveform. This effect explains the diminishing abundance of high frequencies away from a seismic source and determines the upper frequency limit in global seismograms, as seen in the introduction.

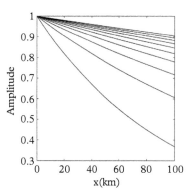

Fig. 2.10 *Seismic attenuation. The quality factor Q is used to describe seismic attenuation (see text). In this graph the amplitude decay due to attenuation only is shown as a function of propagation distance for an initial unity amplitude. The curves illustrate the behaviour for $Q = 10$ (lowest curve, strongest attenuation) to $Q = 100$ (top curve, weakest attenuation) at a frequency of 1 Hz.*

2.3.2 Seismic anisotropy

Today it is widely accepted that most of the Earth's interior—from crust to inner core—shows anisotropic elastic behaviour. Most local observations above active faults, upper mantle body and surface waves, as well as body waves in the inner core cannot be adequately explained without some form of anisotropic symmetry system.

In the most general case, the stress σ_{ij} of an anisotropic body is related to the deformation by

$$\sigma_{ij} = c_{ijkl}\,\epsilon_{kl}, \quad i,j,k,l = 1,2,3, \tag{2.18}$$

where c_{ijkl} is a fourth-order tensor with 81 elastic constants (the Einstein summation convention applies). Due to the symmetry conditions of the elastic tensor $c_{ijkl} = c_{jikl} = c_{klij}$ and further thermodynamical arguments it is possible to reduce this tensor to a 6×6 matrix with the following conventions that $11 \rightarrow 1$, $22 \rightarrow 2$, $33 \rightarrow 3$, $12 \rightarrow 6$, $32 \rightarrow 4$, and $13 \rightarrow 5$ (also known as the Voigt notation). This symmetric matrix is known as c_{pq} and has in general 21 independent elements. The number of independent elements of c_{pq} is reduced if the medium belongs to a certain symmetry system.

For example, the most common system used for seismic wave propagation is hexagonal symmetry. In that case—assuming coordinate axes aligned with the axes of symmetry—the matrix c_{pq} reads

$$c_{pq} = \begin{pmatrix} c_{11} & c_{12} & c_{13} & 0 & 0 & 0 \\ c_{12} & c_{11} & c_{13} & 0 & 0 & 0 \\ c_{13} & c_{13} & c_{33} & 0 & 0 & 0 \\ 0 & 0 & 0 & c_{44} & 0 & 0 \\ 0 & 0 & 0 & 0 & c_{44} & 0 \\ 0 & 0 & 0 & 0 & 0 & \frac{c_{11}-c_{12}}{2} \end{pmatrix} \tag{2.19}$$

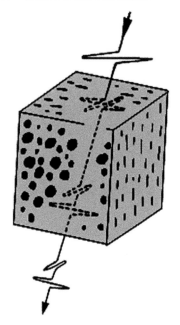

Fig. 2.11 *Shear-wave splitting (bi-refringence). In homogeneous anisotropic material in general there are two quasi-shear waves qS1 and qS2 propagating with different velocities and orthogonal polarizations. This leads to the well-known shear-wave splitting phenomenon. As the orientation of aligned small-scale heterogeneities (e.g. cracks, filled pore space) is usually related to the stress field, the observation of shear-wave polarizations may carry important information about the rock mass at depth.*

with at most five independent elastic parameters. For arbitrary orientation of the symmetry axis, this matrix needs to be rotated and is full.

The consequences of material anisotropy for seismic wave propagation are manifold. First, akin to bi-refringence known from crystal optics, shear waves split into to so-called quasi-shear waves qS_1 and qS_2 that propagate with different velocities and orthogonal polarizations. This is schematically illustrated in Fig. 2.11. There are no longer pure compressional waves polarized in longitudinal direction. They are replaced by quasi-P (qP) waves. Furthermore, wavefronts in homogeneous media are no longer of spherical shape. This is illustrated in an example of a TI (transversely isotropic) medium corresponding to a hexagonal symmetry system. This is a common anisotropic system that results from fine horizontal layering. In this case the axis of symmetry is the vertical axis, and quasi-shear waves are decoupled into pure qSV and qSH waves polarized in vertical and horizontal direction, respectively. In TI media phase velocities are azimuthally isotropic. However, in any vertical plane the phase velocities vary with direction as indicated in Fig. 2.12. The figure was obtained using the weak anisotropy approximation introduced by Thomsen (1986). He introduced three positive parameters ϵ, γ, and δ. The parameter ϵ can be interpreted as the fraction of qP anisotropy, and γ the corresponding fraction for the qSH anisotropy. The phase velocity variations can be obtained by the relations

$$
\begin{aligned}
v_{qP}(\theta) &= v_{P0}\left(1 + \delta \sin^2(\theta)\cos^2(\theta) + \epsilon \sin^4(\theta)\right) \\
v_{qSV}(\theta) &= v_{S0}\left(1 + \frac{v_{P0}^2}{v_{S0}^2}(\epsilon - \delta)\sin^2(\theta)\cos^2(\theta)\right) \qquad (2.20) \\
v_{qSH}(\theta) &= v_{S0}\left(1 + \gamma \sin^2(\theta)\right),
\end{aligned}
$$

where $\theta = 0$ corresponds to vertical direction. The example in Fig. 2.12 was obtained with $v_{P0} = 5$ km/s, $v_{S0} = 2.89$ km/s, $\epsilon = 0.1$, $\delta = 0$, and $\gamma = 0.1$. These values are probably beyond the limit of the approximation and are used here merely to illustrate the principle. Note that the qS-phase velocity surfaces touch each other, in which case the polarization remains undefined (as in the isotropic case in any direction). These directions are called singularities in anisotropy terminology.

There are many ways to initiate the elastic constants for anisotropic material. A comprehensive review of anisotropic symmetry systems is given in Crampin (1984). Anisotropic wave propagation simulated with numerical methods is best verified against quasi-analytical solutions, such as programs based on the anisotropic reflectivity method (Booth and Crampin, 1983).

2.3.3 Poroelasticity

When the rock mass is strongly fractured, and pore space (partially) filled with liquids (water, oil) or gases (air, methane), the elastic equations are no longer sufficient. In this case the stress–strain relation is replaced by an alternative one that was developed using continuum mechanics.[8] The most important effect is that in homogeneous poro-elastic media there are two types of compressional waves (a fast and a slow one) in addition to the classic shear wave (see Fig. 2.13). This result was controversial until such waves were first observed in the eighties. Obviously, poroelastic effects play an important role in reservoir problems, geothermal projects, and volcanology. Some community wave-propagation solvers have optional poroelasticity modules. The most extensive description of poroelasticity can be found in Carcione (2014).

2.4 Boundary and initial conditions

Elastic wave propagation is governed by partial differential equations the solution of which depends on the definition of initial and boundary conditions. In numerical schemes some of these conditions are trivially implemented (like initial conditions), some extremely hard to fulfill (like absorbing boundaries). While not a focus of this introductory text, aspects of boundary conditions are briefly discussed for each numerical method later on. Here, we summarize them from a physical point of view.

2.4.1 Initial conditions

In most cases, everything is at rest when we start. As our solution fields are either displacement $u(\mathbf{x}, t)$ or velocity $v(\mathbf{x}, t)$ this can be expressed as

$$u_i(\mathbf{x}, t = 0) = 0$$
$$v_i(\mathbf{x}, t = 0) = 0. \quad (2.21)$$

We also assume that moments and forces are activated at time $t > 0$ thus

$$f_i(\mathbf{x}, t \leq 0) = 0$$
$$M_{ij}(\mathbf{x}, t \leq 0) = 0. \quad (2.22)$$

As indicated in the section with analytical solutions waves can also be generated by a space-dependent initial condition in displacements, velocities, or stresses, depending on the specific wave equation. Such initial conditions my be described as

$$v_i(\mathbf{x}, t = 0) = v_i^0(\mathbf{x})$$
$$\sigma_{ij}(\mathbf{x}, t = 0) = \sigma_{ij}^0(\mathbf{x}), \quad (2.23)$$

here with velocities or stresses, respectively. A general analytical solution with this initial condition is developed for the velocity–stress formulation in the chapter on the finite-volume method.

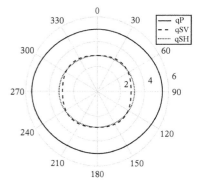

Fig. 2.12 *Seismic anisotropy. Phase velocity variations of qP, qSV, and qSH waves (km/s) in the x–z plane for a hexagonal (transversely isotropic) symmetry system. The axis of symmetry is the vertical axis ($\theta = 0$). See text for details.*

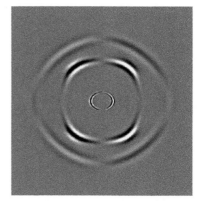

Fig. 2.13 *Poroelastic wave propagation. Snapshot of wave propagation in an anisotropic, poroelastic model of a sandstone material (horizontal motion component). The most important effect of a poroelastic material is the presence of a slow P-wave (centre of image). Figure from de la Puente et al. (2008).*

[8] Poroelasticity goes back primarily to the work of M. A. Biot (1905–1985) in the fifties. During that time he worked as an independent consultant for the Shell Company.

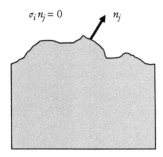

$\sigma_i n_j = 0$ n_j

Fig. 2.14 *Free-surface boundary condition. In the presence of surface topography, the free-surface boundary condition involves the local normal direction n_j to the surface.*

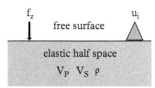

f_z free surface u_i

elastic half space

V_P V_S ρ

Fig. 2.15 *Illustration of a geometrical set-up for Lamb's problem. A vertical point force f_z is activated and recorded at some distance by a receiver recording ground motion components u_i.*

[9] The distinguished British mathematician Sir Horace Lamb (1849–1934) worked in Manchester, Cambridge, and Adelaide (Australia), mainly on problems in fluid mechanics and elasticity.

2.4.2 Free surface and Lamb's problem

The question of which forces act in a specific direction n_j given a space-dependent stress field σ_{ij} leads to the concept of tractions t_i defined as

$$t_i = \sigma_{ij} n_j, \qquad (2.24)$$

where n_j is normalized (i.e. $|\mathbf{n}| = 1$). At the Earth's free surface the tractions perpendicular to it are zero. This is called the *free surface boundary condition*, a condition that leads to the existence of surface waves—a dominant feature in regional and global broadband seismograms. Assuming the z-direction pointing upwards $\mathbf{n} = [0, 0, 1]$, this condition leads to

$$\sigma_{xz} = \sigma_{yz} = \sigma_{zz} = 0. \qquad (2.25)$$

As indicated in Fig. 2.14, this boundary condition depends on the local normal direction in the case of complex surface topography. With some numerical methods this is hard to implement accurately. Thus the involvement of a rugged free surface becomes an important aspect when choosing the right solution strategy.

The so-called Lamb's problem[9] is an important benchmark for numerical solvers that incorporate a free-surface boundary. An analytical solution for the general problem of a point volumetric force acting inside or at the boundary of an elastic half-space was presented by Johnson (1974). The geometrical set-up is illustrated in Fig. 2.15 and an example is shown in Fig. 2.16. As will be discussed in more detail below, the stress-free surface leads to the presence of a Rayleigh-type surface wave. The seismograms show a faint P-wave arrival from propagation along the surface, followed by a Rayleigh surface wave with substantially larger amplitude. These results were obtained with the original Fortran code by Lane Johnson, and some of the results were reproduced from

Fig. 2.16 *Analytical solution to Lamb's problem (vertical displacement). The parameters of the half-space are $v_p = 8$ km/s, $v_s = 4.62$ km/s, and $\rho = 3.3$ kg/m³. The vertical force is input with 10^{15} dyn and the receiver distance is 10 km.* **Left:** *Solution to a step-like source time function.* **Right:** *Solution to a Gaussian-shaped source time function with dominant frequency of 5 Hz.*

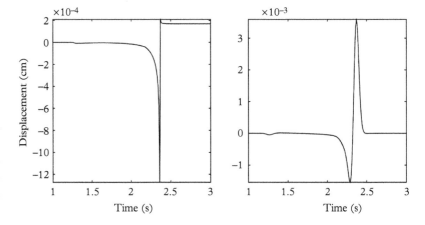

Johnson (1974). The analytical solution is part of the supplementary electronic material.

2.4.3 Internal boundaries

Inside the Earth we frequently have rapid changes of material parameters, often real material discontinuities (e.g. layers). Examples are stratifications in sediments, the crust–mantle (Moho) discontinuity, the core–mantle boundary, or the inner-core boundary. Due to lithostatic pressure such discontinuities can in most cases be treated as perfectly *welded* interfaces. In this case, displacement and tractions are continuous. At the interface between two media 1 and 2 this condition can be expressed as

$$\sigma_{ij} n_j^{(1)} = \sigma_{ij} n_j^{(2)} \tag{2.26}$$
$$u_i^{(1)} = u_i^{(2)}.$$

This is further illustrated in Fig. 2.17 with a curved interface. The good news is that this condition does not have to be explicitly implemented in all the numerical solution strategies we discuss. However, an important question that is still under debate is whether the actual location of such interfaces has to be honoured by a computational mesh, which might be tricky for complex structures (e.g. a sedimentary basin). On the other hand, recent work in the field of *homogenization* suggests that an efficient way of dealing with this problem might be to replace the discontinuity by a smooth structure (see Chapter 11).

There is one situation that might have to be treated explicitly and this is a fluid–solid boundary. Indeed, this is the approach taken by the *specfem3d* developers for global wave propagation to properly simulate the wave effects of the core–mantle boundary (e.g. Komatitsch et al., 2000*b*).

2.4.4 Absorbing boundary conditions

Most seismological applications (global wave propagation is an exception) require the simulation of wave fields in *limited areas* (e.g. reservoirs, continents, fault zones, volcanoes). This implies that at some point seismic waves emanating from a source will hit the boundary of a computational domain. Modelling reality would imply that waves (except at the free surface) are passing undisturbed through that boundary which can then be considered as *absorbing*. It turns out that this is a really tough problem for most numerical methods and is still the topic of ongoing research. Most approaches only solve part of the problem, absorb only in a certain frequency band or propagation direction, and some lead to additional computational instabilities. The art of efficient absorbing boundaries is a field of its own and therefore not covered in this introductory text. Suggestions for further reading are given at the end of this chapter.

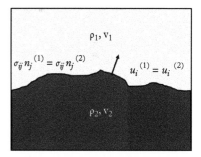

Fig. 2.17 *Material interface. Across an interface with changing geophysical parameters displacement components u_i and traction $t_i = \sigma_{ij} n_j$ are continuous. This is called a welded material interface.*

2.5 Fundamental solutions

What are the basic consequences of the equations of elastic motion?

2.5.1 Body waves

First of all, in the absence of any sources it turns out that there are two body-wave types that propagate in infinite homogeneous elastic media. These are compressional P-waves and transversely polarized shear (S-) waves with velocities

$$v_p = \sqrt{\frac{\lambda + 2\mu}{\rho}} \,, \quad v_s = \sqrt{\frac{\mu}{\rho}}. \tag{2.27}$$

When initializing seismic simulations, often the Lamé parameters need to be given. They are calculated from seismic velocities as

$$\lambda = \rho(v_p^2 - 2v_s^2) \,, \quad \mu = \rho v_s^2. \tag{2.28}$$

Examples for values in the Earth's crust are given in Table 2.2. A useful assumption for crustal velocities is that $v_p = \sqrt{3}v_s$ in which case $\lambda = \mu$. An important concept that we will extensively use to understand numerical approximations to the elastic wave equations is the mathematical description of *plane waves*. The most simple form for a scalar (e.g. acoustic) plane sinusoidal wave in 3D is

$$p(\mathbf{x}, t) = p_0 \sin(\mathbf{kx} - \omega t) = p_0 \sin(k_x x + k_y y + k_z z - \omega t) \tag{2.29}$$

where $p(\mathbf{x}, t)$ is the pressure at position $\mathbf{x}$, p_0 is the maximum amplitude, $\mathbf{k}$ is the wavenumber vector pointing in the direction of propagation, and ω is the angular frequency (see Fig. 2.18). The propagation velocity c is given by

$$c = \frac{\omega}{|\mathbf{k}|} = \frac{\lambda}{T}, \tag{2.30}$$

where $|\mathbf{k}|$ is the modulus of the wavenumber vector, $\lambda = 2\pi |\mathbf{k}|$ is the wavelength, and period $T = 2\pi/f$.

For elastic waves with three motion components, using the common complex exponential form, plane waves can be written as

$$u(\mathbf{x}, t) = \mathbf{A} e^{i(\mathbf{kx} - \omega t)}, \tag{2.31}$$

where $\mathbf{A}$ is the polarization vector, with $\mathbf{A} \| \mathbf{k}$ for P-waves and $\mathbf{A} \perp \mathbf{k}$ for S-waves.

2.5.2 Gradient, divergence, curl

At this point it is useful to muse about the vector field operators gradient ∇, divergence $\nabla\bullet$, and curl $\nabla\times$, and their connections to (simulated) elastic wavefields. The gradient of an elastic wave field is defined as

$$\nabla\mathbf{u}(\mathbf{x}, t) = \partial_j u_i(\mathbf{x}, t). \tag{2.32}$$

Table 2.2 *Examples of crustal seismic velocities and elastic parameters.*

Parameter	Value
v_p	6,000 m/s
v_s	3,464 m/s
ρ	2,500 kg/m^3
μ	3×10^{10} Pa
λ	3×10^{10} Pa

Fig. 2.18 *Plane waves. Illustration of planes of constant phase (e.g. pressure, displacement) travelling with constant speed in the direction of wavenumber vector **k**.*

The separation of this tensor into symmetric and anti-symmetric parts leads to the definition of the symmetric deformation (strain) tensor ϵ_{ij}

$$\epsilon_{ij}(\mathbf{x}, t) = \frac{1}{2}(\partial_i u_j(\mathbf{x}, t) + \partial_j u_i(\mathbf{x}, t)) \tag{2.33}$$

that is an integral part of the elastic wave equation. Note that today it is possible to observe strain components (Fig. 2.19), and/or derive the strain of a seismic wavefield from seismic array measurements. Strain is a linear combination of space-derivatives of the seismic wavefield. Therefore it can be derived by finite-differencing array observations of appropriate dimensions.

In isotropic media, elastic wavefields can be separated into a divergence-free part (S-waves) applying the curl to the wavefield

$$\frac{1}{2}\nabla \times \mathbf{u} = \frac{1}{2}\begin{pmatrix} \partial_y u_z - \partial_z u_y \\ \partial_z u_x - \partial_x u_z \\ \partial_x u_y - \partial_y u_x \end{pmatrix} \tag{2.34}$$

(implicit space–time dependence) and a curl-free part (P-waves) applying the divergence operator

$$\nabla \bullet \mathbf{u} = \partial_x u_x + \partial_y u_y + \partial_z u_z = \epsilon_{ii}, \tag{2.35}$$

where ϵ_{ii} is the trace of the strain tensor corresponding to volumetric change (the Einstein summation rule applies). When simulating seismic wavefields, usually the partial derivatives of the displacement fields are calculated anyway. Therefore, it is easy to output curl or divergence fields. For example, they can be used to separate P- and S-wave fields (e.g. for snapshots with different colouring of P and S).

The combined analysis of displacement field, and its strain components, divergence, and curl offers interesting opportunities for the seismic inverse problem for sources and structures, provided that they can be observed with sufficient accuracy. See the exercises at the end of the chapter for some theoretical examples. Rotational ground motions have been observed relatively recently using ring laser systems (Fig. 2.20, see Igel et al., 2005).

2.5.3 Surface waves

The presence of a free surface with appropriate stress-free conditions (see above) leads to an elastic half-space when the medium below the surface has homogeneous properties. Covering the theoretical background goes beyond the quick introduction this text aims for (see literature at the end of this chapter). In the following we review some of the main properties of surface waves that you are likely to encounter when simulating waves in 3D media with receivers at or near the surface.

Fig. 2.19 *Strainmeter at Piñon Flat Observatory, California. One (of three) leg(s) is shown. A light beam is sent through a tube 700 m in length, reflected at the end, and recombined with a reference beam to measure the change of length Δl. $\Delta l/l$ corresponds to the strain in this direction.*

Fig. 2.20 *Ring laser at Piñon Flat Observatory, California. This instrument measures the rotation (rate) around the vertical axis (vertical component of $\nabla \times v$, v being the ground velocity) by superposition of two counterpropagating mono-frequent laser beams (Sagnac effect).*

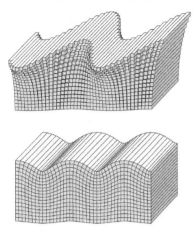

Fig. 2.21 *Surface waves. Illustration of the particle motion of seismic surface waves.* **Top:** *Love waves correspond to SH-waves (horizontally polarized shear waves). Their particle motion is perpendicular to the direction of propagation in the horizontal plane.* **Bottom:** *Rayleigh waves are polarized in the plane through source and receiver (corresponding to P–SV motion). Their particle motion is in general elliptical, and retrograde at the Earth's surface. The amplitude of both wave types decays with depth. From Shearer (2009).*

The presence of the free surface in a homogeneous elastic half-space—assuming incident plane P or SV waves—leads to one additional wave type: Rayleigh waves. They propagate in a horizontal direction, are polarized in the plane through source and receiver, and travel in a homogeneous half-space *without dispersion* with little less than shear-wave speed. Their amplitude decays exponentially with depth. As discussed above, Rayleigh waves in an elastic half-space can be generated by an impulsive vertical force acting on the free surface, providing an excellent reference to check the correct implementation of the free surface boundary condition.

Interestingly, no such waves exist in an elastic half-space for horizontal motion components! However, if we assume a low-velocity layer $v_1 < v_2$ of depth h below the surface, with velocity v_2 in the half-space below, we obtain further solutions—so-called Love waves with a fundamental mode of wave propagation with velocity in between v_1 and v_2. The polarization properties of these surface wave types are illustrated in Fig. 2.21. Love waves are always dispersive.

In layered Earth models Rayleigh waves also show dispersive behaviour. As the Earth's interior (on a global scale)—to first order—can be treated as a layered medium, it is useful to take a look at the velocities expected as a function of frequency. This is illustrated in Fig. 2.22. It is instructive to examine the arrival-time difference between long-period and short-period surface waves at long propagation distances (see exercises). The dispersion curves for Love- and Rayleigh waves illustrate the strong dependence of surface wave velocities as a function of period (or frequency). It is precisely this feature that dominates teleseismic broadband observations as indicated in the initial discussion of observations in Fig. 2.1. As we encounter the phenomenon of *physical* dispersion in the context of surface waves I would like to introduce the concept of *numerical* dispersion, which we will dwell on in more detail later. Numerical dispersion is an unwanted

Fig. 2.22 *Surface wave dispersion. In layered media, Love- and Rayleigh waves are dispersive. Their velocities depend on frequency. Phase and group velocities are shown as a function of frequency for a typical layered Earth model. Figure courtesy of Gabi Laske.*

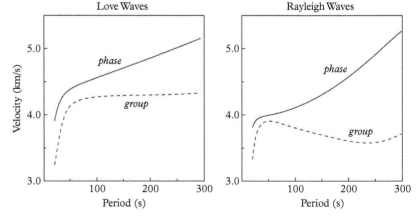

effect of discretizing a wave field and needs to be avoided at all costs. Physical and numerical dispersion are illustrated qualitatively in Fig. 2.23. The graph shows a classic example of Love-wave dispersion (simulated using the finite-difference method), obtained for a surface source in a low-velocity zone. Exact parameters are irrelevant for this demonstration. We note that low frequencies arrive earlier.

Exactly the same source receiver set-up for a higher-frequency source wavelet and an elastic half-space leads to numerical dispersion effects. Here also higher frequencies arrive later. This dispersive effect is entirely artificial and a consequence of a bad choice of simulation parameters. Note that sometimes these effects are hard to distinguish and one of the main goals of this volume is to provide guidelines on how to avoid these errors.

2.6 Seismic sources

In addition to the structural parameters of the Earth model, the physical description of the seismic source parameters will affect the resulting wavefield. In the following, we briefly review the most important concepts relevant for simulation tasks.

2.6.1 Forces and moments

As indicated in the elastic wave equations given above, seismic sources can be injected (1) as stress perturbations using the moment tensor $M_{ij}(\mathbf{x}, t)$ or (2) as external forces $f_i(\mathbf{x},t)$. The latter source type involves energy provided through some external processes (e.g. a hammer hitting the ground, or pressure sources induced by water waves). For seismology and seismic exploration the most important source types can be described using the moment tensor and this will be the focus of this section.

The second-order symmetric moment tensor has units of stress [Pa $= N/m^2$] with elements

$$\mathbf{M} = \begin{pmatrix} M_{11} & M_{12} & M_{13} \\ M_{21} & M_{22} & M_{23} \\ M_{31} & M_{32} & M_{33} \end{pmatrix}, \qquad (2.36)$$

each of which describes a double-couple force system as illustrated in Fig. 2.24. Before we give some examples, let us ask what the values of the moment tensor components mean.

To answer this question we need the concept of the *scalar seismic moment* M_0 that is defined as

$$M_0 = \mu A d, \qquad (2.37)$$

where μ is the shear modulus in the source area, A is the surface area of the rupturing fault plane, and d is the average slip on the fault. This simple equation

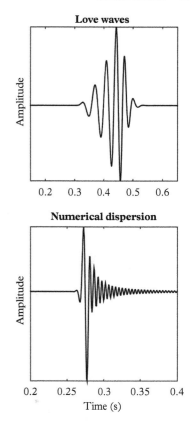

Fig. 2.23 *Physical vs. numerical dispersion. Dispersion is the frequency dependence of propagation velocities.* **Top:** *Physical dispersion of Love-waves for a half-space with a low-velocity zone at the top.* **Bottom:** *Numerical dispersion for the same source–receiver set-up due to the discretization of the wavefield with insufficient number of grid points per wavelength.*

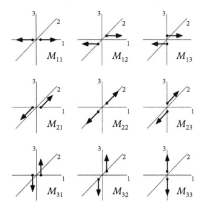

Fig. 2.24 *Seismic moment tensor. Illustration of the double-couple forces corresponding to the elements of the seismic moment tensor. From Shearer (2009).*

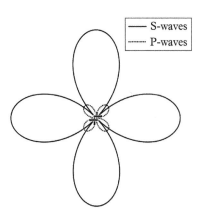

Fig. 2.25 *Far-field radiation pattern of a shear-dislocation source. The figure illustrates the strongly varying radiation of P-waves (dotted) and shear waves (solid) for a double-couple point source (indicated at the centre). Note approximately five times larger peak amplitude of S-waves compared to P-waves.*

is one of the most important results in earthquake seismology, providing the link between radiating wavefield, the size of an earthquake rupture, and—if the earthquake is large enough—geodetically observable displacements at the Earth's surface as a result of the static slip across the fault.

The scalar moment can further be obtained through the following relation:

$$M_0 = \frac{1}{\sqrt{2}} \left(\sum_{ij} M_{ij}^2 \right)^{1/2}. \tag{2.38}$$

The seismic moment determines the energy radiated from the seismic source. It scales the components of the normalized moment tensor that is responsible for the radiation pattern of P- and S-waves. For example, the case of non-zero components $M_{12} = M_{21}$ corresponds to a double-couple source acting across the plane $x = 0$ in y-direction or across the plane $y = 0$ in x-direction. This is the well-known ambiguity between fault and *auxiliary* plane. The resulting double-couple (*dc*) moment tensor is

$$\mathbf{M}^{dc} = \begin{pmatrix} 0 & M_0 & 0 \\ M_0 & 0 & 0 \\ 0 & 0 & 0 \end{pmatrix} = M_0 \begin{pmatrix} 0 & 1 & 0 \\ 1 & 0 & 0 \\ 0 & 0 & 0 \end{pmatrix}. \tag{2.39}$$

Another important moment source is an explosion. The moment tensor corresponds to equal forcing along the axis and has the diagonal form

$$\mathbf{M}^{expl} = M_0 \begin{pmatrix} 1 & 0 & 0 \\ 0 & 1 & 0 \\ 0 & 0 & 1 \end{pmatrix}, \tag{2.40}$$

and the moment M_0 has to be appropriately scaled. An important property of the elastic wave equation is the fact that it is *linear* with respect to the source properties. That means twice the scalar moment leads to twice the observed amplitude (in displacement or velocity). Obviously, a physical source like the double-couple shown above leads to a complicated radiation pattern of seismic body waves. This anisotropic radiation is illustrated for a double-couple point source in Fig. 2.25. Its mathematical form is presented in the next section.

As a first specific example of a simulation in homogeneous media we show snapshots of a displacement wavefield emanating from an explosive point source. The results are shown in Fig. 2.26. When starting with a community simulation code or developing your own code from scratch one should always check such simple model set-ups.

In this example—as expected—the radiation pattern is isotropic when the divergence of the wavefield is shown. However, note that it does not appear to be isotropic when the individual displacement components are visualized. The orientation of the induced motion also leads to polarity reversals depending on the quadrant of the wave field.

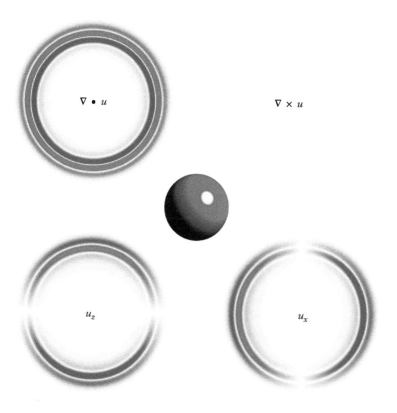

$\nabla \cdot u$

$\nabla \times u$

u_z

u_x

Fig. 2.26 *Seismic point sources. Elastic wave field for a point explosion source at the centre of the homogeneous model.* **Top left:** *Divergence.* **Top right:** *Curl.* **Bottom left:** u_z. **Bottom right:** u_x. *Note that the curl vanishes for an explosion source in homogeneous anisotropic media. Snapshots are calculated with a 2D finite-difference scheme. The central ball illustrates the isotropic radiation pattern, indicating equal extension (or compression) in each direction.*

Because we will later compare with a double-couple source, let us also show the curl of the wavefield that is exactly zero in the case of an explosion source that only generated P-waves. This might be trivial, but in fact there might be numerical artefacts that lead the curl to be non-zero. Therefore, it is useful to check this property.

To fully understand the effects of a seismic source described by the classic moment tensor we must have a look at the analytical solution to the double-couple point source in infinite media. This will be the topic of the next section.

2.6.2 Seismic wavefield of a double-couple point source

Much of seismology, e.g. the understanding of the kinematic properties of seismic sources, the linking of earthquakes to measurable crustal deformation, quasi-analytical solutions to wave-propagation problems, etc., is based on a fundamental analytical solution to the problem of a double-couple point source in infinite homogeneous media. It is an animal of an equation and quite hard to derive. However, if you start a complete 3D elastic wave simulation your solution contains all the elements of this fundamental solution. Therefore, bearing with me for some time on this will prove worthwhile. The focus is to understand the result, not present the full derivation. One could build a whole lecture around this equation! The interested reader is referred to Aki and Richards (2002).

What is the displacement wavefield $\mathbf{u}(\mathbf{x}, t)$ at some distance $\mathbf{x}$ from a seismic moment tensor source with $M_{xz} = M_{zx} = M_0$? This question leads to the following solution, the main aspects of which we will briefly discuss. The displacement $\mathbf{u}(\mathbf{x}, t)$ due to a double-couple point source in an infinite, homogeneous, isotropic medium is (e.g. Aki and Richards, 2002)[10]

$$
\begin{aligned}
u(x, t) = & \frac{1}{4\pi\rho}A^N\frac{1}{r^4}\int_{r/\alpha}^{r/\beta}\tau M_0(t-\tau)d\tau \\
& + \frac{1}{4\pi\rho\alpha^2}A^{IP}\frac{1}{r^2}M_0(t-\frac{r}{\alpha}) + \frac{1}{4\pi\rho\beta^2}A^{IS}\frac{1}{r^2}M_0(t-\frac{r}{\beta}) \\
& + \frac{1}{4\pi\rho\alpha^3}A^{FP}\frac{1}{r}\dot{M}_0(t-\frac{r}{\alpha}) + \frac{1}{4\pi\rho\beta^3}A^{FS}\frac{1}{r}\dot{M}_0(t-\frac{r}{\beta}),
\end{aligned}
\tag{2.41}
$$

with the radiation patterns A^N (near-field), A^{IP} (intermediate-field P-wave), A^{IS} (intermediate-field S-wave), A^{FP} (far-field P-wave) and A^{FS} (far-field S-wave):

$$
\begin{aligned}
A^N &= 9\sin(2\theta)\cos(\phi)\hat{r} - 6\cos(2\theta)\cos(\phi)\hat{\theta} - \cos(\theta)\sin(\phi)\hat{\phi} \\
A^{IP} &= 4\sin(2\theta)\cos(\phi)\hat{r} - 2\cos(2\theta)\cos(\phi))\hat{\theta} - \cos(\theta)\sin(\phi)\hat{\phi} \\
A^{IS} &= -3\sin(2\theta)\cos(\phi)\hat{r} - 3\cos(2\theta)\cos(\phi)\hat{\theta} - \cos(\theta)\sin(\phi)\hat{\phi} \\
A^{FP} &= \sin(2\theta)\cos(\phi)\hat{r} \\
A^{FS} &= \cos(2\theta)\cos(\phi)\hat{\theta} - \cos(\theta)\sin(\phi)\hat{\phi},
\end{aligned}
\tag{2.42}
$$

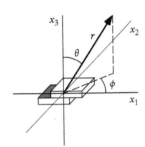

Fig. 2.27 *Spherical coordinate system. Coordinate system used for the analytical solution of a double-couple point source. From Shearer (2009).*

with the coordinate system shown in Fig. 2.27. Furthermore, in these equations r is the distance from the source, ρ, α, β are density, P-velocity, and S-velocity, respectively, and $M_0(t)$ is the source time function. We assume $\int \dot{M}_0 dt = M_0$ where M_0 is the scalar moment. Note that this implies that an arbitrary source time function s(t) can be initialized with

$$
\begin{aligned}
M_0 \int \dot{s}(t)dt &= M_0 \\
\int \dot{s}(t)dt &= 1.
\end{aligned}
\tag{2.43}
$$

The far-field terms of the above radiation patterns were used for the illustrations in Fig. 2.25.

Let us have a closer look at the various term in Eq. 2.41. First, we note that the simplest realistic source time function for an earthquake source is a stress drop occurring over a finite amount of time, the so-called rise time τ_r (see Fig. 2.28).

[10] Dots above symbols denote time derivative.

We start with the situation relevant for regional or global seismology, or if the ratio between propagated wavelengths and fault size is large. This is called the far-field (last line in Eq. 2.41).

In this case, we observe that the displacements for P- and S-waves have the same waveform, the time derivative of the moment source time function. That means that the ground is displaced with the corresponding direction of motion, but goes back to its original position. Another important result is that the far-field amplitude terms decay with $1/r$. This amplitude decay—the geometrical spreading—is the basis for the correction terms in the Richter magnitude scale. Furthermore, the amplitude factors reveal that shear wave displacements are approximately five times larger than P-wave displacements (highly relevant for the damage to buildings during earthquakes).

The important link between seismology and geodesy is made with the near and intermediate terms (top two lines of Eq. 2.41) that contain terms proportional to the moment source time function. This implies that there is a *static displacement* that remains when the source has finished acting, and the seismic wave field has passed through. These are the terms that—for large enough earthquakes—lead to the (sometimes substantial) crustal deformation that can be observed with GPS sensors (or, if possible, by integrating velocity seismograms).

The factors $\mathbf{A}^X$ are merely direction cosines that determine the radiation pattern of the various terms with maximum value 1. It is instructive to plot these terms to understand this directional behaviour (see exercises).

Let us give an example of how to connect earthquake properties with this equation and how it can be realized in a simulation task. The parameters for our earthquake simulation are given in Table 2.3. To obtain the scalar moment M_0 for an earthquake with moment magnitude $M_w = 5$ we use the empirical relationship

$$M_w = \frac{2}{3}(\log_{10} M_0 - 9.1). \tag{2.44}$$

Assuming a typical earthquake stress drop of $\Delta\sigma = 5MPa$ we can determine the fault radius r for a circular rupture through

$$\Delta\sigma = \frac{7M_0}{16r^3}, \tag{2.45}$$

by which the fault surface simply becomes $A_f = \pi r^2$. With shear modulus obtained from the information in Table 2.3 we can determine a static displacement d from the relation $M_0 = \mu A_f d$. In this example we obtain a slip $d \approx 25$ cm and with the assumption of a slip velocity of $v_{slip} = 1$ m/s we obtain a rise time of $\tau_r = 0.25$ s. With this information we can initialize a source time function $M_0(t)$.

A smooth source time function (integrated Gauss function) mimicking a rise time of approx. 0.25 s and the resulting analytical displacement seismograms are shown in Fig. 2.29. The seismograms show a complicated waveform that contain both P- and S-wave signals as well as near- and intermediate field terms that result in permanent displacements.

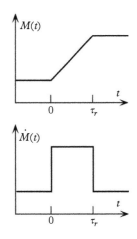

Fig. 2.28 *Source time function. The source time function $M(t)$ (top) and its time derivative $\dot{M}(t)$ with rise time τ_r. This source time function is descriptive of a stress drop occuring over a finite rise time. The rise time determines the slip velocity across the fault plane. From Shearer (2009).*

Table 2.3 *Parameters for model earthquake.*

Parameter	Value
M_w	5
M_0	4×10^{16} Nm
v_s	3,000 m/s
v_p	5,196 m/s
ρ	2,500 kg/m^3
t_{rise}	0.25 s
source	[0,0] km
receiver	[4,4,4] km

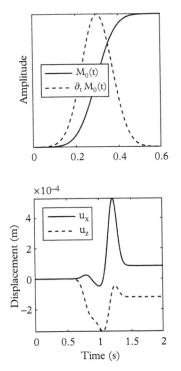

For any earthquake modelling it is useful to check simulation tools for such basic problems, in particular when synthetic seismograms are being compared with observations on absolute scales. We further illustrate the typical simulation results for such a problem in Fig. 2.30. In this 2D calculation, snapshots are shown for the v_x and v_z velocity wave fields emanating from an M_{xz} double-couple point source. Taking the curl and divergence of the wave field separates P- and S-waves. Note the occurrence of nodal lines with vanishing curl, divergence, or velocity components, and the polarity reversal on both sides of these nodal lines. The velocity snapshots also illustrate the dominant amplitude of the S-waves compared to P-waves. Despite the simplicity of this simulation set-up, it is useful to spend some time understanding the details of the results before moving to more complicated Earth and seismic source models!

The elastic wave equation has some very important properties that can be exploited for the simulation of seismic waves, and for both the inversion for source and structure. This is the topic of the following sections.

2.6.3 Superposition principle, finite sources

One of the most important properties of the elastic wave equation is its linearity with respect to the source terms, whether described by volumetric forces or a moment tensor. In fact, this property was already used to derive the geometrical

Fig. 2.29 *Analytical seismograms for a double-couple source in a homogeneous medium.* **Top:** *Source time function $M_0(t)$ (solid line) and its time derivative (dotted line).* **Bottom:** *Displacement for a receiver at 6.93 km distance from a $M_w = 5$ double-couple point source. For parameters see Table 2.3.*

Fig. 2.30 *Seismic point sources. Elastic wave field for a point double-couple source (M_{xz}) at the centre of the homogeneous model.* **Top left:** *Divergence.* **Top right:** *Curl.* **Bottom left:** u_z. **Bottom right:** u_x. *Snapshots are calculated with a 2D finite-difference scheme.*

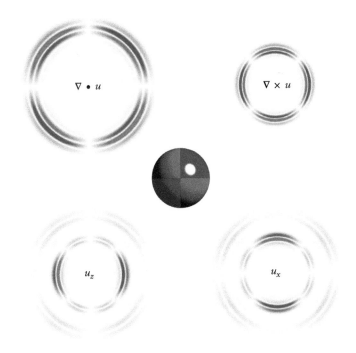

description of an earthquake source by a force double-couple. This linearity is the basis for the *inversion* of the moment tensor and finite-fault properties from observed ground motions. However, note that the inverse problem for finite source behaviour can be strongly nonlinear depending on observed data and source parametrization. With the prospects of solving the wave equation for some discretized Earth model, how can finite sources—at least in principle—be calculated?

A simple example of a realization of the superposition principle in connection with finite sources is given in Fig. 2.31. On a regular computational grid each grid point can be initialized with a time-dependent source. In this particular case, a subgrid of 8×8 points is assembled to form a sub-fault that breaks with the same temporal behaviour. Such a source initialization can be used to simulate waves for a large finite fault with complex kinematic rupture behaviour.

Mathematically, the superposition can be elegantly described in the frequency domain using the convolution theorem. Assuming a numerical solver that returns the Green's function G_{kl}^r in terms of ground velocity components v_l^r at one receiver point r, the complete velocity seismograms can be assembled by

$$v_l^r(\omega) = \sum_{k=1}^{N} \mathrm{slip}_k \exp[-i\omega t_k(c^{rup})] G_{kl}^r(\omega) S(R, \omega), \tag{2.46}$$

where the exponential term is merely a time shift t_k that depends on rupture speed c_{rup}, slip_k is the final slip of source k, and $S(R, \omega)$ is the spectrum of the source time function with R being the rise time. The theory of finite sources is described in detail in Aki and Richards (2002). The source description described above was used in a recent study by Bernauer et al. (2014) to simulate finite source earthquakes. In this case the numerical solver is used to calculate the (normalized) Green's function G_{kl}^r for each source–receiver couple. Once this is done, the above equation can be used to simulate arbitrary finite source scenarios.

For simulation and inversion tasks, the superposition principle offers another interesting opportunity. When sources add to observed seismograms in a linear way, why don't we add up all seismograms (from earthquakes or man-made sources) and use the summed data as *supershot* or *superdata* for the inverse problem for Earth's structure? This process is called *source stacking* or *source encoding* and is indeed a hot topic. Particularly for marine exploration problems where sometimes tens of thousands of sources are used, this is an attractive concept. Without using source stacking every source would require its own simulation which in 3D can be very time consuming. Therefore it is obvious that, when many sources are involved, a substantial amount of computation time could be saved. A recent example which also highlights some of the problems that appear is discussed in Schiemenz and Igel (2013).

In addition to the superposition principle there is another property of the wave equation that might substantially reduce simulation efforts of seismic observations.

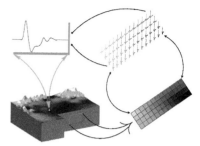

Fig. 2.31 *Finite source simulation by superposition. The superposition principle implies that seismograms for arbitrary finite sources can be obtained by adding up seismograms (numerical Green's functions) from a sufficient number of point sources appropriately timed and scaled to correctly reproduce the kinematic source behaviour. Equal colours denote subfaults and corresponding seismograms summed up to obtain the final seismogram. Figure from Wang (2007).*

2.6.4 Reciprocity, time reversal

The elastic wave equation is symmetric in time. In other words, wave propagation is reversible (we ignore here irreversible effects due to anelasticity). This has tremendous consequences for both forward and inverse modelling. Using the concept of Green's functions G_{ij} corresponding to the solution for a force in direction j recorded in displacement component i, this property can be expressed as

$$G_{ij}(\mathbf{x}, t; \mathbf{x}_0, t_0) \ = \ G_{ji}(\mathbf{x}_0, -t_0; \mathbf{x}, -t), \qquad (2.47)$$

where unprimed and primed variables correspond to source and receiver, respectively. This property holds if the surface around the volume is traction free. This concept is illustrated qualitatively with a simple simulation example in Fig. 2.32. A 1D space is initialized with a model with random velocity perturbations around a constant background model. We simulate acoustic wave propagation for a pressure source at point A recorded at receiver B and vice versa. The figure shows that both seismograms are identical. How can this be used in practice? An impressive example can be given when looking at a (now typical) marine exploration experiment. In many cases marine reservoirs are furnished with thousands of ocean-bottom geophones. To monitor temporal changes of the reservoirs during production, ships cruise several times a year over the area. They send signals hundreds of thousands of times, generating gigantic data volumes.

When faced with the task of simulating such an experiment one can now be clever, and, rather than simulating each source, reverse the problem and simulate each receiver (and record at each source). This leads to a tremendous speeding up of the process! In addition, using the superposition principle one could stack all sources into one supershot (also called source encoding). Unfortunately, some problems arise when doing that and care has to be taken. An application of this approach to exploration data is presented in Schiemenz and Igel (2013).

Reciprocity also allows the concept of *time reversal* (which is just another way of expressing reciprocity). Recorded seismograms from a point source can be reinjected—flipped in time at the receiver locations. The back-propagating wave field will focus in space–time at the original source coordinates. The quality of this focusing will strongly depend on the receiver coverage and, for real observations, on the knowledge of the Earth model. In practice this principle can be used to image the spatio-temporal behaviour of large earthquakes.

A simulation example is given in Fig. 2.33. A 2D finite-difference algorithm is activated with a source at the centre of a square domain. The wave field is recorded on a circular array surrounding the source point. It is recorded long enough such that the entire direct wave field has passed and the receiver points are at rest. In a subsequent simulation the stored seismograms are flipped in time and reinjected at the receiver points as sources. The many sources superimpose and part of the wavefield travels back to the source point and constructively interferes at the correct time (when the reversal wave propagation is carried out in the same

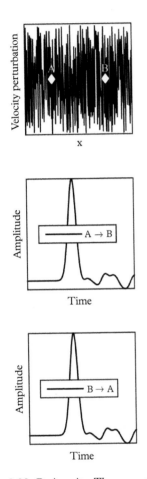

Fig. 2.32 *Reciprocity. The symmetry of the elastic wave equation with respect to time implies that for arbitrary velocity models a seismogram for a source at A recorded at B is the same as a seismogram for the same source at B recorded at A.*

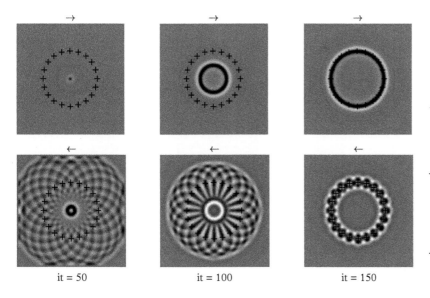

it = 50   it = 100   it = 150

Fig. 2.33 *Time reversal.* **Top row:** *Acoustic wave field at increasing iterations for a point source in the middle recorded in a circular array (crosses). The wavefield is calculated with a 2D finite-difference scheme.* **Bottom row:** *Seismograms recorded at the stations are flipped in time and re-injected as sources at the corresponding receiver locations. The back-propagating wavefield focuses at the original source location. This methodology is for example used to destroy kidney stones.*

model). Examples using 3D simulations for time-reversal studies are given in Larmat et al. (2006) and Kremers et al. (2011).

Structural imaging (reverse-time migration, full-waveform inversion using adjoint techniques) and source imaging are all based on reciprocity. Some aspects of this are further discussed in the application sections at the end of this volume.

2.7 Scattering

If you are planning to use numerical methods to simulate elastic wave propagation you definitely want to simulate waves through a more or less complicated heterogeneous Earth model (otherwise you could use analytical or quasi-analytical tools). How can we characterize heterogeneous Earth models? How do the properties of the Earth model to be simulated affect our choice of computational solutions? These questions leads us to the problem of *scattering*, a topic (again) that could be extended to an entire volume, but we will only scratch at the surface (see recommendations at the end of this chapter).

Any energy propagating in an elastic medium is affected when it encounters a change in medium properties. The most important notions in this context are the so-called *correlation length a* of the medium changes and the (let us say dominant) wavelength λ of the propagating wavefield. The correlation length is a somewhat strange concept and sometimes it might not even exist, but for the moment let us assume that it represents the dominant spatial wavelength of the medium perturbation you want to investigate.[11] Mathematically, the correlation length can be determined by looking at the central part of the autocorrelation function of

[11] At this point in my lecture I usually refer to concert halls. Sometimes concert halls contain columns of a certain diameter. Never sit behind these columns as the wavelength of audible frequencies might well be of that order. In that sense the diameter would be the relevant correlation length.

Fig. 2.34 *Structural heterogeneities. Illustration of heterogeneity classes that impact the choice of specific numerical solvers for seismic wave-propagation problems. The grey scaling indicates variations in seismic velocities (or density, or elastic parameters). **a**: Layered media; **b**: layered media, laterally heterogeneous; **c**: single scattering object; **d**: medium with cavity (boundary conditions apply); **e**: smoothly varying heterogeneities; **f**: heterogeneities on all scales.*

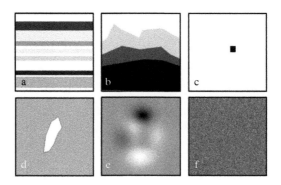

medium perturbations. In Gaussian and exponential media the correlation length a is related to the dominant wavelength of the heterogeneities in the medium.

Let us have a look at some classes of structural heterogeneities and discuss some of the implications for simulations. Examples are given in Fig. 2.34. A layered medium (see Fig. 2.34(a)) is an extremely important class of models. Sometimes it is sufficient to model observations (e.g. simple sedimentary structures, geotechnical problems). For these model classes quasi-analytical schemes (such as the reflectivity method, or normal mode solutions for global wave propagation) exist and one would normally not choose numerical methods for their solution. However, because of these quasi-analytical reference solutions, layered structures represent important benchmark problems against which purely numerical solutions can be tested.

Once layered models become laterally heterogeneous, quasi-analytical methods fail and one has to use numerical solutions. The graph in Fig. 2.34(b) indicates that the material changes are abrupt. It is important to realize that this implies that infinite spatial frequencies make up the model (a fundamental property of δ or step functions). In general, internal boundary conditions are not explicitly implemented; however, this layered model raises the question whether a computational grid needs to follow the interfaces or not (more on this later).

Sometimes one is interested in understanding the scattering behaviour of single objects, as in Fig. 2.34(c). Analytical solutions exist for simple structures; otherwise numerical solutions can be used. If the scattering object inside the medium is an empty cavity (Fig. 2.34(d), e.g. from an explosion, mining structures) special stress-free boundary conditions might apply. Then computational meshes have to be specifically designed to follow the cavity shape. Smoothly varying velocity models as indicated in Fig. 2.34(e) are usually well suited for modelling with numerical tools. Models with all spatial wavelengths involved (Fig. 2.34(f)) might be problematic. For all models, the question of what seismic wavelengths will propagate through the medium is essential. In many cases, it might not be necessary to characterize the medium with high spatial wavenumbers to obtain accurate solutions. In other words, a discontinuous (e.g. layered) medium could be replaced by a smooth version with (almost) no difference to the seismogram obtained for the original model. This is the problem of

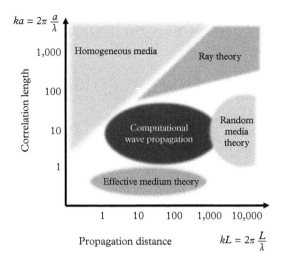

$ka = 2\pi \dfrac{a}{\lambda}$

Correlation length

1,000   Homogeneous media   Ray theory

100

10   Computational   Random
          wave propagation   media
                                theory

1

          Effective medium theory

          1   10   100   1,000   10,000

Propagation distance $kL = 2\pi \dfrac{L}{\lambda}$

Fig. 2.35 *Phase space of scattering problems. Scattering problems characterized by correlation length a and propagation distance L scaled by wavelength λ (after Aki and Richards, 1980). The computational method of choice for a particular seismic wave simulation problem depends strongly on the scattering property a/λ and the propagation distance in terms of wavelengths λ. Numerical methods are required for the strong scattering regime (a ≈ λ) but are restricted to some maximum propagation distance due to limits of computational resources. See text for details.*

homogenization, one of the hottest topics in computational wave propagation today (see Chapter 10 on applications).

Finally, let me present a way to characterize the *phase space* of scattering problems. This concept goes back to Aki and Richards (1980) (my favourite figure in this epic book that did not make it into the second edition). Fig. 2.35 allows us to understand a simulation (or scattering) problem in terms of propagation distance (in terms of wavelengths) and correlation length (scaled by wavelength). In common terms, scattering effects are strong when the wavelength λ and the size of the material perturbation, here characterized by autocorrelation length *a*, are about the same (so around 1 on the vertical axis).

When the scatterers are very small, the medium can be replaced by an *equivalent* homogeneous medium (e.g. homogeneous anisotropic medium due to small cracks). On the other hand, if the medium pertubartions are very long wavelength ($ka >> 1$), they can be ignored and the medium—at least locally—can be considered as a juxtaposition of homogeneous parts.

In an intermediate area ray theory can be applied. In this case $\lambda << a$, the waveform is not altered by the medium perturbations, but the ray path and arrival times are affected. In the scattering regime $\lambda \approx a$, when waves propagate many wavelengths, numerical methods are too expensive and wave propagation can be treated by special theories (e.g. radiative transfer theory). Finally, the domain where numerical simulations play the most important role is the scattering regime $\lambda \approx a$, when the propagation distance is not too large. Anything beyond O(100) wavelengths in 3D is still a challenge today.

It is instructive to review recent (and old) literature and place the simulation tasks into the phase space of Fig. 2.35 (see exercises). Comparing such points for simulation tasks 20 years ago and today, it is a little frustrating that we have not moved very far (but, yes, it is a logarithmic scale!).

2.8 Seismic wave problems as linear systems

In the sections above we have already made use of the concept of Green's functions. Green's functions $G_{ij}(\mathbf{x}, t; \mathbf{x}_0, t_0)$ are the impulse response for a source at $\mathbf{x}_0$ at time t_0 in directon j recorded for ground motion component i at point $\mathbf{x}$. Green's functions contain all information on a *linear system*, and the partial differential equation describing elastic wave propagation (and the Earth for that matter) can be treated as a linear system. While this property is used extensively for the mathematical description of wave propagation and finding analytical solutions, here, we would like to demonstrate the consequences for numerical solutions and simulation studies. There is no such thing as an exact (delta-like) impulse in nature. Therefore, we want to know Earth's response for a reasonable arbitrary source time function, which we denote as $s(t)$.

It is a well-known result that the response to a linear system for an arbitrary input can be obtained by convolving the (arbitrary) input with the Green's function. Thus, in simple mathematical terms

$$u_i(\mathbf{x}, t) = G_{ij}(\mathbf{x}, t, \mathbf{x}_0) \otimes s(t), \tag{2.48}$$

where $\otimes$ denotes convolution, the source time $t_0 = 0$.

According to the convolution theorem, this corresponds to a multiplication in the spectral domain; thus

$$u_i(\mathbf{x}, t) = \mathscr{F}^{-1}[G_{ij}(\omega)S(\omega)], \tag{2.49}$$

where $\mathscr{F}^{-1}$ denotes the inverse Fourier transform and $S(\omega)$ is the (complex) source spectrum.

These relations also hold for the numerical solutions to the elastic wave equations. However, we cannot expect that the Green's function G_{ij} will be evaluated accurately for all frequencies (note that the spectrum of the impulse δ-function is white and contains all frequencies). Denoting the numerical solutions by the ~ sign we obtain

$$\tilde{u}_i(\mathbf{x}, t) = \tilde{G}_{ij}(\mathbf{x}, t, \mathbf{x}_0) \otimes s(t), \tag{2.50}$$

where $\tilde{G}_{ij}(\mathbf{x}, t, \mathbf{x}_0)$ is a Green's function calculated using a numerical solver. Again, this convolution corresponds to a multiplication in the spectral domain; thus

$$\tilde{u}_i(\mathbf{x}, t) = \mathscr{F}^{-1}[\tilde{G}_{ij}(\omega)S(\omega)]. \tag{2.51}$$

We will illustrate this with a simple example using the 1D acoustic wave equation and discuss the potential of this relation. In Fig. 2.36 a simulation problem is set up in a homogeneous acoustic 1D medium. The analytical response of an impulse injected at some source point is a step function shifted by $\Delta\,c$ where Δ is the distance between source and receiver and c is the acoustic velocity.

Initializing a numerical solver (here: a basic finite difference solution to the acoustic 1D wave equation as discussed later in the volume) will result in the numerical Green's function $\tilde{G}(t)$ shown in Fig. 2.36. This raw result is useless as it contains numerical artefacts. However, the fact that our numerical solver is also a *linear system* allows us to perform the convolution *after* the simulation. This is demonstrated in the bottom figure. Both convolving the raw numerical or analytical Green's function with the desired source time function s(t), and directly injecting the source time function s(t) in the simulation lead to the same result, provided that the frequency range of the source time function is chosen such that the final result is accurate (more on this later).

While this might seem a technical aspect, it has tremendous practical consequences. Think of a situation in which you would like to investigate the frequency-dependent effects of waves through a random velocity model. The above properties imply that you can do this with a single simulation, altering the frequency content later, in the convolution step. Other potential applications are numerical Green's function data bases for subfault systems. In principle, by appropriate superposition and convolution, arbitrary finite source scenarios can later be assembled without further simulations. Another powerful example of this concept is the recently published *Instaseis* project providing high-frequency global seismograms almost instantaneously using pre-calculated numerical Green's functions (van Driel et al., 2015*b*).

2.9 Some final thoughts

The material presented in this chapter is intended to motivate you to (1) stick to some simple problems when you start using a numerical solver for wave-propagation problems, and (2) play around with some of the powerful functionalities (e.g. reciprocity, superposition, linearity, convolutions) of numerical simulations before working on more complicated problems. Trust me; this recommendation comes from experience with many students and researchers, who sometimes too quickly dive deep into the numerical simulation realm, and are surprised by the strange results they obtain. Here is a checklist of possible points to consider when trying to understand whether a solver is returning correct answers (often these points can be answered by looking at wavefield snapshots):

- Are P- and S- arrival times correct?
- Does your seismogram correctly reflect your input signal (for example in homogeneous media)? Careful: sometimes there is an integration, sometimes there is a derivative involved.

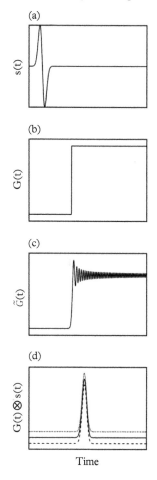

Fig. 2.36 *Green's functions and simulations. Illustration with a numerical approximation of the 1D acoustic wave propagation using the finite difference method. From top to bottom:* **a:** *(Arbitrary) source time function s(t).* **b:** *Analytical Green's function G(t).* **c:** *Numerical Green's function $\tilde{G}(t)$ obtained with a numerical solver.* **d:** *Comparison of analytical solution (solid), numerical Green's function convolved with s(t) (dashed), and numerical solution initialized with s(t) as source time function (shifted for illustration purposes). The results are identical.*

- Is there an analytical solution you can compare your results with (homogeneous medium, Lamb's problem)?
- Is the radiation pattern according to your specific source model (explosion, double-couple)?
- Is the wavefront shape correct?
- Is the relative amplitude of waves (P, S, surface waves) according to expectations?
- Is the particle motion (i.e. polarization) compatible with the wave type?
- Are the polarities of the waves correct in the various directions?

Many of these aspects can be tested with the Python programs provided in the supplementary material.

Chapter summary

- Seismic wave propagation consists of oscillatory phenomena. The wavelengths involved are usually quite small compared to the total physical domain under consideration (planets, continents, reservoirs, volcanoes, rock samples). This implies challenging problems when discretizing the internal structures.
- The relation between wavelength λ and (dominant or maximum) frequency f of the seismic wavefield $c = \lambda f$, where c is phase velocity, is important to understanding the spatio-temporal scales for a seismic simulation problem.
- Seismic wave propagation is governed by a vectorial partial differential equation with the displacement- (or velocity) field as the unknown. Analytical solutions only exist for simple (homogeneous) media.
- Body waves (P- and S-waves) with corresponding polarization and velocities are solutions to the elastic wave equation in homogeneous full space. Surface waves (Rayleigh and Love) are solutions to elastic half- (or layered half-) spaces, respectively.
- The most important physical boundary condition for seismic wave propagation is the stress-free surface boundary condition.
- Various rheologies (e.g. isotropic, anisotropic, viscoelastic, poroelastic) determine the stress–strain relationship.
- Sources of seismic waves can be described with force terms f($\mathbf{x}$,t) or the seismic moment tensor $\mathbf{M}(\mathbf{x}, t)$.
- The superposition principle implies that any finite (or distributed) source can be considered an integral (or sum) over point sources.

- The elastic wave equation is symmetric in time. Sources and receivers can be interchanged. The seismograms are the same.
- The fact that the wave equation can be treated as a linear system has powerful implications for simulation problems.

....

FURTHER READING

- Shearer (2009) provides a very comprehensive introduction to the concepts of seismic wave propagation and earthquake sources at a basic level.
- Stein and Wysession (2003) is another introductory textbook to seismology with a nice section on linear inverse problems.
- Aki and Richards (2002) is the theoretical seismologist's bible. For any in-depth discussion on the theoretical derivations of many results, refer to this book.
- Ben-Zion (2003) provides some of the most important equations for seismology with a focus on results relevant for earthquake physics.
- Kennett (2001) (and the subsequent volume) discusses seismic wave propagation with a strong data point of view.
- Moczo et al. (2014) is a book on the finite-difference method but provides a detailed description of the governing equations for a variety of rheologies. They also cover the problem of absorbing boundary conditions.
- Carcione (2014) provides an in-depth discussion of wave propagation for various rheologies and a discussion of several numerical methods (including absorbing boundary conditions).
- Sato et al. (2012) is a compilation of articles in the field of seismic scattering.
- Rienstra and Hirschberg (2016) develops the analytical solutions for the acoustic wave equation (and others) in a comprehensive way.

....

EXERCISES

Comprehension questions

(2.1) Search for recent or current research projects using seismic methods (e.g. exploration seismics, global seismology, volcanology, laboratory studies, geotechnical problems). Collect information on frequency ranges and size of sensor networks, and discuss consequences for seismic simulation problems.

(2.2)   Search for papers with seismic simulations, and extract information on propagation length and scattering properties of the Earth models. Place these parameters in the phase space of scattering problems shown in Fig. 2.35.

(2.3)   What geophysical parameters make up the Earth model when the isotropic (anisotropic, viscoelastic) wave equation is used?

(2.4)   Seismic velocities are functions of inverse density. Doesn't that mean that the denser the medium, the slower seismic velocities are? Explain!

(2.5)   What is the difference between the *velocity–stress* and the *displacement–stress* elastic wave equation? Is the solution the same?

(2.6)   Describe the various rheologies for seismic wave propagation. How would you rate them in terms of modelling real seismic observations?

(2.7)   What is reciprocity? How can this principle be used in seismic wave problems?

(2.8)   What is time reversal in the context of the wave equation? Find applications in seismology and medicine.

(2.9)   Explain qualitatively the physical model for an earthquake point source. What parameters would you expect in a file that initializes an earthquake simulation? Are the point source properties uniquely defined given seismic observations?

(2.10) Explain the concept of wave dispersion using Love and Rayleigh waves. Describe their dispersive behaviour for various basic Earth models (half-space, layered half-space).

(2.11) What boundary conditions are relevant for seismic simulation problems? Give examples.

(2.12) Explain the vector wave field operators gradient, divergence, and curl for seismic wave simulations (and observations). Why are seismic array measurements relevant in this context?

(2.13) Explain the concept of linear systems, convolution, the convolution theorem, and its relevance for wave simulations. What is the difference between analytical and numerical Green's functions?

Theoretical problems

(2.14) Get information on the PREM model for global Earth structure (see Fig. 2.2) and calculate the maximum and minimum seismic wavelengths (P- and/or S-waves) for frequencies $1.0, 0.1, 0.01$ Hz. Where do they occur?

(2.15) Using the basic form of the 3D isotropic elastic wave equation, derive the 2D version by assuming invariance of all fields in y-direction. Show that you obtain two independent (sets of) equations.

(2.16) Write out all components of the 3D isotropic elastic wave equation in u_x, u_y, u_z in the displacement formulation. Follow the strategy presented in Eq. 2.3.

(2.17) Inject the trial solution $p(x, t) = p_0 e^{i(kx-\omega t)}$ into the source-free 1D acoustic wave equation $\partial_t^2 p = c^2 \partial_x^2 p$. Discuss the solution.

(2.18) Show that $p(x, t) = f(x - ct) + f(x + ct)$ is a general solution to the wave equation $\partial_t^2 p = c^2 \partial_x^2 p$. Discuss the result.

(2.19) Assume two monochromatic plane waves propagating in x-direction: (a) P-wave $u_x = A_x \sin(kx-\omega t)$ and (b) S-wave $u_y = A_y \sin(kx-\omega t)$. Calculate in both cases the elements of stress and strain tensors. Assume that it is possible to observe the z-component of the curl $\nabla \times \mathbf{u}$. The rotation rate around a vertical component is given as the time derivative of the curl applied to the displacement field. How is the vertical component of rotation rate related to the transverse acceleration $\partial_t^2 u_y$ (S-wave)? Would the P-wave contribute to the curl? Discuss the potential of this result.

(2.20) Follow the approach described in the previous exercise. Find a way to obtain phase velocity from collocated measurements of strain and displacement (or velocity or acceleration) from body waves in an infinite space. Which components do you have to combine?

(2.21) The 2003 Hokkaido earthquake (M8.1) led to a maximum horizontal displacement of 1.5 cm for Love waves for an approximately 25-second period recorded in Germany. Estimate the maximum dynamic strain induced by the passing wavefield for a horizontal phase velocity of 5 km/s.

(2.22) Show that for attenuation the relation for the amplitude decay $A(t) = A_0 e^{-\frac{\omega t}{2Q}}$ holds if $\delta = \ln(A_1/A_2)$ relates two subsequent amplitudes, $Q = \pi / \delta$ and the wave propagates one cycle.

(2.23) What is the ratio between maximum S- and maximum P-wave amplitudes in the far field of a homogeneous medium for a double-couple point source? Use Eq. 2.41 and discuss implications for engineering seismology.

(2.24) Estimate the difference of arrival times for Love and Rayleigh waves propagating at various periods [T = 50 s, 200 s] to a distance of 10,000 km. Refer to Fig. 2.22.

(2.25) Show (e.g. graphically) that

$$\delta_a(x) = \frac{1}{\sqrt{2\pi a}} e^{-\frac{x^2}{2a}} \qquad (2.52)$$

converges to a δ-function as $a \to 0$. Show that $\int \delta_a(x) dx = 1$ for any a.

Computational exercises

(2.26) Write a computer program that uses vertical incidence reflection and transmission coefficients (ignore multiples) to calculate Green's functions for a 1D model with a few layers. Apply the convolution model to the Green's function and calculate synthetic seismograms convolving the Green's function with a source time function (e.g. a Gaussian) according to Eq. 2.48. Discuss the results.

(2.27) Use Eq. 2.41 to write a program for *far-field* Green's functions in arbitrary directions. Investigate the radiation pattern and polarization behaviour of body waves.

(2.28) Plot the 3D radiation patterns A^x for P and S far-field energy in Eq. 2.41.

(2.29) Write a program that plots the scalar moment M_0 as a function of energy magnitude M_w (Eq. 2.44).

(2.30) Stress drops $\Delta\sigma$ usually vary between 1 and 10 MPa. Use the relation between stress drop, scalar moment, and (circular) rupture radius to plot the expected radii for varying magnitudes for given stress drop. Carefully check physical units!

(2.31) Write a computer program to check the reciprocity principle with the far-field solutions of Eq. 2.41.

(2.32) Write a computer program that initializes a random 2D velocity perturbation by spatially low-pass filtering random numbers using transform methods.

Waves in a Discrete World

3

In the last chapter we discussed the physics of wave propagation, seismic sources, and some of the phenomena to be expected when seeking solutions to seismic wave-propagation problems. Wave propagation—like most other phenomena in physics—is well described by partial differential equations defined in a continuous world. That is fine, as long as we find analytical solutions to the problems we pose, for example, how waves propagate in infinite homogeneous media.

Obviously, if we want to get synthetic (or theoretical) seismograms for arbitrary Earth models, this approach does not work. We need to find alternative strategies and solve the problem using a (possibly large) computer. This implies that we have to move to the *discrete* world. In other words, anything we describe (e.g. an Earth model with seismic velocities, a displacement field) will have a finite *number of degrees of freedom.*[1]

The fact that we will operate in the discrete world raises a lot of questions. How will we describe space- or space–time-dependent fields in a fully discrete way? What is the impact of the dimensionality (1D, 2D, or 3D) of our problem to its numerical solution? What strategies exist for the initialization of computational meshes? How do we deal with problems in various geometries (e.g. Cartesian, cylindrical, spherical, arbitrary). And finally, how are large-scale problems solved on modern (parallel) computers?

Many of these questions are relevant for the understanding of the specific numerical method applied to the seismic wave-propagation problem. Therefore, we briefly illustrate the concepts behind these issues and indicate their connection with the methods discussed in the following chapters.

3.1 Classification of partial differential equations

The general properties of specific partial differential equations are extremely important for finding efficient numerical solutions. Therefore, let us briefly investigate the properties of our elastic wave equation focusing on the 1D acoustic example. Assuming two independent variables space x and time t and x-dependent variable $c(x)$ (acoustic velocity), the source-free partial differential equation (*pde*) for the acoustic wavefield $p(x, t)$ reads

$$p_{tt} - c^2 p_{xx} = 0, \qquad (3.1)$$

where (only in this section) the lower indices imply differentiation.

[1] Examples: If you parametrize a homogeneous half-space with P- and S-velocities, then that number is 2. Parametrizing a reservoir with $1{,}000 \times 1{,}000 \times 1{,}000$ grid points implies 2×10^9 *d.o.f.* (nothing unusual these days!).

Computational Seismology. First Edition. Heiner Igel.
© Heiner Igel 2017. Published in 2017 by Oxford University Press.

First, we note that this *pde* is *linear*, which is the case when all terms involving *p* (and its derivatives) can be expressed as a linear combination. Another requirement for linearity is that the coefficients (here: $c(x)$ and unity) are independent of *p*. The coefficients may depend on the independent variables (here: *x*). In our problem this will certainly be the case: The coefficients of the wave equation describe Earth's elastic properties and will vary as a function of space. The fact that the wave equation is linear is tremendously important as it allows us—at least under some circumstances—to come up with analytical solutions. We have seen some of those already in the previous chapter and they are very important to verify the accuracy of the numerical solutions. This is in general not the case for nonlinear *pde*s.

Second, *pde*s are classified with respect to the highest-*order* derivative that appears in the equations. In our case this is second order. It turns out that the wave equation can also be written as a first-order system of equations that is formally equivalent (this will be shown later in connection with the finite-volume method) to the advection (or transport) equation, here shown in the homogeneous scalar 1D case

$$p_t + cp_x = 0, \tag{3.2}$$

where space–time dependence is implicit. Furthermore, partial differential equations are classified by analogy with conic sections. Writing the general form of a linear, second-order equation in x, t as

$$Ap_{xx} + Bp_{xt} + Cp_{tt} + Dp_x + Ep_t + Fp = 0, \tag{3.3}$$

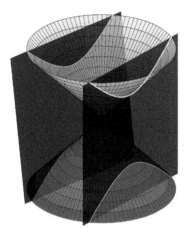

Fig. 3.1 *Hyperbolic conic sections. Partial differential equations are classified by analogy with conical sections. The elastic wave equation is of hyperbolic form. A hyperbola (white lines) is obtained when a vertical plane cuts a cone.*

where the capital letters are the coefficients, the partial differential equation can be transformed (e.g. by a Fourier transform) into an analogous form corresponding to conical sections such that

$$Ax^2 + Bxt + Ctt + Dx + Et + F = 0, \tag{3.4}$$

where the independent variables take on another meaning (e.g. wavenumber and frequency). To classify the *pde* one needs to calculate the *discriminant* $B^2 - 4AC$ to find the category, where

$$\begin{aligned} B^2 - 4AC = 0 &\rightarrow \text{parabolic} \\ B^2 - 4AC < 0 &\rightarrow \text{elliptic} \\ B^2 - 4AC > 0 &\rightarrow \text{hyperbolic.} \end{aligned} \tag{3.5}$$

With these classes it is easy to see that Eq. 3.1 is **hyperbolic** for all possible coefficients c(x). This is also true for the complete vectorial wave equation. The wave equation is *the* classic hyperbolic *pde*. Once initial conditions for the field and its time derivative—for example, $p(x, t = 0)$ and $\partial_t p(x, t = 0)$—are specified, the solution at all times is fixed (in the absence of any further input). The solutions to hyperbolic equations are wavelike and disturbances travel with finite propagation speeds. This distinguishes them from elliptic and parabolic problems, where perturbations of initial conditions or boundaries have an immediate effect everywhere.

What about the first-order advection Eq. 3.2? In fact, the second-order wave equation can be obtained from the first-order advection equation by a few simple steps (see chapter on the finite-volume method). As hyperbolic problems appear everywhere in physics, numerical schemes for their solution can be transferred from other areas of physics (or from engineering) to seismology. Let us have a look at some fundamental concepts concerning numerical solutions to partial differential equations.

3.2 Strategies for computational wave propagation

Our problem is finding numerical solutions to wave equations as indicated by the two examples given above. It is useful to consider the space- and time-derivatives of these equations separately to understand and categorize the methods we will encounter. Luckily, as our wave equations are linear, we can write them as

$$\partial_t^2 p(x, t) = L(p, t) \;\; \rightarrow \; L(p, t) = c(x)^2 \partial_x^2 p(x, t)$$
$$\text{or} \hspace{4cm} (3.6)$$
$$\partial_t p(x, t) = L(p, t) \;\; \rightarrow \; L(p, t) = c(x) \partial_x p(x, t),$$

where $L(.)$ denotes a linear operator, and space–time dependencies are added for clarity. Note that the right-hand sides contain only spatial derivatives, or space-dependent functions like $c(x)$. Assuming that the space-dependent fields p and c can be discretized appropriately in space for a given time, this formulation can be termed a *semi-discrete* scheme.

An important statement at this point is that all the numerical methods presented in this volume—despite their sometimes fundamentally different underlying mathematical concepts—differ primarily in the way the right-hand side of the above equations are treated! With the appropriate initial condition for a wave-propagation problem (everything is at rest), the left-hand side becomes an extrapolation problem that will always be solved by a finite-difference-type approximation. In that sense the right-hand side can be considered a general interpolation problem.

Because one of the main distinctions between the various methods is the way the space-dependent fields are described, we will have a closer look before diving into the details. Durran (1999) distinguishes two basic strategies for spatial discretization: the *grid-point* method and the *series-expansion* method. We illustrate these strategies in Fig. 3.2.

Fig. 3.2 *Basic spatial discretization schemes.* **Top left:** *Grid point approximation of a function* f(x) *(e.g. finite-difference method).* **Top right:** *Function approximation by sum over trigonometric functions (indicated by dotted lines, normalized and shifted). The original discrete function* f(x)*(dashed line) is replaced by an approximation (solid line) that* exactly *interpolates f(x) at the grid points (dots), e.g. pseudospectral methods.* **Bottom left:** *Space is divided into elements. Inside these elements the function* f(x) *(solid line) is approximated by a linear function (dashed line), continuous across element boundaries (e.g. finite-element method).* **Bottom right:** *Space is divided into finite volumes (cells) and the function* f(x) *is approximated by the average value (located at the cell centre indicated by a dot), e.g. finite-volume method. Note the occurrence of discontinuities across cell (volume) boundaries.*

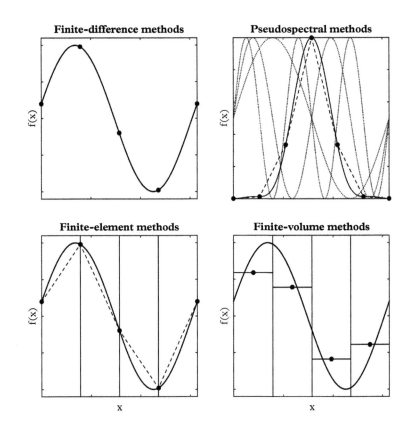

The grid-point method approximates an arbitrary function f(x) at a discrete set of points (Fig. 3.2, top left) and only there. This is the principle used in the finite-difference method. Values are usually not required in between grid points, but if they were, they would have to be obtained by interpolation.

An entirely different strategy is to approximate a function by a sum over some basis functions (e.g. a Fourier series expansion) on the entire domain (Fig. 3.2, top right). For certain grid-series combinations (e.g. regularly spaced grid points and Fourier series) it turns out that an arbitrary function can be *exactly* interpolated (together with its derivatives) at the grid points. That is a cool property also used inside elements in the spectral-element or the nodal discontinuous Galerkin method. When the calculation of derivatives is carried out using Fourier transforms we speak of the Fourier pseudospectral method.

In finite-element-type methods (Fig. 3.2, bottom left) space is divided into elements (possibly of varying size) and in the most basic approach the solution fields are approximated by linear functions inside the elements that are continuous across the element boundaries. Higher-order polynomial approximations (e.g. quadratic functions or Lagrange polynomials) are also possible, leading to more accurate representations.

Another possible discretization is based on defining finite volumes or cells in which the average value of a function is defined (Fig. 3.2, bottom right). Inside the cells higher-order approximations (linear, polynomial) are also possible. However, a fundamental difference is the fact that the approximate fields inside the cells are discontinuous across the cell boundaries. This requires the definition of fluxes as information needs to be exchanged between cells. This discretization scheme is used both in the finite-volume method as well as in the discontinuous Galerkin method.

In the following sections we will investigate what discrete computational models can look like in various dimensions.

3.3 Physical domains and computational meshes

We perceive our world as a 3D space and describe its properties with functions of coordinates x, y, z or r, θ, ϕ. Therefore, it seems natural to discretize a physical domain for a wave-propagation problem in 3D. Obviously, this is computationally the most expensive option. Sometimes a problem can be reduced to 2D (rarely to 1D) with tremendous benefits for computational costs. While the reduction to lower-dimensionality seems trivial there are a few things to be aware of, in particular if one wants to compare synthetic seismograms with observations.

3.3.1 Dimensionality: 1D, 2D, 2.5D, 3D

The bulk of this volume is dedicated to the presentation of various numerical solutions to the wave equation in 1D (with an exception in the chapter on finite differences, with 2D examples). The reason is that once the concepts are understood in 1D they can be extended to higher dimensions in a straightforward way.[2]

The field of computational seismology started some decades ago with 2D simulations, simply because at that time computational resources did not allow 3D calculations. Let us illustrate some of the consequences by looking at the graphs in Fig. 3.3. Nothing much needs to be said about 1D calculations. They have merely tutorial value. However, 2D Earth models can be of substantial use to understanding wave phenomena.

In 2D all space-dependent fields become invariant with respect to one dimension (here: y). In other words, all space derivatives with respect to this variable vanish, leaving two independent variables (here: x and z). Because numerical methods are employed to study heterogeneous structures this implies that all heterogeneities defined in the $x - z$ plane also are invariant in y. For example, a circular velocity perturbation turns into a cylinder, a point scatterer becomes a line scatterer, etc.

However, the most important consequence is the fact that the spatial source function also becomes invariant in y. In 2D a point source defined at some

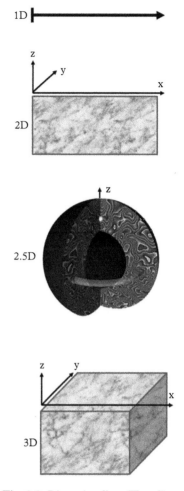

Fig. 3.3 *Dimensionality. The dimensionality of the computational domain is important for the choice of discretization and meshing. In 2D it is important to recognize the invariance of all fields in the third (here y) direction including the source (line source). Fractional dimensions like 2.5D indicate the fact that while the computational domain is 2D the behaviour of the wavefield (e.g. in terms of geometrical spreading) is 3D.*

[2] That's what they always say. Indeed going to higher dimensions may imply far more complicated book-keeping

coordinate $[x_{src}, z_{src}]$ actually becomes a line source. This implies that for any source time function there are always contributions from the line source to the seismogram recorded at any point in the x, z plane.[3] This can be analysed with the 2D numerical schemes available in the supplementary electronic material. Having said that, it is obvious that 2D numerical calculations cannot be compared directly with observations (at least without carefully considering what aspects can be compared; but this goes beyond the scope of this text).

Because for a long time it was impossible to carry out 3D calculations for any realistic Earth structures, there were attempts to find ways around these restrictions. This led to the development of so-called 2.5D schemes (a strange-sounding concept!) in which at least the problem of the incorrect geometrical spreading due to the line source could be fixed. This can be achieved with analytical tools or by moving to other coordinate systems.

An example is given in Fig. 3.3. For global wave propagation—assuming invariance along the lines of constant latitude ϕ—the problem of global wave propagation expressed in spherical coordinates can be reduced to a 2D computational domain in r, θ. When the source is centred at the axis $\theta = 0$ (which is a singularity in the wave equation, but there are ways around that), the correct 3D geometrical spreading is obtained. A powerful recent example of this approach is *Instaseis*, a high-frequency solver for global wave fields based on the spectral-element method (van Driel et al., 2015*b*).

A further issue arising for $>$1D problems is the specific coordinate system used to describe the elastic wave equation. There are fundamental differences when writing the wave equation in cartesian, cylindrical, or spherical coordinates (see exercises). The standard procedure to obtain computational meshes (e.g. regular spaced grid points along the axis) obviously leads to very different meshes. This will be further discussed in the next section, as well as in the Chapter 10 (on applications).

3.3.2   Computational meshes

The generation of computational meshes (or grids; both terms are used synonymously) is a field in itself and a subdomain of computational geometry. Even though in the remainder of the volume we restrict the description of numerical methods mostly to 1D solutions, the very basic concepts of meshes and meshing shall be briefly introduced. In 2D and 3D the choice of the specific mesh on which geophysical parameters are initialized is tightly linked to the numerical method used to approximate the wave equation.

Let us take a simple example to illustrate a fundamental problem when choosing a computational mesh. In many cases seismic velocity models are characterized by surfaces at which there are jumps (discontinuities) in geophysical properties (e.g. seismic velocities or density). Other curved features include the topography at the Earth's surface or internal faults with arbitrary geometrical shapes. Fig. 3.4 illustrates the problem in 2D when the target is to model a curved internal boundary.

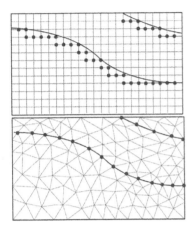

Fig. 3.4 *Honouring (internal) structure. One of the key questions in computational mesh generation is whether it is necessary to honour the geometry of (internal or surface) structures (bottom), or whether it is sufficient to have inaccurate (e.g. blocky) representations of interfaces that do* not *honour their structure (top). (Figure courtesy of Martin Käser.)*

[3] Compared to an impulsive response in a 3D medium, the response of a 2D point (i.e. line) source involves convolution with $1/\sqrt{t}$, thus a vanishing tail. For SH-wave (or acoustic) propagation in a 2D medium waveforms can be converted to 3D seismograms by deconvolution with $1/\sqrt{t}$. In the general case this is not possible. This problem was recently investigated by Forbriger et al. (2014).

When a simple regular grid is used, the boundary cannot be exactly reproduced (*honoured*). It is represented or replaced by a blocky structure. On the other hand, a spatial discretization scheme based on triangles has no problem in following (*honouring*) the internal structure. This simple illustration intuitively suggests that the latter kind of mesh certainly seems better. It turns out that from a computational (and mathematical) point of view the two strategies are actually very different. It is fair to say that the unstructured type of meshes are more difficult to solve. This will become clear when the methods that allow accurate solutions for such meshes (e.g. finite-element-type methods) are being discussed.

The question of the honouring vs. not-honouring strategy is currently a hot topic of research, in particular for seismic wave propagation. This relates to the potential of replacing an Earth model that has discontinuous structures (like the one in Fig. 3.4) with a low-pass version that is smooth but leads (within some small error margin) to the same seismograms. This process is called *homogenization* and will be further discussed in the chapter on applications.

We proceed with a basic classification of meshes.

3.3.3   Structured (regular) grids

Structured or regular grids are subdivisions of (1-, 2-) 3D space characterized by a regular *connectivity*. From a computational point of view this implies that the mesh can be defined uniquely by vectors (1D) or matrices (2D or 3D). Examples are given in Fig. 3.5. In its most basic form a 3D regular mesh consists of brick-like parallel epipeds the corners of which can be addressed with indices i, j, k and physical coordinates (idx, jdy, kdz), where dx, dy, dz are the side lengths of the hexahedral structures. Historically, the numerical solutions for seismic wave-propagation problems (e.g. using finite-difference methods) started with simulations on such regular meshes.

The use of regular meshes works fine, as long as (1) the geophysical model is sufficiently smooth and no complex geometries have to be obeyed, and (2) seismic velocities do not vary too much.[4] Regular grids are also possible when discretizing the wave equation in other coordinate systems (e.g. spherical coordinates, Fig. 3.5(b)). However, the specific characteristics of regular meshes in spherical (or cylindrical) coordinates leads to problems due to the spatially varying grid cells (see section on applications in global seismology).

The problem with honouring (at least smooth) internal or external surfaces can be fixed if an analytical representation of these surfaces can be obtained. Then, grids can be *stretched* by a curvi-linear coordinate transform (Fig 3.5(c)). It is quite obvious that for general surfaces (like Earth's topography) this is not possible. The advantage (compared to the use of unstructured meshes) is the fact that one can stick to vector and matrix descriptions of internal structures.

In seismology (to some extent also in exploration seismics) one can go a long way with relatively smooth Earth models and structures favouring structured mesh approaches. Only in the past decade have strong efforts been undertaken

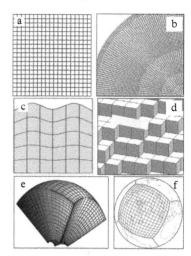

Fig. 3.5 *Structured (regular) grids in 2D and 3D.* **a:** *Regular, equi-spaced 2D grid in cartesian geometry;* **b:** *multi-domain regular 2D grid in spherical coordinates;* **c:** *regular, stretched grid that follows smooth surface;* **d:** *regular 3D cartesian grid with blocky topography surface.* **e:** *regular 3D grid in spherical coordinates (section) with grid points based on Chebyshev collocation points;* **f:** *regular surface grid of sphere meshed with the cubed-sphere approach. After Igel et al. (2015).*

[4] It is hard to give a precise threshold here, but when the velocities vary by an order of magnitude, regular grids are certainly sub-optimal. This situation occurs for example in strong-ground motion problems where surface low-velocity layers have to be honoured, and in global seismology.

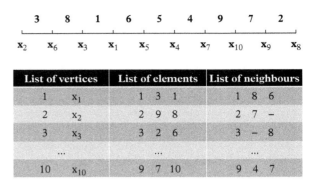

Fig. 3.6 *Connectivity. Space is discretized with nine elements and ten vertices. The connectivity consists of matrices containing the list of vertices, elements, and neighbours (and sides in 2D and 3D).*

List of vertices		List of elements			List of neighbours		
1	x_1	1	3	1	1	8	6
2	x_2	2	9	8	2	7	–
3	x_3	3	2	6	3	–	8
...		...			...		
10	x_{10}	9	7	10	9	4	7

Top number line: 3   8   1   6   5   4   9   7   2 over x_2 x_6 x_3 x_1 x_5 x_4 x_7 x_{10} x_9 x_8

to find efficient numerical methods that work for Earth models defined on unstructured meshes.

3.3.4   Unstructured (irregular) grids

The fundamental difference of unstructured grids is the fact that they can no longer be expressed as a vector (or matrix) on the computer without further specifications. Unstructured grids require the definition of the so-called *connectivity*. What does this connectivity look like? Even though trivial, the 1D example of connectivity matrices shown in Fig. 3.6 illustrates the added complexity of unstructured grids and the substantial larger requirements of computer storage.

Examples are shown in Fig. 3.7. Consider a set of points defined in the 2D plane (e.g. information on geophysical parameters is known at these locations). You want to divide space up into elements (cells) making use of these points. One way of doing this is the so-called Delauney triangulation. There is a (non-unique) way of connecting the points with triangles. Delaunay triangulation breaks up a point set into triangles such that no (other) point is inside the circle defined by each triangle's points. Triangular (or tetrahedral) discretization is the most common meshing strategy for highly complex structures.

Another strategy to subdivide space into elements or cells starting from arbitrary points is the concept of Voronoi cells (Fig. 3.7(b)). Take any two points and determine a line that is equidistant between them (thus orthogonal to the line connecting the points passing through the middle). This is followed by connecting up the intersecting points from the surrounding points. The cells so defined imply that within them each point is closer to the relevant point than to any other. These structures play an important role in many branches of computational geometry and physics (e.g. they are reminiscent of the basalt columns forming through rapid cooling). Voronoi cells can also be used to find interpolation (or differential) weights on unstructured meshes, and to solve nonlinear inverse problems.

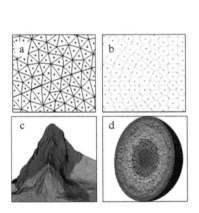

Fig. 3.7 *Unstructured grids in 2D and 3D.* **a:** *Unstructured grid based on Delauney triangulation with cross-cutting interface;* **b:** *Voronoi cells for unstructured grid;* **c:** *tetrahedral mesh for the Matterhorn (mountain in Switzerland);* **d:** *tetrahdral mesh for spherical Earth model. After Igel et al. (2015).*

From a mathematical/computational point of view, regular grids are in general more efficient and converge faster to the correct solutions. The question of which type of mesh is better for a specific geoscientific problem is not easy to answer. Truly unstructured grids are the method of choice for dynamic rupture problems for faults with irregular geometry and for problems with highly complex geometrical features (e.g. boreholes, strong surface topography). More on this in the chapter on applications.

3.3.5 Other meshing concepts

Some of the observations made above indicate that, despite their flexibility, unstructured meshes might lead to substantially more computational work. This raises the question of whether structured and unstructured grids could not be merged and used in those physical domains where they are most efficient. Again this is a topic of current research. An example is shown in Fig. 3.8. This approach is called a hybrid mesh, where a structured mesh is used in the area with little structural complexity and smooth velocity variations. In the region with complex geometry (here the free surface topography) a triangular mesh is used that is more easily generated than a corresponding rectangular mesh that follows the surface.

When physical problems in large computational domains are strongly focused in space (e.g. shock waves, rupture fronts), then it might make sense to densify the grid during run time in the area where things are happening. This is called adaptive-mesh refinement (AMR) and plays an extremely important role in geophysical fluid dynamics. For strongly scattering seismic wave-propagation problems the use of adaptive meshes is usually not advisable as—after relatively short simulation time—energy propagates almost everywhere in the medium and influences the resulting synthetic seismograms (depending of course on source–receiver geometry). It is a matter of current research whether for dynamic rupture problems adaptive mesh refinement should be employed.

Another scheme that is currently being exploited for seismic wave propagation for static and dynamic mesh refinement is the *octree* approach. Octree is a tree data structure. Meshes are refined in a progressive wave. In 3D, subdividing each side by a factor of 2 leads to 8 *children*. An example is shown in Fig. 3.9. If the goal is to densify a mesh around linear structures like the one indicated in the figure (e.g. an interface in a sedimentary layer) the mesh is progressively refined until the desired accuracy of the solution is obtained.

This is useful for static mesh refinements and may also be used for dynamic mesh refinement during simulations. This approach was used to model strong ground motion in combination with the finite-element method by Bielak et al. (2005).

In this section we have encountered a variety of ways to discretize space. However, what we have not yet discussed is how we go from the description of a geophysical model to its computational mesh.

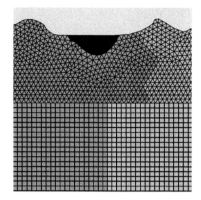

Fig. 3.8 *Hybrid meshes. A structured, regular grid in the lower domain with low degree of geometrical complexity is combined with an unstructured triangular grid at the top of the domain with a complicated free surface. A strong low-velocity domain (black area) is also meshed with an unstructured triangular mesh. From Hermann et al. (2011).*

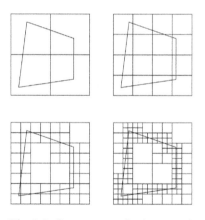

Fig. 3.9 *Octree approach. An octree is a tree data structure. Each node has eight children. Octrees are used to partition a 3D space by recursively subdividing it into eight octants (here shown in 2D). This approach allows efficient refinement of linear or curved internal structures or surfaces. Figure courtesy of Vasco Varduhn.*

3.4   The curse of mesh generation

In a recent meeting, a computational seismologist named his presentation: *Meshing—an underestimated task*. That almost says it all. The efficient generation of computational meshes is (1) often difficult, (2) time consuming, and (3) Earth scientists are usually not trained for it. I think it is fair to say that—while many excellent solvers exist today for the accurate simulation of 3D wave propagation—we are far away from having standard work flows for the generation of computational meshes for wave propagation and rupture problems.

The situation will get worse in the sense that for leading-edge research questions we are using 3D simulation tools to investigate the response of fairly complicated models (you won't get the Crafoord Prize[5] with 1D simulations). Now that community codes like *specfem3d* or *SeisSol* exist that take as input externally generated meshes, researchers who want to use those codes need to deal with mesh generation. Here, we will only introduce some fundamental concepts that help to understand the issues involved.

Let us take an example. Grenoble is situated in a beautiful valley in the French Alps (and also hosts one of the best labs in seismology in Europe). Like most of the alpine valleys, Grenoble is built on fairly thick layers of alluvial sediments that might resonate when seismic waves enter. With some potential for sizeable earthquakes in the vicinity, it is worth investigating the ground motion expected for some realistic scenario. As the velocity structure of the area is fairly well known, a few years ago a community benchmark project was set up (Chaljub et al., 2010*b*). The goal was to compare solutions from various numerical solvers for this challenging problem.

What is the input for such a geophysical model? First, there is the surface topography that—given the alpine setting of Genoble—has to be taken into account. Surface topography is usually given in terms of a digital elevation model (DEM) that consists of a regular surface grid in appropriate coordinates with the elevation in vertical direction. Second, the location of the interfaces of the sedimentary layers might be given in a similar form, or (even better) as parametric surfaces. Third, the location of internal fault surfaces might be given, on which (finite) seismic sources have to be activated. How (on Earth) can we create a computational mesh with this information?

Mesh generation has two major steps: (1) geometry creation, and (2) mesh generation. *Geometry creation* involves the definition of surfaces bounding the mesh volume. This can be the free surface, internal interfaces, or the domain boundaries. Often, this is the most time-consuming process. For example, when arrays of grid points (or even unstructured points) describe surfaces a parameteric description has to be found (e.g. using spline functions). Often this involves simplifying and subdividing geometric structures manually. Finally, surfaces have to be joined up to create a closed volume.

The *mesh-generation* process takes as input the created geometry and subdivides the entire volume into grid cells. At this point, the decision that has to be taken, according to which solver is going to be used, whether to use hexahedra

[5] There is no Nobel Prize for Earth sciences; the Crafoord Prize comes closest.

Fig. 3.10 *Hexahedral mesh. Computational mesh for the Grenoble basin code verification exercise. The mesh is based on curved hexahedral elements. The mesh is refined towards the centre of the model (sedimentary basin) where seismic velocities are substantially lower. Interfaces are not honoured. This mesh was used for spectral-element simulations. From Igel et al. (2015).*

or tetrahedra as basic element structure (no other element types are currently in use for large-scale 3D simulations in seismology). In Fig. 3.10 an example of a hexahedral mesh for the Grenoble valley model is shown. Hexahedral meshes cannot usually be obtained fully automatically, and require substantial manual interaction during the mesh-generation process. Once the mesh has been created and grid points are defined, the geophysical parameters can be initialized and boundary conditions can be associated with the surfaces.

Concerning automatic mesh generation, tetrahedral elements have substantial advantages. Once the geometry is defined, tetrahedral meshes of high quality can usually be obtained in an automated way, considerably speeding up the overall meshing work flow. However, this comes at the price of substantially higher computational costs for the solver. This will become clear when discussing the various numerical methods.

An example of a tetrahedral mesh for a volcano structure is shown in Fig. 3.11. Unstructured grids have several advantages. In this case, as the interest is in modelling waves right under the volcano summit, the structure of interest can be embedded in a half sphere in an elegant and efficient way.

There are a number of (mostly commercial) programs available to solve the geometry- and mesh-generation workflow for hexahedral meshes.[6] To my knowledge, the most commonly used meshing software at the present time is CUBIT (with a linked open source library called GEOCUBIT providing specific functionalities for Earth sciences)

If you are planning a simulation task involving mesh generation from scratch, the following rough estimate of timing might be useful. Table 3.1 indicates how much time is spent for a meshing and simulation problem. Of course this might vary substantially depending on the specific geometries to be meshed. It is a warning, however, that the effort for meshing should not be underestimated.

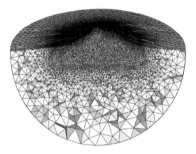

Fig. 3.11 *Tetrahedral mesh of Merapi volcano, Indonesia. The volcano is embedded in a hemispherical mesh with densifiction in the summit area where the mesh follows the topography. From Igel et al. (2015).*

Table 3.1 *Timing estimates for meshing and simulation problems*

Human time	Simulation workflow	CPU time
15%	Design	0%
80% (weeks)	Geometry creation, meshing	10%
5%	Solver	90%

Source: E. Casarotti, personal communication

3.5 Parallel computing

When you download a community code for 3D wave propagation (e.g. *specfem3d* or *SeisSol*) it will be a parallel implementation of the specific numerical method. What is parallel computing? Why do we need it? How does it work? What

[6] Examples are: CUBIT, TRELIS, GMESH, GID, ABAQUS, ANSYS, HEXPRESS, and others.

programming languages are used to parallelize codes? Does it matter which computer it will be running on? No doubt, one could write entire books answering these questions! The goal here is to provide you with some basic concepts and terminology that should allow you to understand the documentation provided with current parallel community codes. Some oversimplifications are unavoidable. The computational scientists may excuse.

The original approach for computing based on the von Neumann[7] model is *serial*, that is, one operation is done at a time operating on single data. This computing model obviously has its limitations. First, it is limited by the time it takes to do one operation step (a cycle). The continuing pursuit was on until recently to increase the so-called clock rate. This raised in particular the temperature (and energy consumption) of the processing units (until laptops started catching fire!).

For many problems (e.g. cryptography, simulation of physical phenomena, Monte Carlo sampling) that require large computational resources, (1) the same operation has to be carried out on different data (*data parallelism*) or (2) different tasks need to be carried out in parallel to obtain a final result (*task parallelism*). Therefore, the efforts to develop computer hardware that works in parallel started early. Computer scientist Michael Flynn came up with a terminology that allowed the classification of computer models. It is still very useful today, and is illustrated in Fig. 3.12.

The classic serial model (SISD) was extended to allow a single instruction to be applied to a large amount of data (SIMD) distributed in different processors. An example is a matrix that is multiplied by a real number. In this case, no information exchange between processors is needed (this is called *embarrassingly parallel*). However, if you would like to know the maximum value of the matrix, information needs to be exchanged (see exercises). One of the first such (massively) parallel computers dedicated to the Earth sciences was installed in 1990 at the Institut de Physique du Globe in Paris in connection with the visionary work of Albert Tarantola and Peter Mora in the Geophysical Tomography Group ($G^T G$).

The SIMD architecture can be viewed as memory that is distributed and accessed by a front-end computer. This simple architecture works very efficiently for image processing and also for some numerical simulation approaches. Recently this type of architecture has been revived through the use of GPU technology. It appears that history played a substantial role in the evolution of parallel computing. The early architectures were strongly supported by military applications (e.g. cryptography), allowing the development of machines (like the Connection Machine CM2) with relatively limited domains of application. Other support came from the geophysical exploration industry which saw opportunities to dramatically speed up seismic processing and imaging. The end of the Cold War led to fundamental changes in the funding of computer technology and the requirement for parallel computers with more flexibility. Support from military sources faded.

At least to some extent this accelerated the development of the so-called MIMD computer models where basically each processor can perform different

SISD	MISD
Single Instruction Single Data	**Multiple Instruction Single Data**
Serial computer	*Cryptographic decoding*
SIMD	MIMD
Single Instruction Multiple Data	**Multiple Instruction Multiple Data**
GPU cluster, CM2	*Supercomputers PC cluster*

Fig. 3.12 *The Flynn taxonomy. Classification of computer architectures.*

[7] We will encounter John von Neumann (1903–1957) again when analysing numerical solutions.

tasks in parallel. Such parallel computers are a lot more flexible and can improve computational efficiency for most conceivable problems. This initiated a route through technical developments in terms of hardware and programming software that is best characterized as *disruptive technology*. But more on this later.

Let us have a look at the consequences of parallel computing for simulations of dynamic physical systems.

3.5.1 Physics and parallelism

Nature works in parallel. The same physical laws (as far as we know) are valid everywhere in space. The partial differential equations describing processes like elastic wave propagation or geophysical fluid dynamics are statements of this parallel nature. Take the example of (source-free) 1D wave propagation

$$\partial_t^2 p(x, t) = c^2(x) \partial_x^2 p(x, t), \tag{3.7}$$

where $p(x, t)$ is the pressure, and $c(x)$ is the velocity model. Writing this equation using finite-difference approximations to the partial derivatives (see next chapter for the derivation) leads to

$$p(x, t + dt) = c^2(x) \frac{dt^2}{dx^2} \left[p(x + dx, t) - 2p(x, t) + p(x - dx, t) \right]$$
$$+ 2p(x, t) - p(x, t - dt), \tag{3.8}$$

where dt and dx are temporal and spatial increments. This mathematical statement can be phrased as:

The (immediate) future $t + dt$ of a physical system—here the pressure $p(x, t + dt)$ at some point in space x—depends (only) on the values in its immediate neighbourhood $p(x \pm dx)$, the presence $p(x, t)$, and the recent past $p(x, t - dt)$.

This illustrates (maybe clearer than the original partial differential equation) that the spatio-temporal interaction of elastic wave (and many other) phenomena is of a *local* nature. This property has tremendous implications for the parallelization of this type of problem (at least when finite-difference type approximations are used): Assuming that space-dependent fields are distributed across many processors (today $O(10^6)$), it appears that—when communication is only necessary between neighbouring processors—this is very efficient (see Fig. 3.13). Clever networking schemes can alleviate problems with non-local communication.

This degree of *locality* of a numerical algorithm has important consequences and might differ substantially depending on the specific mathematical approximations used. In fact, the need for non-local communication has hampered the use of some (very elegant and accurate) methods for wave-propagation problems

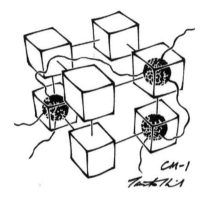

Fig. 3.13 *Parallel processing. Design study following a form-follows-function approach for one of the first parallel computers, the Connection Machine CM-1 of Thinking Machines Corp. with connected cubes indicating processors that communicate with each other. A photo of famous physicist Richard Feynman (who designed the communication scheme) wearing a T-shirt with this logo was used by Apple^{TM} in their 'Think Different' campaign in the nineties. The actual design of the CM-1 was based on this logo; apparently the only time that a computer was designed after a T-shirt. Design concept by and figure courtesy of Tamiko Thiel.*

(e.g. pseudospectral methods) on parallel hardware, because the communication overhead is too high.

3.5.2 Domain decomposition, partitioning

How can we make best use of parallelization for our wave-propagation problem? All numerical methods discussed in this volume (and most 3D wave simulation codes currently used in the seismological community) are based on time–space domain numerical solutions to the wave equations. This implies that space-dependent fields (e.g. displacements, stresses, strains, elastic parameters, seismic velocities, density) are mapped on parallel hardware by *domain decomposition* using the *distributed memory concept*. That means the tasks only see local memory.

For regular 1–3D grids this is usually a trivial task, as shown in Fig. 3.14, top. The discretized spatial domain, specified as vector or matrix shape accordingly, is subdivided into *n* areas of equal size, *n* being the number of available processors. A further, less obvious task is that of deciding which processor gets which chunk of the subdivided volume. In Fig. 3.14, processors are illustrated by different colours and various strategies for their allocation exist. As indicated above, communication between processors might well depend on the physical distance between processors, which means, depending on the communication requirement of a specific problem, the mapping may have a strong impact on the computational efficiency.

This problem is much harder for unstructured grids, with the added complexity that each element might require a different amount of operations per time step (e.g. when space-dependent fields are described with different polynomial order, this is called p-adaptivity). Fig. 3.14 (bottom) is an example of a mesh partitioning and mapping onto various processors, optimizing the *load balancing* for the volcano mesh. With load balancing one tries to keep all processors happily active until some synchronization between the program parts is carried out.

If you ask how this can be done efficiently, and how portable such codes are, you are asking the right questions. Even for regular meshes, optimizing load balancing and overall performance has become an issue that is best left to computational scientists. That is one of the reasons why we are increasingly using community codes that have been developed with the help of computational scientists. That makes it more difficult to understand what is going on inside (just like with every new generation of cars).

Well, one reason for writing this volume is to provide a look under the bonnet.

3.5.3 Hardware and software for parallel algorithms

We have learned that we make use of parallel computers by distributing our meshes (at best equally) across the available processors. Most likely when you read this, you will already have encountered some programming language that is used on parallel hardware such as C, C++, Fortran, or Python. The remaining

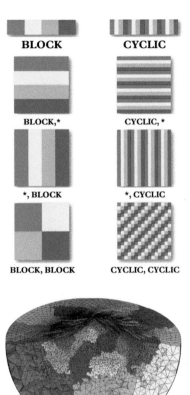

Fig. 3.14 *Domain decomposition. When space-dependent fields are discretized the question is which processor gets which chunk of data. Even for regular meshes in 1D or 2D (**top**) this is not obvious. Here each colour represents a specific processor (or node). Partitioning volumes for unstructured grids in 3D (**bottom**), optimizing the load balance of processors (or nodes), can be a challenging task.*

questions concern (1) how to tell a program to run in parallel and (2) how your parallel coding depends on the specific hardware.

Before describing the situation as it is today, allow me to go back to the time when the first (SIMD-type) parallel computers appeared on the market.[8] Due to their relatively simple architecture, high-level extensions to standard languages like Fortran (e.g. Connection Machine Fortran, CMF) were provided with the software that made parallel code development relatively easy.

With the increasing complexity of hardware (e.g. the CM-5 was an MIMD machine), and the variety of vendors of parallel machines, high-level programming solutions for simple parallel problems (like domain decomposition) vanished, despite attempts like High-Performance Fortran (HPF), Cray Fortran (CRAFT), and others. What survived was the more flexible approach based on the Message-Passing Interface (MPI), which allows solutions for most types of parallel problems, albeit with a substantially bigger overhead for parallelization of serial programs. The situation, as far as programming was concerned, got even more complicated with the appearance of hybrid architectures and GPU clusters.

In a panel discussion a few years ago with several representatives of parallel computer vendors, it was acknowledged that this is a very unfortunate situation for domain scientists who are used to developing their own software solutions. The consequence was a paradigm shift in the development of efficient parallel simulation software that is still ongoing. Today, efficient parallel software development is only possible (and yes, it is fun) by permanent interaction and collaboration with computational scientists.

3.5.4   Basic hardware architectures

To understand the existing programming models for parallel algorithms it is necessary to introduce the basic hardware models. In fact, due to recent developments (e.g. in GPU technology) computational scientists often speak of hardware–software co-design. That means that sometimes hardware is designed to be optimal for a specific software problem (e.g. image processing) and vice versa (e.g. simulation software is programmed such that it is optimal on specific hardware).

Some basic parallel hardware concepts are illustrated in Fig. 3.15. These are:

Shared Memory: All processors have direct access to common physical memory. This is a model where parallel tasks see the same state of memory and can access the same logical memory locations no matter where physical memory exists.

Distributed Memory: Architecture with distributed memory implies network-based access to physical memory that is not common. A computer program only sees local memory. To access memory from other machines (processors) specific communication is necessary (e.g. message passing).

Hybrid Distributed–Shared Memory: A parallel computer that consists of distributed *nodes*, each of which is a parallel shared-memory model with a certain

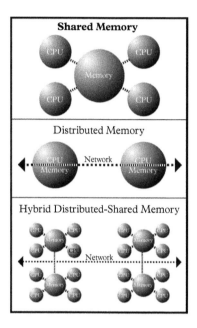

Fig. 3.15 *Basis parallel architectures.* ***Top:*** *Shared memory architecture.* ***Middle:*** *Distributed memory architecture.* ***Bottom:*** *Hybrid model with several shared-memory nodes.*

[8] In fact, in early 1990, three weeks after I started my PhD at the Institut du Physique du Globe in Paris, the Connection Machine CM-2 was delivered, and our research group had almost exclusive access to it. The programming environment was a dream compared to today. Domain parallelization was basically automatic. The programming overhead for parallelization was almost nil. In addition, application engineers assisted with the scientific code development.

Fig. 3.16 *Parallel supercomputer. The hybrid supercomputer SuperMUC located at the Leibniz Supercomputing Centre in Munich, Germany, with 241,000 cores, which together delivered almost 7PFlop in 2015. SuperMUC uses a revolutionary form of warm water cooling. The buildings are heated reusing this thermal energy.*

number of processors. At the time of writing this is a popular model for flexible supercomputers (e.g. the SuperMUC at the Leibniz supercomputing centre in Munich, Germany, Fig. 3.16). While, in principle, hybrid computers can be programmed with message-passing concepts, in practice code that can exploit them most efficiently needs to be optimized at both the shared-memory and distributed-memory levels.

As indicated above, the industry of parallel computing started with relatively homogeneous assemblies of processors connected by an appropriate network (like the Connection Machine CM-2 or CM-5). With the increasing power of personal computers (PCs) in the '90s, and the evolution of free operating systems like Linux, the concept of *parallel clusters* (mostly Linux-based) emerged. Local area networks (LANs) could then in principle be used as parallel computers. Today, departmental-scale clusters with $O(10^3 - 10^4)$ processors are still attractive alternatives for meso-scale simulation tasks to using supercomputer resources (with potentially long queueing times).

Soon after, the cluster concept was taken to a higher level, linking up computers that were not (more or less) physically collocated. This was the birth of the *GRID* initiatives linking many (really geographically) distributed, heterogeneous computational resources at local, national, and international levels. It is fair to say that *cloud computing* developed out of the concepts of GRID Computing, providing on-demand resources. With a few exceptions, GRID and cloud computing have not (yet) played an important role for seismic simulations, but this might well change in the near future.

3.5.5 Parallel programming

What programming models are used today to parallelize seismic simulation software? Most solvers implemented on classic *CPU*-based supercomputers or computer clusters use the Message-Passing Interface standard (MPI, http://www.mpi-forum.org). As there are a massive number of tutorials and information on the Web, we will just present the basic underlying concept. MPI consists of libraries that can be called from Fortan, C, and C++ programs. Recently, MPI implementations with Python have also been developed (e.g. pyMPI).

Consider the following MPI Fortran code example:

```
! My first MPI program
program main
  use mpi
  integer   error
  integer   id
  integer   np
! Initialize MPI.
  call MPI_Init ( error )
! Get the number of processes.
  call MPI_Comm_size ( MPI_COMM_WORLD, np, error )
```

```
! Get the individual process ID.
  call MPI_Comm_rank ( MPI_COMM_WORLD, id, error )
! Print a message.
  write (*,*) "The overall number of processors is ",np
  write (*,*) "I am processor ",id
! Shut down MPI.
  call MPI_Finalize ( error )
  stop
end
```

In this Fortran90-MPI program the MPI library is called and the program is initialized for *np* processors. After compilation into an executable program like main.x, the program can be executed using the *mpirun* command as
  ≫ mpirun -np 4 main.x
which will run the executable independently on each of the four processors in this case, with the result

```
The overall number of processors is 4
I am processor 0
The overall number of processors is 4
I am processor 2
The overall number of processors is 4
I am processor 3
The overall number of processors is 4
I am processor 1
```

where the sequence is indeed random, depending on the clock cycle in which the output is performed in each processor. This simple example illustrates the flexibility of the MPI concept. Identical programs run independently on any processor of the computer cluster until there is a statement to exchange information between processors. Tasks for individual processors can be given using *if* statements that identify the processor-*id*. The interested reader is referred to the available online material for further examples. For shared-memory models, additional libraries exist such as openMP (http://www.openmp.org), which exploit the specific architecture. As indicated above for hybrid models, the shared- and distributed-memory programming models might have to be combined to achieve optimal performance of a specific parallel computer code.

How do you measure whether a parallel code is performing well? This question leads to the concept of *scaling* or *scalability*. Scalability refers to the speeding up of a program with increasing resources. A simple equation that describes scalability is

$$\text{speed-up} = \frac{1}{\frac{P}{n} + S},\tag{3.9}$$

where P is the fraction of the code that can be parallelized, S is the serial fraction, and n is the number of processors. This is illustrated in Fig. 3.17. It is interesting

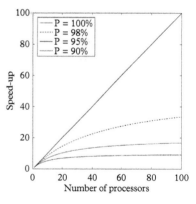

Fig. 3.17 *Speed-up of parallel software. The speed-up is given as a function of the fraction of the code that is parallelizable P and the serial part S (see text for details).*

to note how adding more processors does not lead to much more speed-up for even high percentages of parallelizability. For many problems (like seismic wave calculations) usually the parallel fraction increases with problem size, improving the speed-up for large problems. It is easy to understand that I/O might substantially deteriorate the performance of parallel applications. Therefore parallel I/O is a hot topic, particularly for data-rich problems (see next section).

There are two classes of scaling. *Strong scaling* investigates how the run time varies with the number of processors for a fixed size problem. *Weak scaling* refers to the run time when additional processors with the same problem size are added. A recent example of strong scaling for the *SeisSol* code based on the discontinuous Galerkin method is shown in Fig. 3.18. This remarkable scaling behaviour as well as the obtained peak performance (1.09PFlops) was the result of several years of performance optimization (Breuer et al., 2014) which was recognized by being nominated as a finalist of the Gordon Bell Prize in 2014.[9]

Finally, it is worth noting that—similar to the efforts in the car industry to reduce the fuel per 100 km and CO_2 emissions—supercomputing is going green. *Green computing* refers to the attempt to develop environmentally sustainable computing. Supercomputers burn energy comparable to small cities. For example, the analysis of the impact of convergence order, CPU clock frequency, vector instruction sets, and chip-level parallelism on the execution time, energy consumption, and accuracy of the obtained solution for the *SeisSol* software led to a reduction of the computational error by up to five orders of magnitude, while increasing the accuracy order of the numerical scheme from 2 to 7; and all this while consuming no extra energy (Breuer et al., 2015).

As the number of processors inside supercomputers, Linux clusters, PCs, tablets, and even your smartphone increases, scaling to large processor numbers

[9] The spectral-element code *specfem 3d* won the Gordon Bell Award for Best Performance at the SuperComputing 2003 conference in Phoenix.

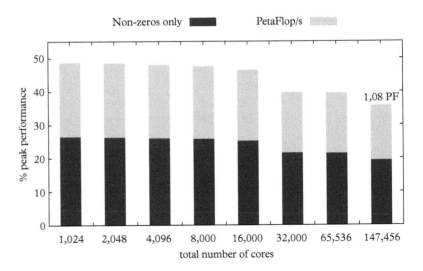

Fig. 3.18 *Strong scaling. Example of strong scaling behaviour of the* Seis-Sol *seismic simulation software based on the discontinuous Galerkin method. The simulation of wave propagation through a volcano is discretized with about 10^8 tetrahedra. Flops refers to floating point operations per second including zero operations. The* non-zeros *values refer to the scaling for calculations not involving any zeros. The obtained absolute peak performance is quite sensational for a real application code leading to the code being a finalist for the Gordon Bell Prize in 2014. Figure courtesy of Michael Bader and Alexander Breuer.*

is an extremely important issue. When requesting resources on national or international supercomputer infrastructure you have to demonstrate that your program scales. Again, this is an issue that requires interaction with computational scientists in most cases.

3.5.6   Parallel I/O, data formats, provenance

Let us make a simple calculation. You are running a simulation on a supercomputer for an Earth model discretized on a grid with $1,000 \times 1,000 \times 1,000 = 10^9$ grid points (nothing unusual). You might want to output some snapshots, for example of the seismic wave field, to analyse the results. How would you extract this information on a supercomputer with data-distributed architecture? How would you store such a file? How would you post-process it? Would you transfer it to your local institutional PC and work with? Well, one file in double precision would have 8 GBytes. When looking at snapshots you usually have hundreds or thousands of them, so you quickly have TBytes of data. This is not exorbitant compared to other fields like astronomy where sometimes Pbytes of data are generated from observations. Nevertheless, it is obvious that you are likely **not** to be able to easily handle, transfer, or post-process this data on usually serial (or mildly parallel) hardware locally.

The question of how to input (e.g. an Earth model) or output (e.g. snapshots) from a parallel computer leads to *parallel I/O* formats. Obviously, for computers with distributed memory and many cores the worst option would be to send all data to one main processor and then output it serially. Parallel I/O allows data to be input and output in parallel to a storage device, implying that, outside the supercomputer, parallel storage facilities are available. Libraries like *MPI* allow parallel I/O, and the interested reader is referred to the relevant documentation. In many fields the increase of data volumes has led to domain-specific parallel data formats that adapt to the needs of a specific scientific domain, allowing efficient *data exchange* and *data sharing*. The main goals are: (1) handling large volumes; (2) coping with complex data (e.g. time series with difference formats, sampling rates); (3) provision for heterogeneous data (e.g. binary seismic data, earthquake information, earth model infomation); (4) providing options for simulation data (in addition to observations); and (5) access to provenance information (e.g. what happened to the data previously, which solver was used for synthetic data).

An attempt to provide a novel data format for seismology is currently under way (called Adaptible Seismic Data Format ASDF, Krischer et al., 2016), that addresses some of these problems. It is based on the *hierarchical data format* approach (HDF) with its current implementation *HDF5*. HDF5 is a standard format for binary data with a large amount of tools and support. The seismology-specific adaptation consists of a *container* that stores three fundamental categories of information: data (source and waveforms), metadata (attributes for a piece of waveform data), and provenance (description of how a piece of waveform data was generated). The contents are illustrated in Fig. 3.19.

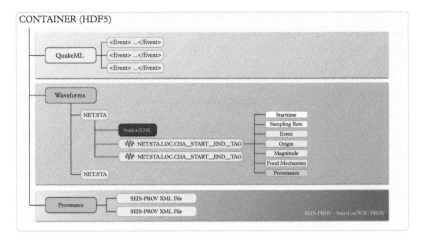

Fig. 3.19 *Adaptable Seismic Data Format (ASDF, http://www.seismic-data.org). A file with seismic data containing not only the seismic traces themselves but additional station information, and information on earthquake parameters, synthetic seismograms, and provenance (let's not forget the data itself!). The format is based on the* hierarchical data format *approach. Figure courtesy of Lion Krischer.*

A data format like this has the potential to (1) substantially reduce time-to-research if accepted by the community, and (2) make results more reliable and reproducible. There is a multitude of formats for seismic observations alone.[10] Until recently, that multitude was complemented by another, of conversion programs that allowed the movement between these formats. In the past few years the situation has been improved substantially by the open-source project ObsPy (http://www.obspy.org, Krischer et al., 2015*b*; Beyreuther et al., 2010). The Python library ObsPy has incorporated the various data formats and allows stable operations on most available seismic data.

This section has touched on the *big data* discussion that is currently ongoing in many fields. It is a rapidly evolving field and we should expect new developments along the lines discussed above that substantially improve the practice of our everyday scientific work.

3.6 The impact of parallel computing on Earth Sciences

Earth Sciences and seismology in particular (with a few exceptions such as earthquake physics) can be considered *data-rich*. The analysis of seismic data, mostly using approximate theories to explain observations (e.g. ray theory, ray tomography, and wave propagation in Cartesian and spherical layered media), has dominated seismology in the first decades of its relatively short history. The situation today, with both digital data and the capacity to compare observations with synthetic data based on 3D simulations has already begun to, and will continue to, substantially improve our understanding of the structure of the Earth's interior on all scales, and will thereby help us understand better how our planet works.

[10] SAC, MiniSEED, GSE1, GSE2, SEISAN, Q, SEG-Y.

This implies that most seismological research projects involve the use of 3D simulation software. The common practice until a few years ago for a PhD student or postdoc was to basically start from scratch, write a solver (e.g. using the finite difference method), do research with it, and submit a dissertation (and leave). In many cases maintaining software for later use was extremely difficult. This approach no longer works today, leading to the paradigm shift that I alluded to above: Most leading-edge problems require larger-scale simulations working on huge data sets, requiring substantial post-processing resources (e.g. visualization, filtering, etc.) that can rarely be developed and maintained by (small) research groups. As discussed, efficient IT solutions require a substantial amount of development time. Therefore the ongoing developments to openly distribute simulation software for Earth sciences (e.g. CIG, http://www.geodynamics.org) with sufficient documentation and training material, as well as the development of community platforms such as VERCE (http://www.verce.eu) or EPOS (http://www.epos-ip.org), should be welcomed.

However, this situation creates new problems. While during the time of *heroic coding* of PhD students and postdocs at least the researcher knew exactly what was under the bonnet, now we are mostly dealing with black boxes. Today, community simulation codes are fairly stable and portable, and can be easily downloaded and run without knowledge of numerical methods or parallel computing. The danger is that incorrect initializations can lead to erroneous results that are hard to distinguish from correct solutions.

A main goal of this volume is to turn the black box at least into a fairly transparent box in which you have some idea of what is going on inside the codes. It is my firm belief and the basic concept for this volume that if you manage to write a 1D simulation code **from scratch** in whatever language (Matlab, Python, Fortran, C, Julia), and investigate how it works, you have come quite a long way to understanding how a 3D code works, and you are much less likely to commit errors when dealing with complex simulation tasks. The substantial electronic material that we provide with this volume should help you in taking this path.

Chapter summary

- (Most) 3D wave-propagation simulation software is based on the discretization of an Earth model on a structured or unstructured mesh.

- The generation of meshes in particular for complicated Earth models is a challenging task.

- Large-scale simulations in seismology require parallel computing, that is, hardware that allows the performance of several tasks at the same time, and software that allows the programming of the task and/or data distribution.

- Most (time–space domain) seismic simulation codes are based on domain decomposition. Space-dependent fields are subdivided and distributed in parallel memory.

- Efficient parallelization of an algorithm often requires careful identification of the parallelizable part and specific coding with respect to the targeted parallel hardware.
- The efficiency of parallel software is characterized by its performance (in terms of percentage of peak performance) and its scalability (speed-up when more processors are used).
- Parallel code development in the Earth sciences should be done in close collaboration with computational science.

FURTHER READING

- There is a huge amount of information in tutorials available online on parallelization using MPI. A good starting point is http://www.mpi-forum.org.
- A very comprehensive treatment of parallel computing is given in Pacheco (2011).
- Ismail-Zadeh and Tackley (2010) contains a slightly more elaborated section on parallel programming in Chapter 9 with a view to problems in geophysical fluid dynamics.
- Durran (1999) is an excellent introduction to numerical methods for wave-like phenomena and dissipative flows.

EXERCISES

Comprehension questions

(3.1) Describe concepts that show how to represent space-dependent functions in a discrete way.
(3.2) Explain the concept of 2D, 2.5D, and 3D simulations. What problems can arise for <3D simulations when comparing with observations?
(3.3) Explain the concepts of structured and unstructured meshes. Give examples.
(3.4) Illustrate the differences between regular meshes in various coordinate systems: Cartesian, cylindrical, spherical. What are the consequences for simulation problems?
(3.5) What is a *cubed sphere*?
(3.6) Explain the concept of Delauney triangulation and Voronoi cells.

(3.7) Discuss pros and cons of structured vs. unstructured grids.

(3.8) What are adaptive meshes? Can they be used in seismology?

(3.9) Give reasons why the generation of meshes is relevant for seismological problems. Give examples.

(3.10) What are the basic models for parallel computers?

(3.11) What is the most common model of parallelization for seismic wave propagation and why?

(3.12) Explain the concepts of strong and weak scaling.

(3.13) Find some current supercomputers on the internet and extract the main specifications (e.g. number of processors, memory, peak performance, etc.).

(3.14) Class exercise: Every student gets an integer number starting with 0. This number denotes the processor. Each student writes four numbers on a page. Perform the following tasks:

  • Single-Instruction-Multiple-Data: The tutor tells all students to multiply each number by 5.

  • Embarrassingly parallel problem: Add the first three numbers and subtract the fourth.

  • Circular shift operation: Pass the first number to your right neighbour. If there is none, pass it to the far left neighbour.

  • Global reduce: Find the maximum value of all initial 4 numbers of all processors. Processor 0 speaks out the maximum value loudly.

  • Global distribute: The tutor gives 2 numbers to processor 0. Processor 0 distributes these 2 numbers to all processors.

  • Extend these exercises to your liking.

Theoretical problems

(3.15) Classify the following partial differential equations in terms of elliptical, hyperbolic, or parabolic problems:

$$u_{xx} + 2cu_{xt} + c^2 u_{tt} = 0$$
$$xu_{xx} - 4u_{xt} = 0 \qquad (3.10)$$
$$u_{xx} - 6u_{xt} + 12u_{tt} = 0.$$

(3.16) Use Eq. 3.9 to find out what fraction of the code needs to be parallel to achieve a speed-up of 10,000 for 20,000 processors.

(3.17) You want to simulate a physical domain of size (or side length) 1,000 km with a grid distance of 1 km. Estimate the required memory of one space-dependent field (double precision) in 1D, 2D, and 3D. Compare with the RAM of your smartphone, laptop, and with the specifications of current supercomputers. Discuss the results.

Programming exercises

(3.18) Write a program (e.g. Matlab, Python) that generates arbitrary point clouds. Triangulate them using the Delauney method. Calculate and visualize the corresponding Voronoi cells. Find appropriate libraries to carry out the tasks.

(3.19) Distributed data: Write a small parallel program using (e.g. Fortran/MPI or pyMPI). Define a matrix A(2,000,2,000) and distribute it on n processors. Initialize it with random numbers and extract minimum and maximum values. Perform operations on the matrix in a loop and time the operations. Compare to the serial case $n = 1$ and distributed task parallelism.

(3.20) Task parallelism: Write a small parallel program using (e.g. Fortran/MPI or pyMPI). Load a seismogram trace. In one processor calculate filtered seismograms (e.g. looping through low-pass filters with various corner frequencies). In a second processor perform an equal number of subsequent auto-correlations. Time the parallel and serial codes and compare the results. Note: If you use Python the ObsPy library offers tools for seismic data processing.

(3.21) Search for open-source mesh generators (e.g. *MeshPy*), invent some simple geometries, and generate *vtk* (visualization toolkit) files and visualize them (e.g. with *paraview*).

(3.22) Install the *ObsPy* Python library (http://www.obspy.org). Follow the tutorials and investigate the potential to access and process observed and/or simulated seismic data.

(3.23) Use *ObsPy* to download data from any seismic station you are interested in for the M9.1 Tohoku-Oki earthquake of 11 March 2011. Save the data using the *ASDF* format (http://www.seismic-data.org). Explore the provenance options.

Part II

Numerical Methods

The Finite-Difference Method

Without doubt, the finite-difference method is conceptually the simplest method presented in this volume. Historically it was the first numerical method that was widely used in seismological research. It is also justified to say that for decades it was the workhorse for many research applications. Despite its 'brute force' reputation, it is important to note that a well-designed finite-difference algorithm still is capable of beating some other—mathematically more sophisticated—techniques with better reputations.

It all depends on the specific seismological problem. A major advantage of mathematical simplicity is that an algorithm is quickly adapted to a specific problem. In my view, that is the main reason that some of the best research in seismology involving seismic wave calculations in heterogeneous media has been achieved with this approach. This involves in particular applications in exploration geophysics, strong ground motion problems (see Fig. 4.1), dynamic rupture simulations, and inverse problems.

For students who are interested in understanding partial differential equations, the finite-difference method offers an efficient and fast way to develop numerical approximations that allow the investigation of some of the main characteristics of the problem. In the exercises section of this chapter, some examples that go beyond seismology are given.

This chapter is structured as follows. After a brief section on the history of the finite-difference method in seismology we proceed with the introduction of the fundamental mathematical concepts. The method is first applied to the acoustic wave equation in 1D and 2D and the algorithm is analysed analytically. This leads to some of the most important concepts in numerical methods relevant to most other methods that are presented in this volume, such as stability, numerical dispersion, and convergence. The finite-difference approximation of the 1D elastic wave equation in the velocity–stress form is presented, leading to the staggered-grid concept in both space and time. At the end of the chapter, several specific issues are discussed, including the implementation of boundary conditions, other non-standard implementations, and recent developments.

4.1 History

A historical view is always a compromise. For the purpose of this volume I only highlight a few milestones in the history of finite differences in seismology.

Fig. 4.1 *Snapshot of horizontal ground motion using a 3D finite-difference method. The 1992 M5.3 Roermond earthquake in the Cologne area, Germany, is simulated using a 3D structure of the sedimentary basin. Red and blue colours denote positive and negative horizontal ground velocity, respectively. The low-velocity basin structure amplifies motion compared to the surrounding bedrock and substantially prolongs shaking. From Igel et al. (2015).*

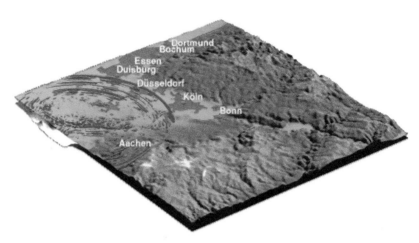

Fig. 4.2 *Image of the massively parallel supercomputer CM-2 of Thinking Machines Corp. that was used in the early nineties for such tasks as seismic wave simulations at the Institut de Physique du Globe in Paris. The external hard drive (the* data vault, *seen centre-right) was the size of a pub bar and had 20 GByte memory (compare with microSD cards today!). The CM-2 had 65,536 processors and a total RAM (rapid access memory) of 512 MBytes. On such machines some of the first parallel finite-difference simulations were performed by Peter Mora. Figure courtesy of Tamiko Thiel.*

An extensive review of the historic developments is given in the excellent book by Moczo et al. (2014).

The first application of the finite-difference method to elastic wave propagation can be attributed to Alterman and Karal (1968) who—interestingly enough at around the same time that the reflectivity method was developed by Fuchs and Müller (1971)—approximated the elastic wave equation in cylindrical coordinates. This allowed wave-propagation simulations in layered media. Boore (1970) used the finite-difference method to simulate Love waves and was (to my knowledge) the first to show snapshots of seismic wave fields. Alford et al. (1974) carried out a thorough analysis of the finite-difference approximation to the acoustic wave equation, comparing numerical results with analytical solutions. This was followed by applications to the elastic wave problem by Kelly et al. (1976).

The now widely used concept of staggered grids was introduced to solve the problem of rupture propagation (Madariaga, 1976; Virieux and Madariaga, 1982), and later adapted to the problem of elastic SH and P-SV wave propagation in 2D (Virieux, 1984; Virieux, 1986), respectively. High-order operators improving accuracy were presented by Levander (1988). He also introduced the concept of stress imaging to implement the free-surface boundary condition.

The evolution of numerical methods applied to wave propagation was tightly linked to the development of computational hardware. A major step towards realistic simulations was possible with the invention of parallel computers (an example is shown in Fig. 4.2). The CM-2 was tailor-made for finite-difference algorithms as it favoured near-neighbour communication. Increasing computer power led to the extension of the method to 3D, with initial applications by Frankel and Vidale (1992), Graves (1993), Olsen and Archuleta (1996), and Pitarka and Irikura (1996). Other rheologies such as viscoelastic behaviour (Day and Minster, 1984; Emmerich and Korn, 1987; Robertsson et al., 1994) and anisotropic material (Igel et al., 1995) were incorporated shortly thereafter.

Finite-difference methods were applied to the problem of global wave propagation with formulations in spherical coordinates, first with the axisymmetric approximation (Igel and Weber, 1995, 1996; Chaljub and Tarantola, 1997) and for 3D spherical sections (Igel et al., 2002). The implementation of frictional boundary conditions in 3D finite-difference algorithms (Olsen et al., 1997) had a strong impact in the field of dynamic rupture analysis. Nielsen and Tarantola (1992) presented a finite-difference scheme for wave propagation in which a threshold criterion would allow any node to fail (i.e., break or rupture).

Despite the difficulties of implementing (e.g. free-surface) boundary conditions, methods were developed and tested to allow the simulation of strong topographies, for example in connection with volcano seismology (Ohminato and Chouet, 1997) and for other rheologies (Robertsson and Holliger, 1997). Moczo et al. (2002) developed strategies for the accurate simulation of waves through strongly heterogeneous media.

At an early stage finite-difference simulations were incorporated in full waveform inversion schemes, initially in 2D, as with Crase et al. (1990), and later in 3D (Chen et al., 2007). Particularly for exploration-type problems, the finite-difference method is now widely used for 3D inversion as body wave propagation with finite differences is highly efficient. For a review see Virieux and Operto (2009).

4.2 The finite-difference method in a nutshell

The finite-difference method is the classic example of a grid method. For space–time dependent problems, such as the equation(s) describing the propagation of seismic waves, space and time are both discretized (usually) on regular space–time grids. This implies (unlike series expansion methods) that the values are only known at these points. In their simplest form, the partial derivatives in the original equations are replaced by the well-known finite differences that use the function values at adjacent grid points to approximate the function's derivatives.

The acoustic wave equation in 1D with constant density

$$\partial_t^2 p(x, t) = c(x)^2 \partial_x^2 p(x, t) + s(x, t) \tag{4.1}$$

with pressure p, acoustic velocity c, and source term s contains two second derivatives that can be approximated with a difference formula such as

$$\partial_t^2 p(x, t) \approx \frac{p(x, t + dt) - 2p(x, t) + p(x, t - dt)}{dt^2} \tag{4.2}$$

and equivalently for the space derivative. Injecting these approximations into the wave equation allows us to formulate the pressure p(x) for the time step $t + dt$ (the future) as a function of the pressure at time t (now) and $t - dt$ (the past). This is

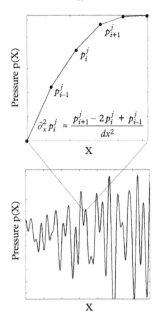

Fig. 4.3 *Principle of the finite-difference method illustrated with a simulation of acoustic waves in 1D through a medium with random velocity perturbations.* **Bottom:** *Snapshot in space of the pressure field $p(x,t)$.* **Top:** *Close-up of a detail of the wave field with the grid points indicated by dots. The finite-difference method is based on the calculation of partial differentials with formulas as indicated in the figure using values in the neighbourhood $p^j_{i\pm 1}$ appropriately weighted. The indices i, j denote discrete space and time levels, respectively. Here a three-point operator is used to approximate the second derivatives in space (and time).*

---

[1] *Implicit schemes are algorithms where the state at $t + dt$ depends on values at $t + dt$ leading to a system of equations that needs to be solved at each time step. Such schemes are sometimes more stable but are not necessarily more accurate. For the wave equation implicit schemes play a minor role and are therefore not discussed here.*

called an explicit scheme allowing the *extrapolation* of the space-dependent field into the future only looking at the nearest neighbourhood.[1] Actually, it is this near-neighbour communication that makes the finite-difference method so efficiently parallelizable on modern supercomputers.

The schema is illustrated in Fig 4.3 showing a snapshot of a 1D wavefield. The values at three adjacent grid points are used to calculate the second derivatives necessary to extrapolate to the next time step.

The finite-difference method is very popular and widely used, primarily because of its simplicity and the easy adaptation to parallel hardware. A disadvantage of the finite-difference method is the relative difficulty with implementation of sufficiently accurate and stable boundary conditions, especially the non-planar free surface. Therefore the finite-difference method plays an important role in exploration geophysics where surface waves are often considered noise. Recent developments (see (Moczo et al., 2014)) indicate that accurate implementations of the free surface with modified operators or hybrid schemes are possible.

4.3 Finite differences and Taylor series

Our first problem is to find a way to calculate derivatives of functions that are defined on regularly spaced grid points. The finite-difference method is based on the definitions of the derivative of a function $f(x)$ as the limit of the distance between two functional values defining the slope approaches zero. As is obvious from the definitions below, there are three different ways of introducing the derivative; the forward derivative,

$$d_x f(x) = \lim_{dx \to 0} \frac{f(x + dx) - f(x)}{dx}, \qquad (4.3)$$

the centred derivative

$$d_x f(x) = \lim_{dx \to 0} \frac{f(x + dx) - f(x - dx)}{2dx}, \qquad (4.4)$$

and the backward derivative

$$d_x f(x) = \lim_{dx \to 0} \frac{f(x) - f(x - dx)}{dx}. \qquad (4.5)$$

The equal sign and the same naming is justified here as in the limit the derivatives are equal (provided that the function $f(x)$ is continuous and smooth around x). The *finite* in the difference method originates in not taking the limit but keeping a finite dx. As a consequence we obtain three definitions; the forward difference denoted

$$d_x f^+ \approx \frac{f(x + dx) - f(x)}{dx}, \qquad (4.6)$$

the central difference

$$d_x f^c \approx \frac{f(x+dx)-f(x-dx)}{2dx}, \tag{4.7}$$

and the backward difference

$$d_x f^- \approx \frac{f(x)-f(x-dx)}{dx}. \tag{4.8}$$

The *approximate* sign is important here as the derivatives at point x are not exact. One of the most fundamental problems in numerical analysis is always to quantify *how* accurate numerical derivative operations are,[2] and we will devote a substantial part of this chapter to this question. For the moment we will restrict ourselves to understanding the accuracy of finite-difference-based derivatives in the space domain. Let us have a look at the definition of Taylor series:[3]

$$f(x+dx) = f(x) + f'(x)dx + \tfrac{1}{2}f''(x)dx^2 + O(dx^3), \tag{4.9}$$

where $O(dx^3)$ denotes the remaining error term with the leading order 3. Subtraction with $f(x)$ and division by dx leads us right away to the definition of the forward derivative given above

$$\frac{f(x+dx)-f(x)}{dx} = f'(x) + O(dx), \tag{4.10}$$

but also provides us with the leading order of the error term, which in this case is 1. Also written as $O(dx)$. Surprisingly, using the same approach, adding the Taylor series for $f(x+dx)$ and $f(x-dx)$ and dividing by $2dx$ leads to

$$\frac{f(x+dx)-f(x-dx)}{2dx} = f'(x) + O(dx^2) \tag{4.11}$$

with an error term $O(dx^2)$! This implies that a centred finite-difference scheme converges more rapidly to the correct derivative on a regular grid, if one reduces the grid spacing dx. At this point some important conclusions can be drawn. First, it does matter which of the above formulations one uses to approximate a derivative. Second, this does not imply one or the other finite-difference approximation is always the better one. This will become clear when playing with these definitions and various simple partial differential equations. The message is that one always has to verify the overall accuracy of a solution scheme. It might even be that one of the approximations never leads to an accurate solution![4]

4.3.1 Higher derivatives

What about higher derivatives? The partial differential equations we are dealing with often (not always) have second (seldom higher) derivatives. Let us start

[2] In his most famous speech Abraham Lincoln said: "Whatever differs from this to the extent of the difference (…)". Much of numerical analysis is about the *extent of the difference*. How much inaccuracy can we afford in the solution of our problem?

[3] Space derivatives:
$\partial_x f(x) \to f'(x)$
$\partial_x^2 f(x) \to f''(x)$

[4] Examples are problems with advection terms, e.g. flow problems. Here, for the advection term so-called upwind derivatives are used, which are realized by non-centred derivatives.

by using the definitions for the first derivatives developed above. Taking the derivative of those terms mixing a forward and a backward definition leads to

$$\partial_x^2 f \approx \frac{\frac{f(x+dx)-f(x)}{dx} - \frac{f(x)-f(x-dx)}{dx}}{dx}$$
$$= \frac{f(x+dx) - 2f(x) + f(x-dx)}{dx^2} \tag{4.12}$$

and the order of the leading error term can be determined in a straightforward manner (see exercises). Let us introduce a more general way of determining the weights with which the function values have to be multiplied to obtain derivative approximations. This schema will prove useful for finding more accurate differential operators.

We write down the Taylor series for 2 grid points at $x \pm dx$, include the function at the central point as well, and multiply each by a real number.

$$af(x+dx) = a\left[f(x) + f'(x)dx + \tfrac{1}{2!}f''(x)dx^2 + \cdots\right]$$
$$bf(x) = b\left[f(x)\right] \tag{4.13}$$
$$cf(x-dx) = c\left[f(x) - f'(x)dx + \tfrac{1}{2!}f''(x)dx^2 - \cdots\right].$$

Our goal is to find solution(s) for coefficients a, b, c such that the sum over the weighted functional values leads to approximations of the second derivative. First, we sum up the above equations, drop higher-order terms, and rearrange, to obtain

$$af(x+dx) + bf(x) + cf(x-dx) \approx$$
$$f(x)\,[a + b + c]$$
$$+dxf'\,[a \quad\quad - c] \tag{4.14}$$
$$+\tfrac{1}{2!}dx^2 f''\,[a \quad\quad + c].$$

Looking at these equations it is easy to see that to obtain an approximation of the second derivative we require

$$a + b + c = 0$$
$$a \quad\quad - c = 0 \tag{4.15}$$
$$a \quad\quad + c = \frac{2!}{dx^2}.$$

This is a linear system of equations that we can solve with standard linear algebra tools. Casting this in matrix form we get

$$\begin{pmatrix} 1 & 1 & 1 \\ 1 & 0 & -1 \\ 1 & 0 & 1 \end{pmatrix} \begin{pmatrix} a \\ b \\ c \end{pmatrix} = \begin{pmatrix} 0 \\ 0 \\ \frac{2!}{dx} \end{pmatrix} \tag{4.16}$$
$$\mathbf{A} \qquad\qquad \mathbf{w} \;=\; \mathbf{s}$$

defining **A** as the system matrix, **w** as the vector with the differential operator (weights), and **s** as the vector specifying the desired solution. It turns out the square matrix so defined is invertible and we find the operator weights by

$$\mathbf{w} = \mathbf{A}^{-1}\mathbf{s} \tag{4.17}$$

and obtain

$$
\begin{aligned}
a &= \frac{1}{dx^2} \\
b &= -\frac{2}{dx^2} \\
c &= \frac{1}{dx^2}
\end{aligned}
\tag{4.18}
$$

which is the approximation to the second derivative in its lowest-order form, as presented in Eq. 4.12.[5] It is instructive to investigate the behaviour of these derivative approximations for various functions and explore the accuracy as the number of grid points per wavelength decreases (see exercises).

The way we derived the differential operator by formulating a linear system of equations is actually very powerful. We can use it to derive more accurate operators by looking at grid points further away from the point at which the derivative is calculated.

4.3.2 High-order operators

What happens if we extend the *domain of influence* for the derivative(s) of our function $f(x)$? Does it allow us to improve the accuracy of the derivative approximations? From the solution scheme above we expect to have to define a square linear system with as many unknown coefficients as Taylor expansions. As an example, let us search for a five-point operator for the second derivative. We seek

$$
f''(x) \approx
$$
$$
af(x + 2dx) + bf(x + dx) + cf(x) + df(x - dx) + ef(x - 2dx), \tag{4.19}
$$

and sum up the corresponding Taylor series up to order 4 obtaining a linear system of equations for the coefficients

$$
\begin{array}{llllll}
a & +b & +c & +d & +e = 0 \\
2a & +b & & -d & -2e = 0 \\
4a & +b & & +d & +4e = \dfrac{1}{2dx^2} \\
8a & +b & & -d & -8e = 0 \\
16a & +b & & +d & +16e = 0,
\end{array}
\tag{4.20}
$$

[5] Note that the weights for the second derivative are symmetric re. the central point, whereas the weights for the first derivative are antisymmetric. In other words the information at the central point is not contributing to the first derivative!

after multiplying each line in order to obtain integer coefficients. Using matrix inversion we obtain a unique solution

$$a = -\frac{1}{12dx^2}$$
$$b = \frac{4}{3dx^2}$$
$$c = -\frac{5}{2dx^2}$$
$$d = \frac{4}{3dx^2}$$
$$e = -\frac{1}{12dx^2}.$$

(4.21)

The leading error term for the second derivative using these difference weights is $O(dx^4)$ (see exercises). This accuracy improvement is indeed substantial and from a practical point of view such a five-point operator (or equivalently four-point operator for the first derivative) should always be preferred.

In principle—using the linear system above—we could also seek the corresponding weights for a first or third derivative (or an interpolation) by simply changing the solution vector on the right-hand side. We note—and for two-sided operators this is general—that the modulus of the coefficients decreases with increasing distance from the central point (see illustration in Fig. 4.4).

As will be explained below, one of the most widely used concepts in connection with finite-difference solutions to the elastic wave equation is so-called *grid staggering*. With this method—either in space or time—the derivative is calculated half-way between two points on a regular spaced grid. We have just documented that the number of points used for the calculation of numerical derivatives (we also say the length of the finite-difference operator) is directly related to the accuracy of the derivative. Obviously, this comes at the price of having to undertake more floating-point operations. Finding the right balance between computational effort and mathematical accuracy is at the heart of numerical analysis. It also depends on the specific hardware a computer program is implemented on.

As far as finite-difference operators are concerned, it turns out that in the case of first space derivatives four-point operators are the most widely used (and five-point operators for second derivatives). The accuracy improvement from second to fourth order is substantial. Using longer operators usually does not pay off. The mathematical approach presented in this section can also be used to develop one-sided high-order differential operators for boundaries (see exercises). It is instructive to compare high-order Taylor operators with those obtained using Fourier transforms (see chapter on the pseudospectral method).

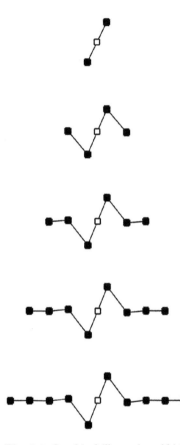

Fig. 4.4 *Graphical illustration of high-order Taylor operators for the first derivative. The derivative is defined in between the central grid points (white square). The differential weights (black squares) rapidly decrease with distance from the central point of evaluation.*

4.4   Acoustic wave propagation in 1D

Let us now proceed to find numerical solutions to the wave equation. The finite-difference method allows a fairly simple analysis of the basic properties of numerical solutions for wave equations that are quite general. Therefore, we start with the simplest wave equation—the constant density acoustic wave equation—in 1D. In terms of physics, this equation might describe pressure waves in a gas or stationary fluid. A snapshot example of acoustic waves in a homogeneous medium is shown in Fig. 4.5. With slight modifications this scalar wave equation is also descriptive of the vibrations of a string.

Using dense notation and omitting the spatial and temporal dependencies the scalar acoustic wave equation in Cartesian coordinates can be written as

$$\partial_t^2 p \;=\; c^2 \partial_x^2 p + s, \tag{4.22}$$

imposing pressure-free conditions at the two boundaries as

$$p(x)\,|_{x=0,L} \;=\; 0. \tag{4.23}$$

The following dependencies apply:

$$
\begin{aligned}
p &\rightarrow p(\mathbf{x}, t) &&\text{pressure}\\
c &\rightarrow c(\mathbf{x}) &&\text{P-velocity}\\
s &\rightarrow s(\mathbf{x}, t) &&\text{source term.}
\end{aligned}
$$

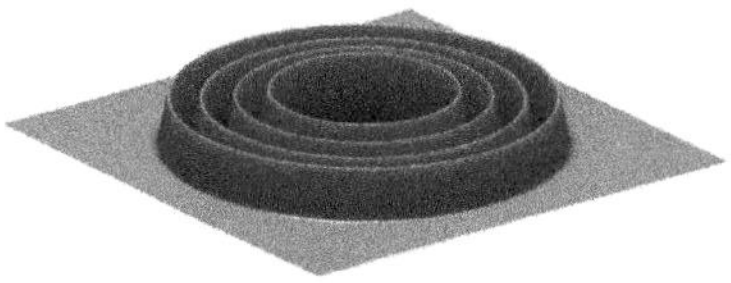

Fig. 4.5 *Acoustic wave simulation. Pressure waves radiate isotropically from the central point. A sinusoidal source time function was applied.*

As a first step we discretize space and time with constant increments dx and dt. Thus

$$
\begin{aligned}
x_j &= j\,dx, & j &= 0, j_{max}\\
t_n &= n\,dt, & n &= 0, n_{max}.
\end{aligned} \tag{4.24}
$$

The choice of these space and time increments is crucial and we will discuss this in more detail when we are able to forecast the consequences for our wave-propagation problem. We now make the step from the continuous description of the partial differential equation to a discrete description using the indices introduced above. From now on, the upper index will correspond to the time discretization, and the lower index (or indices) will correspond to the spatial discretization, for example

$$
\begin{aligned}
p_j^{n+1} &\rightarrow p(x_j, t_n + dt)\\
p_j^{n} &\rightarrow p(x_j, t_n)\\
p_j^{n-1} &\rightarrow p(x_j, t_n - dt)\\
p_{j+1}^{n} &\rightarrow p(x_j + dx, t_n)\\
p_j^{n} &\rightarrow p(x_j, t_n)\\
p_{j-1}^{n} &\rightarrow p(x_j - dx, t_n).
\end{aligned} \tag{4.25}
$$

This discretization implies that, when describing the space-dependent fields on the computer, we will initialize arrays with the corresponding dimensions (in 1D

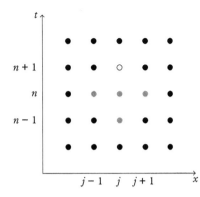

Fig. 4.6 *Illustration of the space-time discretization scheme of the finite-difference algorithm for the 1D acoustic wave equation. The x-axis corresponds to space, the y-axis to time. The open circle denotes the point $p(x_j, t_{n+1})$ to which the state of the pressure field is extrapolated. Such a space-time operator is also called a* stencil.

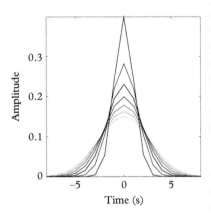

Fig. 4.7 *Dirac delta function. Discretized Gauss functions converging towards the Dirac delta function as the halfwidth is decreasing. The integral remains constant and is unity (see Eq. 4.32).*

column vectors) and are able to address the values at certain locations and their neighbours by means of their indices.

With the definitions of the derivatives introduced in the previous section we can replace the partial derivatives in Eq. 4.22 to obtain

$$\frac{p_j^{n+1} - 2p_j^n + p_j^{n-1}}{dt^2} = c_j^2 \left[\frac{p_{j+1}^n - 2p_j^n + p_{j-1}^n}{dx^2} \right] + s_j^n. \tag{4.26}$$

Note that the right-hand side is defined at the same time level n, at which information around the grid point j is used to calculate the spatial derivatives (see Fig. 4.6). The calculation of the time derivatives on the left-hand side requires information from three different time levels. Assuming that the information at the levels n (the presence) and $n-1$ (the past) is known, we can solve for the unknown future field p_j^{n+1}

$$p_j^{n+1} = c_j^2 \frac{dt^2}{dx^2} \left[p_{j+1}^n - 2p_j^n + p_{j-1}^n \right] + 2p_j^n - p_j^{n-1} + dt^2 s_j^n. \tag{4.27}$$

In practice, we will loop over time and, at each time level n, calculate the space derivatives using the information from neighbouring grid points (i.e. loop over space). The initial condition of our wave simulation problem is such that everything is at rest at time $t = 0$:

$$p(x, t = 0) = 0, \ \partial_t p(x, t = 0) = 0. \tag{4.28}$$

Waves radiate as soon as the source term $s(x, t)$ starts to act. Let us have a closer look at the spatial and temporal form of the source term. In many situations (seismic exploration, earthquake seismology) sources of seismic waves are treated as points (later we will consider the case of finite-sized sources). In general, a geophysical problem requires the source point to be at a certain location that might not coincide with a grid point. For the sake of simplicity let us assume for the moment that the source acts directly at a grid point with index j_s.

What about the temporal behaviour of the source? In fact, from a physical point of view it is often useful to calculate the Green's function, that is, the response to an impulse in both time and space of the form

$$s(x, t) = \delta(x - x_s) \, \delta(t - t_s), \tag{4.29}$$

where x_s and t_s are source location and source time, respectively, and $\delta(.)$ corresponds to the Dirac delta function (see Chapter 2 for analytical solutions). While the injection of a spatial point source can in principle be realized (see section below), a delta function in time cannot. A delta function contains all frequencies (white spectrum) and we cannot expect that our numerical algorithm will be capable of providing accurate solutions in this case (see discussion of this problem in Chapter 2). Therefore, we will operate with a band-limited source time function

$$s(x, t) = \delta(x - x_s) f(t) \tag{4.30}$$

where the temporal behaviour $f(t)$ is chosen according to our specific physical problem.

The correct implementation of sources in a numerical solver such that the results converge to the analytical solution requires some care. Referring to the discussion of the analytical solutions in Chapter 2 a delta-like point source can be implemented, provided that its spatial integration leads to unity. This can be achieved with appropriately scaled functions that converge to the delta function. One possibility is the boxcar function

$$\delta_{bc}(x) = \begin{cases} 1/dx & |x| \leq dx/2 \\ 0 & \text{elsewhere} \end{cases} \tag{4.31}$$

scaled by the width of the box (here: the grid point distance dx, in general the grid cell volume). For an illustration see Fig. 2.8. This function solves the scaling issue for the spatial source function. What about time?

Another function that, in the limit $a \to 0$, converges to a delta function is the Gaussian

$$\delta_a(t) = \frac{1}{\sqrt{2\pi a}} e^{-\frac{t^2}{2a}} \tag{4.32}$$

with the required integration property (see exercises in Chapter 2). An illustration is given in Fig. 4.7. It is important to test numerical schemes against the analytical solutions with respect to proper source scaling (see supplementary material).

Let us discuss an example. Suppose we want to simulate acoustic wave propagation in a 10 km column (e.g. the atmosphere) and assume an air sound speed of $c = 343$ m/s. We would like to *hear* the sound wave so it would need a dominant frequency of at least 20 Hz (at the bottom of the audible frequency range). For the purpose of this exercise we initialize the source time function $f(t)$ using the first derivative of a Gaussian function (because we are aware that in 1D the resulting signal is an integral of the source time function and we want a Gaussian waveform)

$$f(t) = -8 f_0 (t - t_0) e^{-\frac{1}{(4f_0)^2}(t - t_0)^2}, \tag{4.33}$$

where t_0 corresponds to the time of the zero-crossing, and f_0 is the dominant frequency. The source time function and its amplitude spectrum are illustrated in Fig. 4.8.

At this point we have to choose the spatio-temporal setup of the simulation and this is fundamental to the solution of all wave-propagation problems, independent of the specific numerical methods. The following questions have to be answered carefully before starting any simulation:

- What is the minimum spatial wavelength that propagates inside the medium?

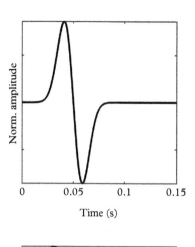

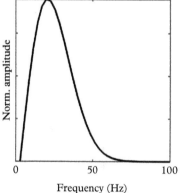

Fig. 4.8 *Source time functions:* **Top:** *Frequently used source time functions in wave propagation are Gaussian functions (and their derivatives). Here, a source time function (first derivative of a Gaussian) with a dominant frequency of 20 Hz is shown.* **Bottom:** *Amplitude spectrum of the source time function given above.*

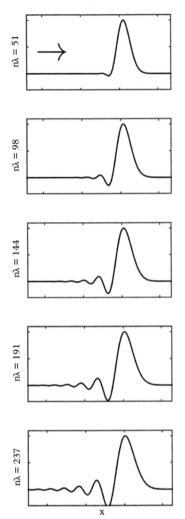

Fig. 4.9 *Simulation results (snapshots in space) for the 1D acoustic wave equation at various propagation distances (given as number of dominant wavelengths propagated). In theory, the Gaussian shaped signal should propagate undistorted for ever. However, the finite-difference discretization results in the signal becoming dispersive and it disintegrates with time (or distance). The propagation direction is indicated by an arrow.*

- What is the maximum velocity inside the medium?
- What is the propagation distance of the wavefield (e.g. in dominant wavelengths)?

In order to answer these questions it is sufficient to look at the relation between frequency and wavenumber

$$c = \frac{\omega}{k} = \frac{\lambda}{T} = \lambda f, \tag{4.34}$$

where c is velocity, T is period, λ is wavelength, f is frequency, and $\omega = 2\pi f$ is angular frequency. We have chosen a source time function with a dominant frequency of $f_0 = 20\,\text{Hz}$. From Fig. 4.8, however, we can see that a substantial amount of energy in the wavelet is at frequencies above 20 Hz.

For the given velocity the corresponding wavelengths are $\lambda = 17\,\text{m}$ and $\lambda = 7\,\text{m}$ for frequencies 20 Hz and 50 Hz, respectively. For this exercise we choose a grid increment of $dx = 0.5\,\text{m}$ which would result in about 34 points per spatial wavelength for the dominant frequency. The time increment will be set to $dt = 0.0012\,\text{s}$ corresponding to $\approx$40 points per dominant period. Restrictions of the choice of the space–time discretization will be discussed in the next section.

In the following a Python code fragment is presented with the core of the finite-difference algorithm:

```
# Time extrapolation
for it in range(nt):
    # calculate partial derivatives (omit boundaries)
    for i in range(1, nx - 1):
        d2p[i] = (p[i + 1] - 2 * p[i]\
            + p[i - 1]) / dx ** 2
    # Time extrapolation
    pnew = 2 * p - pold + dt ** 2 * c ** 2 * d2p
    # Add source term at isrc
    pnew[isrc] = pnew[isrc] +  dt ** 2 * src[it] / dx
    # Remap time levels
    pold, p = p, pnew
```

Here, *nt* is the maximum number of time steps, $nx = 20{,}000$ is the number of grid points for *x*, *isrc* is the grid point at which the source is injected, and *src[it]* is the source time function scaled by the grid increment. For the sake of simplicity the acoustic velocity c is kept constant. The results of this simulation example at various time steps (i.e. propagation distances in terms of dominant wavelengths) are shown in Fig. 4.9.

In theory, the integral of the injected source time function (thus a Gaussian) should propagate for ever without distortion. This is not what we observe in our numerical simulation. After propagating about 100 wavelengths, the waveform begins to disintegrate and forms a distinctive tail. This phenomenon is called numerical dispersion and is an intrinsic feature of almost all numerical approximations to wave equations. The velocity of the wavefield becomes frequency dependent as a function of the discretization. Of course this phenomenon needs to be avoided. The choice of the right set-up is the central task for any seismic simulation experiment. In the next section we shed light on these phenomena using analytical tools.

How can we analyse our numerical approximation to the wave equation and compare with the analytical solution in a quantitative way? This question leads us to the so-called von Neumann analysis.[6]

4.4.1 Stability

We start with the definition of a plane complex harmonic wave for pressure p propagating in x-direction with wavenumber k and angular frequency ω

$$p(x, t) = e^{i(kx - \omega t)}. \tag{4.35}$$

In our finite-difference approximation to the wave equation we introduced the spatio-temporal discretization as

$$\begin{aligned} x_j &\to jdx \\ t_n &\to ndt \end{aligned} \tag{4.36}$$

which we will use in the plane-wave formula such that

$$\begin{aligned} p_j^n &\to e^{i(kjdx - \omega ndt)} \\ p_{j+1}^n &= e^{i(k(j+1)dx - \omega ndt)} \\ &= e^{ikdx} e^{i(kjdx - \omega ndt)} \\ &= e^{ikdx} p_j^n \\ p_j^{n+1} &= e^{-i\omega dt} p_j^n. \end{aligned} \tag{4.37}$$

With this discrete plane-wave trial solution we can enter the (source-free) finite-difference approximation of the acoustic wave equation (Eq. 4.22), replacing all terms in Eq. 4.27 with the terms above to obtain

$$e^{i\omega dt} + e^{-i\omega dt} - 2 = c^2 \frac{dt^2}{dx^2} (e^{ikdx} + e^{-ikdx} - 2), \tag{4.38}$$

where we divided by the term p_j^n on both sides. Using the definition

$$\cos x = \frac{1}{2}(e^{ix} + e^{-ix}) \tag{4.39}$$

we get

$$\cos(\omega dt) - 1 = c^2 \frac{dt^2}{dx^2} (\cos(kdx) - 1), \tag{4.40}$$

and with the trigonometric relation

[6] John von Neumann (1903–1957) was a mathematician of Hungarian origin who is considered (amongst other things) as one of the fathers of quantum mechanics, game theory, and computational science.

$$\sin \frac{x}{2} = \pm \sqrt{\frac{1 - \cos x}{2}} \tag{4.41}$$

finally arrive at

$$\sin \left(\omega \frac{dt}{2}\right) = c \frac{dt}{dx} \sin \left(k \frac{dx}{2}\right). \tag{4.42}$$

Note that here we assume true physical wavenumber resulting in a grid-dependent frequency (and thus numerical wavenumber). From this relation one of the most important conclusions in numerical analysis can be drawn. The equation only has real solutions in general when the connecting term has the following property

$$\epsilon = c \frac{dt}{dx} \leq 1, \tag{4.43}$$

as the *sine* terms on both sides have values in the interval [−1, 1]. If this condition is not met the solution will explode and is called unstable. This stability condition is also called the Courant–Friedrichs–Lewy (CFL) criterion (named after three scientists who described it in a paper in 1928).[7] This result has important consequences:

- The space–time discretization cannot be arbitrarily chosen but depends on the medium properties (here: velocity c).

- The CFL criterion describes the *conditional stability* that leads to convergent behaviour of the solution.

- As the space discretization (increment dx) is often imposed by considering the smallest seismic velocities in the medium and the highest frequencies (i.e. the shortest wavelengths), the CFL criterion determines the time increment dt and thus the number of time steps to achieve a certain simulation length.

- The actual value (in this case 1) that has to be respected depends on the number of space dimensions and the overall algorithm.

- It is important to note that the fulfilment of the CFL criterion is necessary but not sufficient to guarantee an accurate simulation!

Equation 4.42 allows us to investigate the relationship between wavelength and frequency in our numerical realm in more detail.

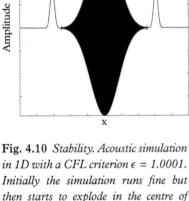

Fig. 4.10 *Stability. Acoustic simulation in 1D with a CFL criterion $\epsilon = 1.0001$. Initially the simulation runs fine but then starts to explode in the centre of the domain with exponentially growing numbers. Try this out with computer practicals. It is fun!*

4.4.2 Numerical dispersion

Following equation 4.42 the angular frequency ω can be expressed as

$$\omega = \frac{2}{dt} \sin^{-1} \left[c \frac{dt}{dx} \sin k \frac{dx}{2} \right]. \tag{4.44}$$

[7] The stability criterion is *the* most important concept to remember. Its principle is essential for the planning of simulation tasks. Note that it is dimensionless and is essentially the ratio of a physical velcity c and a *grid* velocity dx/dt.

The phase velocity c that—in theory—should be identical to the acoustic velocity can be obtained by dividing the above equation by wavenumber k

$$c(k) = \frac{\omega}{k} = \frac{2}{kdt} \sin^{-1}\left[c\frac{dt}{dx}\sin k\frac{dx}{2}\right]. \tag{4.45}$$

The relationship explains the observations we documented above in our first numerical example. The phase velocity is no longer independent of wavenumber as was the case in the original acoustic wave equation. Our numerical approximation leads to (unphysical!) dispersion.

This is illustrated in Fig. 4.11 where the phase velocity is shown as a function of the number of grid points per wavelength.[8] This figure has an important message. Unless a sufficient number of grid points per (dominant or minimum) wavelength is used, the wavefield is strongly dispersive and leads to inaccurate results.

Note that a curious situation occurs if you set the CFL limit $\epsilon = 1$ in the acoustic 1D example. In that case you would not observe any dispersion! While this sounds interesting, it is not really significant for practical applications (remember we are using numerical methods to be able to simulate wave propagation through heterogeneous material, so the CFL limit varies in space).

In many publications—and particularly when new numerical approaches appear—there are statements about the number of grid points to be used for accurate simulations. By themselves, such statements are not meaningful as the accuracy of a simulation also depends on the overall propagation distance. The longer the propagation distance, the more the errors accumulate, and the more grid points per wavelength have to be used.

4.4.3   Convergence

We introduced finite-difference operators with the idea that when the finite difference becomes infinitesimally small the analytical derivative is recovered. Does that hold for our numerical approximation of the acoustic wave equation? We can answer this question by careful inspection of Eq. 4.45 using

[8] This is more instructive than showing the wavenumber. The Nyquist wavenumber $k_N = \pi/dx$ corresponds to 2 grid points per wavelength, dx being the grid increment.

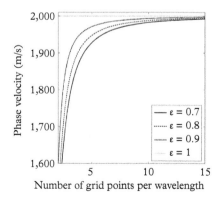

Fig. 4.11 *Numerical dispersion. The numerical phase velocity is shown as a function of number of grid points per wavelength for various CFL criteria. The left limit corresponds to two points per wavelength (Nyquist wavenumber). The correct propagation velocity is 2,000 m/s. As the number of grid points per wavelength increases the correct velocity is recovered.*

$$\sin x \approx x \qquad \text{for small } x$$
$$\sin^{-1} x \approx x \qquad \text{for small } x. \tag{4.46}$$

It is easy to show that as the spatial increment dx and the time increment dt converge to zero the original dispersion relation

$$\lim_{\substack{dt \to 0 \\ dx \to 0}} c(k) = \frac{\omega}{k} = c_{\text{exact}} \tag{4.47}$$

is recovered, demonstrating the *convergence* of the numerical algorithm (homogeneous case) to the analytical solution.

4.5 Acoustic wave propagation in 2D

Even though the same principles should apply in higher dimensions we will consider here the acoustic wave equation in 2D as it allows us to investigate the behaviour of the errors as a function of propagation direction. Also, in practice the 2D acoustic algorithm (see supplementary material) already allows many useful practical exercises with interesting model set-ups.

In 2D the constant-density acoustic wave equation is given by

$$\partial_t^2(x, z, t) = c(x, z)^2 (\partial_x^2 p(x, z, t) + \partial_z^2 p(x, z, t)) + s(x, z, t), \tag{4.48}$$

where the z-coordinate is chosen because in many applications the $x - z$ plane is considered a vertical plane with z as depth coordinate. In accordance with the above developments we discretize space–time with

$$p(x, z, t) \rightarrow p_{j,k}^n = p(ndt, jdx, kdz). \tag{4.49}$$

Using the three-point operator for the second derivatives in time leads us to the extrapolation scheme

$$\frac{p_{j,k}^{n+1} - 2p_{j,k}^n + p_{j,k}^{n-1}}{dt^2} = c_j^2 (\partial_x^2 p + \partial_z^2 p) + s_{j,k}^n, \tag{4.50}$$

where on the r.h.s. the space and time dependencies are implicitly assumed and the partial derivatives are approximated by

$$\partial_x^2 p = \frac{p_{j+1,k}^n - 2p_{j,k}^n + p_{j-1,k}^n}{dx^2}$$
$$\partial_z^2 p = \frac{p_{j,k+1}^n - 2p_{j,k}^n + p_{j,k-1}^n}{dz^2} . \tag{4.51}$$

Note that for a regular 2D grid $dz = dx$ (see Fig. 4.12). Let us set up an example and investigate the behaviour of the wavefield. We want to simulate P-wave

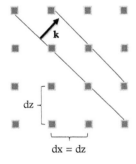

Fig. 4.12 *Regular grid with horizontal increment dx and vertical increment dz equal. This is not a requirement but differences in these increments result in stronger directional dependencies of the errors. In a regular grid propagation at an angle of 45° with respect to the coordinate axes is most accurate (see text).*

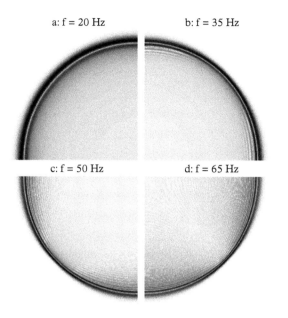

a: f = 20 Hz   b: f = 35 Hz

c: f = 50 Hz   d: f = 65 Hz

Fig. 4.13 *Numerical anisotropy: Snapshot of acoustic wavefield for various dominant frequencies keeping all other parameters constant: a: 20 Hz; b: 35 Hz; c: 50 Hz; d: 65 Hz. While at 20 Hz an isotropic wave field can be observed, at higher frequencies (fewer points per wavelength) numerical dispersion appears first in the direction of the grid axes. This propagation direction dependent effect is called numerical anisotropy.*

propagation in a reservoir scale model with maximum velocity c_{max} = 5 km/s and minimum velocity c_{min} = 3 km/s. The dominant frequency is chosen to be 20 Hz (from the discussion of the source time function above we expect energy up to 50 Hz to be present in the waveforms). The dominant wavelength is λ_{dom} = c/f_{dom} = 150 m. For this exercise we simulate a spatial domain of 5 km × 5 km and use a grid point distance dx = 10 m resulting in 15 grid points per wavelength for the dominant frequency. We examine the behaviour of the wavefield looking at the snapshots resulting from a source injected at the centre of the model.

The results are shown in Fig. 4.13 after a propagation time of $\approx$ 0.5 s. For acoustic wave propagation we expect an isotropic radiation pattern, that is, identical waveforms propagating in all directions. While this is indeed the case for the example with 20 Hz, for higher frequencies we can observe that the wavefield becomes anisotropic in the sense that in certain directions the wavefield deteriorates faster. This effect is called *numerical anisotropy*. In order to understand it we shall take another look at the numerical dispersion equation, but this time in 2D.

4.5.1 Numerical anisotropy

Can we understand the observed artificial anisotropic behaviour using our analytical approach? We start with the description of a plane harmonic wave propagating in 2D with wavenumber vector **k** pointing in the direction of propagation

$$p(x, z, t) = e^{i(kx-\omega t)} = e^{i(k_x x + k_z z - \omega t)} \tag{4.52}$$

and use the discretization of our 2D problem to obtain

$$\hat{p}_{j,k}^n = e^{i(k_x j dx + k_z k dx - \omega n dt)}. \tag{4.53}$$

Substituting this formula with the pressure field of the source-free 2D acoustic wave equation (Eq. 4.50) and following the same steps as done for the 1D numerical dispersion analysis leads to the following relation for the numerical phase velocity in 2D (assuming a regular grid $dz = dx$)

$$c^{num}(k_x, k_z) = \frac{2}{\lambda dt} \sin^{-1}\left[\frac{dt^2}{dx^2} c^2 \left(\sin\left(\frac{k_x dx}{2}\right) + \sin\left(\frac{k_z dx}{2}\right) \right) \right]. \tag{4.54}$$

This relation can be analysed as a function of propagation direction noting that

$$\mathbf{k} = \begin{pmatrix} k_x \\ k_z \end{pmatrix} = \begin{pmatrix} |\mathbf{k}| \cos\alpha \\ |\mathbf{k}| \sin\alpha \end{pmatrix} \tag{4.55}$$

and the results are shown in Fig. 4.14. The error of the phase velocity (compared to the true acoustic velocity) is shown as a function of propagation direction and varying number of grid points per wavelength. The anisotropic behaviour is striking. For $n_\lambda = 5$ the error varies from 1.8% to 5%, with the most accurate propagation direction at an angle of 45° to the coordinate axes. This directional behaviour is independent of the number of grid points per wavelength, but of course the absolute error decreases with increasing n_λ.

Again there is a strong message in this observation and the analytical result. Seismic observations provide evidence for physical anisotropy on almost all scales. While the effects on the travel times are usually small, any errors we are committing in our numerical simulation map into errors in our resulting Earth model (in terms of tomographic inverse problems). Therefore, we must be careful to choose simulation parameters to ensure that we avoid numerical anisotropy.

4.5.2   Choosing the right simulation parameters

With the 2D acoustic algorithm we now possess a simulation tool that, despite its simple physics, allows us to investigate many wave-propagation phenomena relevant for seismology. Therefore, we can use this example to highlight the preparatory thinking that has to be done prior to any simulation task. We illustrate this with an example from earthquake seismology: the simulation of fault-zone trapped waves (see Fig. 4.15). Note that the scalar acoustic wave equation is mathematically identical to the SH-wave-propagation problem (assuming constant density). So, in the case of fault-zone trapped waves that are predominantly observed for SH-type ground motions, this is a useful basic physical model.

Let us investigate the effects of a narrow 200 m-wide fault zone with a 25% velocity decrease. In accordance with observations the target frequency is 10 Hz (dominant, maximum 30 Hz). We expect that for a seismometer sitting at the top of the fault zone a seismogram length of $t_{max} = 3.5$ s will be sufficient to observe trapped waves. The parameters of the physical model are summarized in

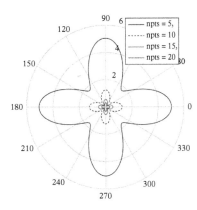

Fig. 4.14 *Numerical anisotropy. The error of the phase velocity is shown as a function of propagation direction (in %) for varying numbers of grid points per wavelength. The directional dependence is the same but the error increases as the number of grid points per wavelength decreases.*

Table 4.1 *Earth model set-up for the fault zone simulation example.*

	Description	Value
f_0, f_{max}	Dominant, maximum frequency	10 Hz, 30 Hz
c_{min}, c_{max}	Min., max. velocity	2250 m/s, 3,000 m/s
x_{max}, z_{max}	Min., max. extension	10 km, 10 km
t_{max}	Seismogram length	3.5 s

Table 4.1. To map this physical model on a computational grid and initialize a simulation we must answer the following questions:

- What is the smallest wavelength propagating in the medium?
  Answer: With the f_{max} = 30 Hz and the smallest velocity c_{min} = 2250 m/s the shortest wavelength is given by $\lambda_{min} = c_{min}/f_{max} = 75$ m.

- How many grid points per smallest wavelength are required by the numerical method (given the wave-propagation distance that needs to be covered)?
  Answer: The dominant wavelength in the low-velocity medium is $\lambda_{dom} = c_{min}/f_{dom} = 225$ m. Thus we expect to propagate more than 20 wavelengths to the surface. As we use a five-point operator we choose 20 points per dominant wavelength, resulting in about 7.5 points per smallest wavelength.

- What is the grid spacing that needs to be implemented?
  Answer: With 20 points per dominant wavelength we obtain $dx = \lambda_{dom}/20 = 11.25$ m grid spacing.

- What is the size of the physical domain and how many overal grid points are required?
  Answer: The spatial extent of the 2D model is 10 km × 10 km. In each dimension we thus need $10,000$ m/dx ≈ 900 grid points, leading to 900^2 overall grid points.

- Will the seismogram(s) be influenced by artificial boundary reflections (i.e., is it necessary to implement absorbing boundaries or increase the model size)?
  Answer: With the model set-up as chosen—the fault zone in the middle of the model—we might be able to avoid reflections from the domain boundaries, when seismograms are extracted only at the centre of the domain (above the fault zone). For receivers near the boundaries or very long time series, absorbing boundaries would be necessary.

- What is the maximum velocity in the model, and the resulting time step (given the grid increment and the CFL criterion)?
  Answer: The maximum velocity in the model is c_{max} = 3,000 m/s. Assuming a CFL value of $\epsilon = 0.7$ we can determine the time step required for a stable simulation as $dt = \epsilon \, dx/c_{max} = 0.0026$ s.

Fig. 4.15 *Landers, California. View East along a road crossing the epicentral area of the M7.3 Landers earthquake in 1992 with a horizontal surface slip of several metres. In many places trapped waves can be observed right above fault zones. Trapped waves were extensively investigated with a finite-difference method by Jahnke et al. (2002).*

- What is the overall number of time steps to be propagated?
  Answer: For a desired simulation time of $t_{max} = 3.5$ s the number of time steps required is $t_{max}/dt \approx 1{,}300$.

- How much core memory (RAM) will the simulation approximately require?
  Answer: For a simple estimate we focus on the space-dependent fields that will constitute the largest part of the memory allocation. Those fields are: (1) the velocity model, (2) the pressure field at three different time levels, and (3) two temporary fields containing the second space derivatives. Assuming double precision floating point numbers (8 bytes per number) this will require approximately $6 \times 900^2 \times 8$ bytes $\approx$ 40 MBytes. So you should be able to run this simulation easily on your smartphone!

Results for a simulation with these parameters are shown in Fig. 4.16. The snapshot at a simulation time of $t = 2$s indicates a highly focused wavefield propagating inside the fault zone towards the surface. The earthquake source is located at the left side of the low-velocity zone at 5 km depth. In the host medium (high velocity) head waves develop at the edges of the fault zone. Within the fault zone there are delayed and amplified fault-zone (trapped) waves propagating towards the surface. In any case this simple but quite realistic structural heterogeneity has a dramatic effect on the wave field recorded at the surface. The short scale high amplitudes measured directly above the fault zone have considerable relevance with respect to the shaking hazard in seismically active regions, but are usually ignored.

For many wave-propagation problems the questions just raised (usually of course for 3D problems) should be asked as part of planning a research project involving simulations. You want to be realistic in terms of what specifically is feasible in terms of memory and CPU time available to you. Do you want to carry out a few extremely highly resolved simulations or are you targeting tens of thousands of low-resolution simulations (e.g. for inverse problems or parameter space studies)? Answers to these questions are also required when applying for large computational resources at the supercomputer centres.

There are several issues that we have not addressed here, mainly because they go beyond this introductory level. Some examples are: (1) What rheologies are necessary for your specific problem? (2) What is the degree of heterogeneity (do you need to use parameter-averaging schemes)? (3) What is the wavenumber spectrum of your Earth model and which numerical model suits best (e.g. strongly heterogeneous random models with short-wavelength structures might be less suitable for spectral methods)? (4) Will your problem require parallel implementation? (5) How are maximum/minimum propagation velocities related? Does it make sense to have a uniform grid or do you need to have space-dependent grid increments (or variable operator accuracy in the medium)? Some of these issues will be discussed when presenting research problems with the various methods at the end of this volume.

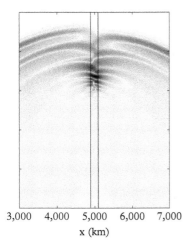

3,000   4,000   5,000   6,000   7,000

x (km)

Fig. 4.16 *Snapshot of pressure amplitude for a source located at the (left) edge of a fault zone model (limited by vertical lines) with a 25% velocity decrease. The vertical extension (equivalent to the source depth) is 5 km. The snapshot indicates the occurrence of head waves (e.g. top left) and the development of a dispersive wavefield trapped inside the low-velocity zone (i.e., fault-zone trapped waves). Such phenomena are now widely observed above faults in seismically active areas with high deformation rates (e.g. San Andreas Fault, North Anatolian Fault).*

4.6 Elastic wave propagation in 1D

Let us move towards a physical description of waves that is closer to seismic wave propagation. In Part I we introduced the stress–strain relation as $\sigma_{ij} = \lambda \epsilon_{kk} \delta_{ij} + 2\mu \epsilon_{ij}$, where λ, μ are the Lamé parameters and σ_{ij} and ϵ_{ij} are the stress and strain tensors, respectively. When considering 1D wave propagation in x-direction and particle motion in a horizontal direction y this relation reduces to

$$\sigma_{xy} = \sigma_{yx} = 2\mu\epsilon_{xy} = 2\mu\tfrac{1}{2}(\partial_x u_y + \partial_y \underbrace{u_x}_{=0}) = \mu\partial_x u_y, \qquad (4.56)$$

where u_y is the only non-zero horizontal displacement component (and σ_{xy} correspondingly the only non-zero stress component). The propagation of elastic waves on strings can be described with this 1D relation. Despite the fact that we are dealing again with a scalar wave equation it will allow us to introduce the concept of grid-staggering that is a key element in many 3D finite-difference algorithms used in research today. But first let us apply the concepts of central finite differences to the resulting wave equation.

4.6.1 Displacement formulation

The 1D stress–strain relation presented in Eq. 4.56 leads to the 1D elastic wave equation in which density $\rho(x)$ and shear modulus $\mu(x)$ are allowed to vary in space (under some smoothness constraints)

$$\rho\partial_t^2 u = \partial_x(\mu \, \partial_x u) + f, \qquad (4.57)$$

where the space–time dependence of unknown field u and force f is implicitly assumed. This is called the displacement formulation. The spatial discretization around an arbitrary grid point i located inside the medium is illustrated in Fig. 4.17. All space-dependent fields are defined at these regular-spaced grid locations at time level j through

$$u_i^j = u(idx, jdt). \qquad (4.58)$$

We proceed by using the definition of centred finite differences replacing the partial differentiation in space and time.

In space we obtain the derivative of the displacement u' with reference to space coordinate x as (omitting the superscript for time)

$$u' = \partial_x u = \frac{u_{i+1} - u_{i-1}}{2dx}, \qquad (4.59)$$

where the derivative is defined at location i. Note that the function value u_i is not used for the evaluation of the derivative because of the anti-symmetry of the

$$i-2 \qquad i-1 \qquad i \qquad i+1 \qquad i+2$$

Fig. 4.17 *Spatial discretization and indexing in 1D.*

difference operator. To obtain the first term of the right-hand side of Eq. 4.57 we multiply by the shear modulus μ defined at location i. We indicate this locality with a vertical bar and subscript i

$$\mu\partial_x u\big|_i = \mu\,u'\big|_i = \mu_i \frac{u_{i+1} - u_{i-1}}{2dx}. \tag{4.60}$$

The next step is to take the derivative of this term by evaluating the difference at positions $i+1$ and $i-1$, again using a central difference formula, to finally obtain

$$\begin{aligned} \partial_x(\mu\partial_x u)\big|_i &= \frac{\mu u'\big|_{i+1} - \mu u'\big|_{i-1}}{2dx} \\ &= \frac{\frac{\mu_{i+1}(u_{i+2}-u_i)}{2dx} - \frac{\mu_{i-1}(u_i-u_{i-2})}{2dx}}{2dx} \\ &= \frac{\mu_{i+1}u_{i+2} - \mu_{i+1}u_i - \mu_{i-1}u_i + \mu_{i-1}u_{i-2}}{4dx^2} \end{aligned} \tag{4.61}$$

for the stresses at grid point i.

Approximating the left-hand side of the wave equation with a centred scheme for the second time derivative at time level j for grid point i leads to

$$\rho_i \frac{u_i^{j+1} - 2u_i^j + u_i^{j-1}}{dt^2} = \frac{\mu_{i+1}u_{i+2}^j - \mu_{i+1}u_i^j - \mu_{i-1}u_i^j + \mu_{i-1}u_{i-2}^j}{4dx^2} + f_i^j \tag{4.62}$$

and the final extrapolation scheme for the displacement-stress 1D elastic wave equation using a central difference scheme

$$\begin{aligned} u_i^{j+1} &= \frac{dt^2}{4\rho_i dx^2}\left[\mu_{i+1}u_{i+2}^j - \mu_{i+1}u_i^j - \mu_{i-1}u_i^j + \mu_{i-1}u_{i-2}^j\right] \\ &\quad + 2u_i^j - u_i^{j-1} + \frac{dt^2}{\rho_i}f_i^j. \end{aligned} \tag{4.63}$$

In this formulation we calculate directly derivatives of the elastic parameters (that might be discontinuous in the Earth). Also, we employ a central difference scheme with a $2dx$ grid spacing ignoring the information of the field at the central location. This situation can be improved by solving another form of the wave equation and using the staggered-grid concept. While instructive for introducing finite-differences to wave equations, the centered schemes in this formulation are inefficient for realistic problems.

4.6.2   Velocity–stress formulation

As indicated at the beginning of this chapter, the error of a finite-difference approximation depends on the size of the increment dx employed in the derivative approximation. In the present scheme the leading error is a quadratic function of

dx. If we were able to reduce the size of this increment by a factor of 2 the error of the scheme would be four times smaller. Defining

$$\partial_t u = v$$
$$\sigma = \mu \partial_x u, \tag{4.64}$$

where v is velocity, $\sigma = \sigma_{xy} = \sigma_{yx}$ representing the only non-zero stress component, and implicitly assuming space–time dependencies leads to the wave equation as a coupled system of two first-order partial differential equations

$$\rho \partial_t v = \partial_x \sigma + f$$
$$\partial_t \sigma = \mu \partial_x v. \tag{4.65}$$

Note that we are not taking second derivatives any more, nor are we directly calculating derivatives of the material parameters. Our unknowns are the discrete velocity and stress values

$$v_i^j = v(idx, jdt), \tag{4.66}$$

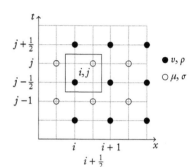

Fig. 4.18 *Spatio-temporal grid staggering in 1D. The vertical indexing corresponds to time and the horizontal indexing to space.*

again defined on a regular spaced grid in time and space. We proceed by replacing the partial differentials with centred finite-difference approximations to the first derivative. However, these are not defined at the grid points of the function but in between (i.e., $i \pm \frac{1}{2}$). Remember that the difference operator is antisymmetric, which implies that the information at the location of the derivative is not used anyway. The *grid staggering* is illustrated in Fig. 4.18.[9] The following computational scheme does the trick:

$$\frac{v_i^{j+\frac{1}{2}} - v_i^{j-\frac{1}{2}}}{dt} = \frac{1}{\rho_i} \frac{\sigma_{i+\frac{1}{2}}^j - \sigma_{i-\frac{1}{2}}^j}{dx} + \frac{f_i^j}{\rho_i}$$

$$\frac{\sigma_{i+\frac{1}{2}}^{j+1} - \sigma_{i+\frac{1}{2}}^j}{dt} = \mu_{i+\frac{1}{2}} \frac{v_{i+1}^{j+\frac{1}{2}} - v_i^{j+\frac{1}{2}}}{dx}, \tag{4.67}$$

leading to the extrapolation scheme

$$v_i^{j+\frac{1}{2}} = \frac{dt}{\rho_i} \frac{\sigma_{i+\frac{1}{2}}^j - \sigma_{i-\frac{1}{2}}^j}{dx} + v_i^{j-\frac{1}{2}} + \frac{dt}{\rho_i} f_i^j$$

$$\sigma_{i+\frac{1}{2}}^{j+1} = dt\, \mu_{i+\frac{1}{2}} \frac{v_{i+1}^{j+\frac{1}{2}} - v_i^{j+\frac{1}{2}}}{dx} + \sigma_{i+\frac{1}{2}}^j. \tag{4.68}$$

First of all, we note that the space-dependent properties of the (Earth) model ρ, and μ are not defined at the same locations. Also, stress and velocity are staggered in space and time. Yet, the scheme is consistent in the sense that we

[9] For any staggered finite-difference scheme it is useful to draw the spatial scheme in this way in order to make sure the indexes are properly addressed in the computer program. On first encounter it may take a while to digest these schemes. Please note that the definition of indexing is not unique. The results are the same but it has to be consistent.

are always multiplying or adding terms that are defined at the same location or time level. This is an inherent property of staggered elastic finite-difference schemes, with some exceptions (e.g. when the medium is anisotropic or wave equations in other coordinate systems are being solved). In the following, we will present a simulation example using this scheme and investigate its dispersion characteristics.

4.6.3   Velocity–stress algorithm: example

Table 4.2 *Simulation parameters for 1D velocity–stress simulation*

Parameter	Value
ρ	$2,500\,\text{kg/m}^3$
μ	$5 \times 50\,\text{GPa}$
v_S	$4,500\,\text{m/s}$
x_{max}	$1,000\,\text{km}$
dx	$1,000\,\text{m}$
dt	$0.18\,\text{s}$
f_0	$1/15\,\text{Hz}$

Let us illustrate the velocity–stress algorithm with an example. We initialize the homogeneous model with the parameters given in Table 5.3. The example has dimensions encountered in regional seismology with physical domain of 1,000 km, a shear-wave-propagation velocity of 4,500 m/s (representative of an upper mantle shear velocity or a near-surface Love wave velocity), and a dominant period of 15 s. The Python code fragment shown below represents the core of the 1D velocity–stress finite-difference algorithm.

```
# Time extrapolation
for it in range(nt):
    # Stress derivative
    for i in range(1, nx-1):
        ds[i] = (s[i+1] - s[i])/dx
    # Velocity extrapolation
    v = v + dt/rho*ds
    # Add source term at isx
    v[isx] = v[isx] + dt*src[it]/(dx*rho[isx])
    # Velocity derivative
    for i in range(1, nx-1):
        dv[i] = (v[i] - v[i-1])/dx
    # Stress extrapolation
    s = s + dt*mu*dv
```

In this code fragment, which presents the time extrapolation loop, *it* is the time level, and *s*, *v* are respectively unknown stress, and velocity values. Note that velocity and stress, as well as the given Earth model parameters (ρ, μ) are vectors (allowing heterogeneous models to be initialized). The source is injected as a force term at grid point *is*. We stress here that code lines like the velocity extrapolation are not mathematical statements. In this case the updated velocity field is allocated to the same vector (thus overwritten) as only two time levels are required in this formulation.[10] In the computer code one does not work with index fractions as in the mathematical algorithm. The spatial loops avoid the points beyond the boundaries. Boundary conditions are briefly discussed at the end of this chapter.

Results of the simulation are shown in Fig. 4.19. In the case of a homogeneous medium, we expect theoretically the source wavelet (the first derivative of

[10] Remember the displacement-stress formulation requires three time levels to be evaluated and stored.

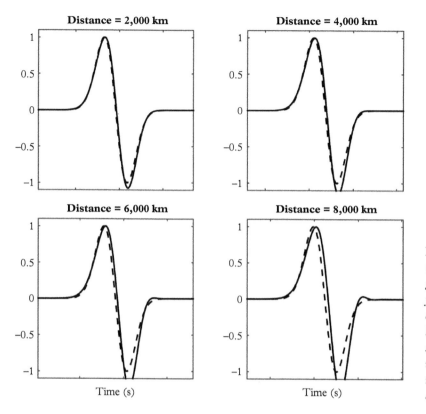

Distance = 2,000 km

Distance = 4,000 km

Distance = 6,000 km

Distance = 8,000 km

Time (s)

Time (s)

Fig. 4.19 *Simulation example with the velocity–stress formulation. Seismograms are shown at various distances from the source for a dominant period of 15 s. The numerical results (solid line) are compared with the analytical solution (dashed line). Note the increasing difference between analytical and numerical solution with propagating distance.*

a Gaussian) to propagate undisturbed for ever. Of course this is not the case in our numerical world. We observe, as in the acoustic case, that the waveform breaks down as a consequence of numerical dispersion. Let us quantify again this dispersion behaviour for this frequently–used elastic grid-staggering scheme and discuss the consequences.

4.6.4 Velocity–stress: dispersion

Applying the same procedure as given in the section on numerical dispersion of the acoustic problem above we obtain the condition for stable calculations

$$\sin\left(\frac{\omega dt}{2}\right) = \pm\sqrt{\frac{\mu}{\rho}}\,\frac{dt}{dx}\sin\left(\frac{kdx}{2}\right), \tag{4.69}$$

where the physical parameters μ and ρ are assumed constant. Note that this is the same relation as obtained in the acoustic case.

For the numerical phase velocity as a function of wavenumber (wavelength, or frequency, or number of points per wavelength) we obtain

$$c^{num} = \frac{\omega}{k} = \frac{\lambda}{\pi\,dt}\sin^{-1}\left(c_0\frac{dt}{dx}\sin\frac{\pi\,dx}{\lambda}\right), \qquad (4.70)$$

where $k = 2\pi/\lambda$ was used. Energy propagates with group velocity, so the accuracy of simulations should be checked against group velocity c_g, defined as

$$c_g = \frac{d\omega}{dk} = \frac{c_0\cos\frac{\pi\,dx}{\lambda}}{\left[1-\left(c_0\frac{dt}{dx}\sin\frac{\pi\,dx}{\lambda}\right)^2\right]^{\frac{1}{2}}}. \qquad (4.71)$$

These relations allow us now to investigate the behaviour of the propagation velocities as a function of frequency. It is instructive to translate the frequency axis into the number of grid points per spatial wavelength using $c = \omega/k$ and the corresponding dx of our simulation example. The results are illustrated in Fig. 4.20. For any regularly spaced field the Nyquist wavenumber corresponds to 2 points per wavelength (the left end of the horizontal axis). As can be seen in the figure the phase, and even more so the group velocity, substantially deviate from the theoretical propagation velocity as the number of grid points becomes smaller.

It is instructive to superimpose the spectrum of the source time function (a first derivative of a Gaussian) and translate it into the same coordinate system

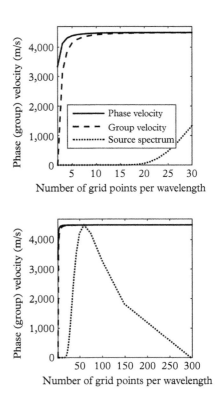

Fig. 4.20 *Bottom: Dispersion curves for the 1D velocity–stress staggered-grid finite-difference scheme as a function of number of grid points per spatial wavelength. Phase and group velocity are given as solid and dashed lines, respectively. The normalized spectrum of the source time function (simulation example with dominant frequency $f_0 = 1/15\,Hz$ is superimposed.* **Top:** *Detail towards the short-wavelength end of the spectrum.*

(amplitude normalized). For the simulation example discussed in the previous section there is substantial energy in the pulse between 20 and 300 grid points per wavelength while the dominant frequency is sampled with ≈66 points. Given these relatively large numbers it may appear surprising how bad the results are. However, an important aspect to consider is always the propagation distance. Also, we used here the lowest-order finite-difference operator. The results can be substantially improved using a four-point operator for the derivative calculations.

4.7 Elastic wave propagation in 2D

4.7.1 Grid staggering

Despite the focus of this volume on the presentation of various numerical schemes in the 1D case, we present here some basic aspects of grid staggering in 2D because (1) it contains the fundamental aspects of grid staggering for the stress–strain relation for higher dimensions, and (2) the extension to 3D is straightforward. Furthermore, this scheme is one of the most widely used numerical approximations in seismological research today. Here we only discuss the stress–strain relation. The complete 2D or 3D staggered grid algorithm is presented in other books or review papers (see suggestions at the end of this chapter).

Let us recall the (time-derivative of the) stress–strain relation

$$\partial_t \sigma_{ij} = \lambda \partial_t \epsilon_{kk} \delta_{ij} + 2\mu \partial_t \epsilon_{ij} \tag{4.72}$$

and rewrite it in 2D using the definition of the strain tensor to obtain for each component

$$\partial_t \sigma_{xx} = (\lambda + 2\mu)\partial_x v_x + \lambda \partial_z v_z$$
$$\partial_t \sigma_{zz} = (\lambda + 2\mu)\partial_z v_z + \lambda \partial_x v_x \tag{4.73}$$
$$\partial_t \sigma_{xz} = \mu(\partial_x v_z + \partial_z v_x),$$

where v_x and v_z are the two components of the velocity vector. The first derivatives of the velocity field with respect to both spatial coordinates need to be evaluated. Following the concepts described above, this can be achieved using the staggering scheme presented in Fig. 4.21.

Here, the diagonal elements of the stress tensor are defined at the same locations, while both velocity components and the off-diagonal stress component are at staggered locations, shifted along the coordinate axes. This leads to a consistent scheme in which the first derivatives only need to be calculated at these staggered grid locations. However, note that to evaluate the stresses and extrapolate the velocity field, the physical parameters ρ, λ, μ have to be known at different locations inside one grid cell. In the case of a heterogeneous medium, this has to be taken

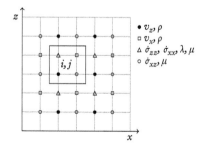

Fig. 4.21 *Spatial grid staggering for the 2D velocity–stress formulation. Dots denote time derivatives.*

into account and, unless the variations of the parameters are known analytically, the values will have to be interpolated to the staggered grid locations.

4.7.2   Free-surface boundary condition

For simulation problems involving the Earth's surface, the implementation of the free-surface boundary condition is crucial. As discussed in the introductory sections on wave propagation, assuming z as the vertical direction, the traction at the surface is zero, which implies that stress components

$$\sigma_{xz} = \sigma_{yz} = \sigma_{zz} = 0 \tag{4.74}$$

vanish. It is instructive to reduce this problem to 2D and demonstrate how it can be implemented in a basic way. This approach goes back to Levander (1988). From the stress–strain relation we know that

$$\begin{aligned}
\partial_t \sigma_{zz} &= 0 = \lambda \partial_x v_x + (\lambda + 2\mu) \partial_z v_z \\
\partial_t \sigma_{xz} &= 0 = \mu (\partial_x v_z + \partial_z v_x).
\end{aligned} \tag{4.75}$$

The procedure is illustrated in Fig. 4.22. The medium is extended beyond (above) the interior domain for as many points as required by the length of the finite-difference operator. One index level (here i) is defined as the location of the free surface. Here, we impose the stresses to be zero. If the stresses are extended beyond the free surface in an anti-symmetric way, the stress-free condition is fulfilled. The velocities are imposed to be symmetric such that the vertical gradients vanish. The horizontal derivatives in the above equation can be calculated as usual. This implementation is not unique. There are other options, which were discussed in Gottschämmer and Olsen (2001).

   This approach is a low-order implementation of the free surface which often is not accurate enough when dealing with surface waves that propagate many wavelengths. Finding more accurate solutions for the free-surface problem led to alternative strategies such as one-sided approximations (Kristek et al., 2002; Moczo et al., 2004) and hybrid solutions, exploiting the fact that in finite-element simulations the free-surface boundary condition is implicitly fulfilled (Galis et al., 2008). These schemes are presented in detail and compared with each other in Moczo et al. (2014).

Fig. 4.22 *Stress imaging. The stress-free boundary condition can be implemented numerically by adding an artificial domain outside the medium and applying symmetry conditions to stresses and velocity (see text for details).*

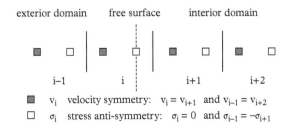

4.8 The road to 3D

The extension of a 2D *acoustic* finite-difference scheme as presented here to 3D is very easy: just add another dimension to all fields. This is a useful exercise (e.g. using the 2D schemes presented in the supplementary material) and 3D acoustic simulations can be run on PCs providing the models are not too large. A review of 3D acoustic finite-difference methods was presented by Etgen and O'Brien (2007).

The extension of staggered-grid schemes is also straightforward, but requires care concerning the proper implementation of differential operators and boundary conditions. There are many papers that present entire algorithms that can be used to develop 3D codes. Examples are Graves (1996), Igel et al. (1995) (anisotropic case), Pitarka (1999) introducing heterogeneous grids, and Kristek and Moczo (2003) the viscoelastic case. The book by Moczo et al. (2014) contains most conceivable finite-difference schemes in detail and is the best source for developers. The classic 3D staggered grid is shown in Fig. 4.23.

The following sections aim at winding you down (or hyping you up) with some interesting further aspects and developments that might raise your interest. Some of the topics covered here go beyond the introductory level but might be important for the implementation of competitive algorithms.

4.8.1 High-order extrapolation schemes

This volume has a clear focus on explaining the mathematical approaches of the various numerical techniques concerning the space-dependent discretization

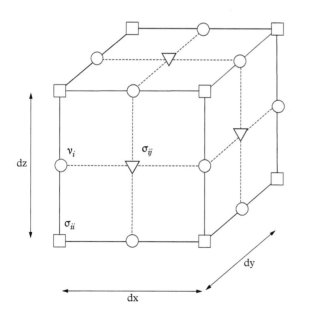

Fig. 4.23 *3D staggered finite-difference grid (velocity stress). The off-diagonal stresses σ_{ij} are separated by the diagonal elements σ_{ii} and the velocity components v_i by half a grid distance* (dx, dy, dz).

of wave propagation. Therefore—with the exception of the finite-volume and the discontinuous Galerkin methods—the time extrapolation scheme is of lowest order. The accuracy of all schemes can be improved by applying high-order extensions. Most community software packages do not go beyond second-order extrapolation schemes.

As a first example we illustrate the predictor–corrector scheme for a general first-order system (e.g. advection equation, velocity–stress formulation), which can be described by

$$\partial_t q(x, t) = L(q, t) = c \, \partial_x q(x, t) + s(x, t), \tag{4.76}$$

where q is the solution field, c is velocity, and L is a linear operator. The first-order Euler scheme corresponds to

$$q^+ = q(x, t + dt) = q(x, t) + dt L(q, t), \tag{4.77}$$

which in many cases is not accurate enough. A far better scheme is the so-called predictor–corrector method (or Heun's method; it also belongs to the family of Runge–Kutta methods) that is obtained from the Euler method using the trapezoidal rule. With the definitions above we obtain

$$\begin{aligned} k1 &= L(q, t) & \text{predictor} \\ k2 &= L(q + dt q^+, t + dt) & \text{corrector} \\ q(x, t + dt) &= q(x, t) + \frac{1}{2} dt (k1 + k2), & \end{aligned} \tag{4.78}$$

with marked improvement compared to the Euler scheme. We make use of this approach in the chapters on the finite-volume and the discontinuous Galerkin methods for implementation examples and results.

As a second example we present a powerful concept that goes back to the work of Lax and Wendroff (1960), sometimes also referred to as the Cauchy–Kowaleski procedure. We start with the Taylor expansion of a space–time dependent function (e.g. acoustic wavefield)

$$\begin{aligned} q(x, t + dt) &= q(x, t) + dt \partial_t q(x, t) + \frac{dt^2}{2} \partial_t^2 p(x, t) + \dots \\ &= \sum_{j=0}^{N} \frac{dt^j}{j!} \partial_t^j p(x, t). \end{aligned} \tag{4.79}$$

We know that q(x,t) is the solution to the advection problem (Eq. 4.76). Thus the following relation also holds

$$\partial_t^{j+1} q(x, t) = c \, \partial_x \left[\partial_t^j q(x, t) \right], \tag{4.80}$$

indicating that we can calculate time derivatives of q(x,t) of any order in a recursive way making use of the advection equation. In other words we replace time derivatives by space derivatives. This approach was used to develop the Arbitrary high-orDER (ADER) schemes for the finite-volume and discontinuous Galerkin methods (e.g. Titarev and Toro 2002; Dumbser and Munz, 2005a). This formulation also works for the second-order wave equation (e.g. Dablain 1986; Igel et al., 1995). Further solution schemes not discussed here are the Crank–Nicolson scheme, the Newmark scheme, and high-order Runge–Kutta schemes.

4.8.2 Heterogeneous Earth models

The finite-difference method belongs to the class of numerical approaches that is primarily used on regular equally spaced grids, multi-domain equally spaced grids, or space-tree-based solutions with local refinements. Therefore, in cases where interfaces (i.e. material discontinuities) are not aligned with the coordinate axes there is a problem (see Fig. 4.24 for an illustration). The geometry of the interfaces is not accurately modelled. In part this restriction motivated the development of other methods such as Galerkin-type approaches (e.g. finite/spectral elements) for seismic wave propagation.

However, there are ways to improve the situation by modifying the elastic parameters and density around the interface, replacing them using appropriate averaging schemes. Muir et al. (1992) applied equivalent medium theory to this problem, replacing the isotropic parameters with anisotropic parameters. An alternative approach for smoothly varying media was presented by Moczo et al. (2002) using volume and arithmetic averaging of elastic moduli and density for isotropic media and for viscoelastic media Kristek and Moczo, 2003.

Yet another strategy to cope with strongly heterogeneous media is to make the grid density space-dependent. Examples are presented in Jastram and Tessmer (1994). More recent results are reported in Moczo et al. (2010a) comparing numerical solutions for an earthquake ground motion problem, indicating the benefits of discontinuous finite-difference grids. Other technical developments include the introduction of spatially varying time steps for very heterogeneous problems (Tessmer, 2000). The results presented by Moczo et al. (2010b) indicate that care has to be taken when the ratio between P- and S-velocity exceeds certain values.

A more general treatment of the strongly heterogeneous Earth models leads to the problem of homogenization (e.g. Capdeville et al., 2010a; Capdeville et al., 2010b), discussed in Chapter 11.

4.8.3 Optimizing operators

Many attempts have been made to develop better, *optimal* finite-difference operators specifically for the elastic wave-propagation problem, that go beyond the more general improvements of the time-extrapolation schemes described above. A very clever approach is to artificially make errors in the space derivatives that

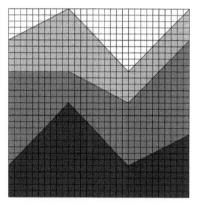

Fig. 4.24 *Internal interfaces. Interfaces that do not coincide with grid lines are not properly accounted for by regular spaced grids. There are several strategies to improve the situation, by assigning equivalent medium properties to the grid points adjacent to the interfaces (see text for details).*

Conventional $(1/dt^2)$

	x−dx	x	x+dx
t+dt		1	
t		−2	
t−dt		1	

Optimal $(1/dt^2)$

	x−dx	x	x+dx
t+dt	1/12	10/12	1/12
t	−2/12	−20/12	−2/12
t−dt	1/12	10/12	1/12

Conventional $(1/dx^2)$

	x−dx	x	x+dx
t+dt			
t	1	−2	1
t−dt			

Optimal $(1/dx^2)$

	x−dx	x	x+dx
t+dt	1/12	−2/12	1/12
t	10/12	−20/12	10/12
t−dt	1/12	−2/12	1/12

Fig. 4.25 *Optimal operators. The conventional second-order finite-difference operators for the second derivative are compared with the optimal operators developed by Geller and Takeuchi (1998) (see text for details).*

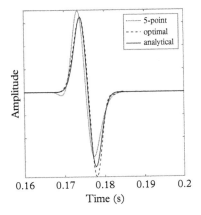

Fig. 4.26 *Simulation with optimal operators. Synthetic seismograms for the 1D acoustic case are compared. In this example the energy misfit with the analytical solution for the classic five-point operator is 10.1% and for the optimal operator 1.2%.*

[11] In many publications it is stated that *n*th order schemes are used. Often such statements refer to the accuracy in space only. This rarely translates to the overall accuracy of the space–time-dependent solutions. Only careful convergence analysis allows quantification of the true convergence order.

compensate for the errors committed by the time extrapolation, to obtain a truly high-order scheme.[11] This approach was taken, for example, by Emmerich and Korn (1987).

Geller and Takeuchi (1995) developed criteria against which the accuracy of frequency-domain calculation of synthetic seismograms could be optimized. This approach was transferred to the time-domain finite-difference method for homogeneous and heterogeneous schemes by Geller and Takeuchi (1998). Let us have a look at an *optimal* operator and compare with the classic scheme. The space–time stencils are illustrated in Fig. 4.25. Note that by summing up the optimal operators one obtains the conventional operators. This can be interpreted as a smearing out of the conventional operators in space and time. The optimal operators lead to a locally implicit scheme, as the future of the system at $(x, t + dt)$ depends on values at time level $t + dt$, that is, the future depends on the future. That sounds impossible, but it can be fixed by using a predictor–corrector scheme based on the first-order Born approximation.

The optimal operators perform in a quite spectacular way. With very few extra floating point operations an accuracy improvement of almost an order of magnitude can be obtained. The results shown in Fig. 4.26 were obtained by coding the algorithm presented by Geller and Takeuchi (1998). The optimal scheme performs substantially better than the conventional scheme with a five-point operator. The details about the simulation set-up are not important here. The message is simple: While finite-difference methods often have a brute-force reputation it is important to note that a well-written finite-difference algorithm is highly competitive when compared to other high-order schemes such as the spectral element method. In the light of this it is surprising that the optimal operator concept does not seem to be widely used.

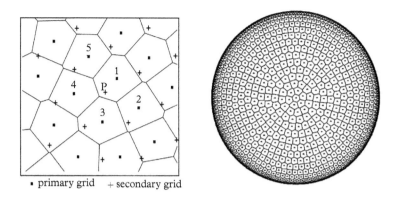

• primary grid   + secondary grid

Fig. 4.27 *Finite differences on unstructured grids. Unstructured grids can be treated using Delauney triangulation and Voronoi cells.* **Left:** *Akin to staggered-grid schemes, unstructured primary grids (e.g. stresses) can be interleaved with a secondary grid (e.g. velocities), allowing the calculation of (first) derivates for a velocity–stress wave-propagation scheme. The differential weights can be calculated using natural neighbour coordinates or finite-volume concepts.* **Right:** *Voronoi cells for a spherical shape with grid densification near the boundary.*

4.8.4   Minimal, triangular, unstructured grids

Can the finite-difference method be applied to triangular, or unstructured grids? This question dominated my own research when working towards solutions for global wave propagation. Magnier et al. (1994) presented a very elegant algorithm for equilateral triangular grids (hexagonal structures). By interleaving two such grids a staggered-grid-type scheme was developed allowing the efficient solution of the elastic wave equation in 2D. Such grids have the neat property that the error is isotropic due to the hexagonal symmetry of the spatial grid. Unfortunately no such isotropic grid exists in 3D! Later it turned out that the difference weights derived for the minimal grids are essentially first-order finite-volume weights.

But what happens when the triangles are not equilateral? Can we increase the influence domain and use more points for the derivative calculation like in the classic finite-difference method? These questions led to the search for general ways to calculate differential weights for unstructured grids. At around that time, Braun and Sambridge (1995) had imported the concepts of natural neighbour coordinates from computational geometry to geophysics. The fact that this paper appeared in *Nature* is a sign of how important this problem was considered by the scientific community.

The differential operators proposed by Braun and Sambridge (1995) were investigated and applied to the 2D elastic wave-propagation problem by Käser et al. (2001) and Käser and Igel (2001), with applications to spherical structures, and media with curved internal boundaries or complicated free surfaces. An illustration of the natural neighbour concepts with Voronoi cells is shown in Fig. 4.27. The results with this approach indicated that the operators are not accurate enough for the elastic wave-propagation problem and not easily extended to high order. Eventually this frustration led to the adaptation of the more sophisticated discontinuous Galerkin method to the wave-propagation problem (Käser and Dumbser, 2006). It is fair to say that, for problems involving unstructured tetrahedral grids, today this is the method of choice.

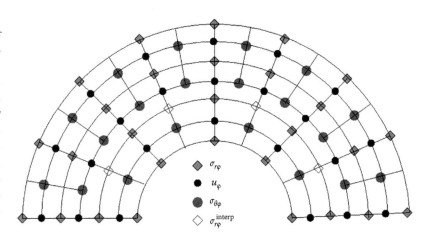

Fig. 4.28 *Schematic illustration of multidomain axisymmetric mesh for staggered-grid finite-difference calculations with grid refinement towards the surface. Note that, in addition to the grid densification due to the spherical grid, velocities increase with depth in the mantle. This leads to small time steps due to the CFL criterion. At the domain boundaries additional interpolations are necessary to calculate the space derivatives. From Igel and Gudmundsson (1997). Reprinted with permission.*

4.8.5 Other coordinate systems

How about other coordinate systems? In fact, some of the first papers applying the finite-difference method to wave-propagation problems were using non-Cartesian formulations of the wave equation. Alterman and Karal (1968) used cylindrical coordinates, and the same group developed a scheme for spherical coordinates (Alterman et al., 1970) targeting global wave propagation. This was heroically forward looking, given the tiny computational resources that existed at the time.

Because working in other coordinate systems is still very interesting today, let us have a quick look at the acoustic wave equation in spherical coordinates (the same concepts apply to cylindrical coordinates). Assuming a standard spherical coordinate system r, θ, ϕ and invariance in ϕ (so-called zonal or axisymmetric model) we can write the wave equation using the Laplace operator as

$$\partial_t^2 p = c^2 \left[\frac{1}{r^2} \partial_r (r^2 \partial_r p) + \frac{1}{r^2 \sin \theta} \partial_\theta (\sin \theta \partial_\theta p) \right] + s, \tag{4.81}$$

where $p(r, \theta)$ is pressure, $c(r, \theta)$ is acoustic velocity, and $s(r, \theta)$ is a source term.

This looks much more complicated than the Cartesian case, and for the finite-difference-based numerical solution there are dramatic consequences. First, the equation contains singularities and is not defined along the axis $\theta = 0$. Second, regular discretization along the coordinate axes r and θ leads to increasing grid point distance for increasing radius. This is illustrated in Fig. 4.28. For global wave propagation this is the opposite of what we need. In the mantle, seismic velocities increase with depth; therefore we would prefer to increase grid spacing with depth.

Nevertheless, it is possible to devise staggered-grid finite-difference schemes for this case (Igel and Weber, 1995, 1996; Chaljub and Tarantola, 1997) that led to some of the first numerical investigations of scattering effects in laterally heterogeneous mantle structure (e.g. Toyokuni and Takenaka 2006; Jahnke et al.,

Fig. 4.29 *SH waves in the Earth's mantle. SH-wave propagation based on a staggered-grid finite-difference scheme for the spherically symmetric Earth model PREM. The source is at 600 km depth and the dominant period is 25 s. Note the reflections from the core–mantle boundary, and the increasingly complex wavefield near the Earth's surface from multiply reflected S-phases (building up Love surface waves).*

Fig. 4.30 *Wave propagation in spherical sections using finite differences. The singularities in the wave equation formulated in spherical coordinates can be avoided by appropriate positioning in the spherical domain (e.g. around the equator). Snapshot of elastic wave propagation for an explosive source at 600 km depth. Note the P–S conversions at the core–mantle boundary. From Igel et al. (2002). Reprinted by permission.*

2008; Thorne et al., 2013*b*). An example of some of the first simulations is shown in Fig. 4.29 for global SH-wave calculations. At that time the visualization of such wavefields, despite their limited scientific value, allowed an unprecedented view of what happens when waves pass through the Earth's interior. Even today I find these visualizations fascinating. They have enormous educational value; a fact that was recently recognized by Thorne et al. (2013*a*), who provided numerous openly accessible animations.[12]

In principle 3D wave propagation in spherical coordinates (spherical sections excluding poles) using finite differences is possible. Igel et al. (2002) presented an anisotropic scheme and used the method to investigate wave effects of sources located inside subduction zones. An illustration of this approach is shown in Fig. 4.30.

4.8.6 Concluding remarks

In its simplest form the finite-difference method offers a straightforward introduction to the world of numerical methods. For many physics problems a first approximate solution can be obtained relatively quickly, and it can help us to develop a deeper understanding of the necessary properties of any solution. Therefore it plays an important role for the development of simulation methods in many fields of science.

The finite-difference method continues to play an important role in Earth science today. As indicated, with relatively minor extensions to the basic solution schemes, finite-difference simulations can be very competitive with other methods, in particular for problems concerning strong ground motion, local scale wave propagation, and seismic exploration problems (see Chapter 10 on applications for specific examples).

Chapter summary

- Replacing the partial derivatives by finite differences allows partial differential equations such as the wave equation to be solved directly for (in principle) arbitrarily heterogeneous media.

---

[12] http://web.utah.edu/thorne/animations.html.

- The resulting space–time discretization leads to unphysical phenomena such as numerical dispersion that can only be avoided by sampling with enough grid points per wavelength.
- The accuracy of finite-difference operators can be improved by increasing the number of grid points (i.e. longer operators) used to approximate the derivatives. The weights for the grid points can be obtained using Taylor series (or spectral methods).
- Plane-wave (or von Neumann) analysis of the approximative scheme leads to a stability criterion that restricts the choice of the space–time discretization.
- In 2D and 3D the error of wave propagation becomes anisotropic. In regular-spaced grids the most accurate direction is at 45° to the grid axes.
- The finite-difference method—despite usually low-order implementations—remains an attractive numerical scheme for many applications in seismology, even for problems that require accurate surface waves, provided that the free surface gets special treatment.
- In principle finite-difference-type operators are possible on unstructured grids but only with low-order accuracy.
- Finite-difference approximations to the wave equation in cylindrical or spherical coordinates are possible, with restrictions due to the intrinsic singularities.

....

FURTHER READING

For the finite-difference method, some good references are:

- The recent book by Moczo et al. (2014) is the most complete work on the finite-difference method applied to elastic wave propagation to date. It provides many different algorithms and discusses pros and cons of various implementation strategies. There are also detailed sections on different rheologies (viscoelasticity) and boundary conditions (free surface, internal material interfaces, absorbing boundaries).
- The book by Fichtner (2010) on modelling and inversion of seismic waves has a section on finite differences and spectral elements with some additional features such as spherical coordinates.
- The Society of Exploration Geophysicists (SEG) has published two special volumes (Kelly and Marfurt 1990; Robertsson et al., 2012) with many classic papers on the various numerical methods applied to wave propagation.

- A more mathematical treatment of finite-difference (and other) methods applied to wave-propagation problems can be found in Durran (1999).

. . . .

EXERCISES

Comprehension questions

(4.1) Characterize problems that necessitate the use of numerical methods such as finite differences.

(4.2) Are finite-difference-based approximations of partial-differential equations unique (give arguments)?

(4.3) What strategies are there to improve the accuracy of finite-difference derivatives? Give the procedures in words.

(4.4) What is stability in connection with finite-difference algorithms. Give the relevant condition for the 1D wave-propagation problem.

(4.5) What is convergence?

(4.6) What is the difference between physical and numerical dispersion?

(4.7) Which propagation direction is most accurate on a rectangular (square) grid? Can you suggest any reasons why this might be so?

(4.8) Give strategies to check whether a finite-difference simulation is accurate for (a) a homogeneous medium, and (b) a strongly heterogeneous medium.

(4.9) Explain why staggered grids appear to be useful for the elastic wave equation.

(4.10) What is the difference between phase- and group velocity? To be on the safe side, which velocity should be accurately modelled, and why?

(4.11) Are finite-difference methods easily parallelized using domain decomposition? Do processors need to communicate with each other? Make an illustration for a 2D problem.

(4.12) Explain why for Earth models with large variations in seismic velocities, varying the grid cell size is highly desirable. What is the problem with having to have a global time step dt, though (i.e. one dt for all grid cells)?

Theoretical problems

(4.13) Show that

$$\frac{f(x + dx) - 2f(x) + f(x - dx)}{dt^2}$$

is an approximation for the second derivative of $f(x)$ with respect to x at position x. Hint: Use Taylor series

$$f(x + dx) = \sum_{n=0}^{\infty} \frac{f^{(n)}(x)}{n!} dx^n,$$

where $f^{(n)}(x)$ is the nth derivative of $f(x)$. What is the leading order of the error term?

(4.14) Derive the numerical dispersion equation (Eq. 4.44) for the 1D acoustic wave equation using the von Neumann analysis.

(4.15) Use Taylor's theorem to approximate the derivative of $f(x)$ with the functional values given by $f(x + dx/2)$ and $f(x - dx/2)$. What is the order of accuracy? You are not happy with this accuracy and would like to have a higher-order approximation. Calculate the derivative weights, if you also use information at points $x + 3/2dx$ and $x - 3/2dx$.

(4.16) Generalize the procedure of the previous exercise and derive equations for the system matrix A (Eq. 4.16) for centred and staggered-grid finite difference operators of arbitrary length.

(4.17) You want to estimate the derivative of a function $f(x)$ near a boundary using the high-order finite-difference method. Develop the required system matrix and calculate the one-sided derivative weights for operators of arbitrary length. Hint: Define the derivative at $f(x + dx/2)$ and search for weights at $f(x)$ and $f(x + ndx)$. Discuss the results.

(4.18) The source-free advection equation is given by

$$\partial_t u(x, t) = v \partial_x u(x, t),$$

where $u(x, t = 0)$ could be a displacement waveform at $t = 0$ (an initial condition) that is advected with velocity v (this will become important in Chapters 8 and 9 on finite volumes and the discontinuous Galerkin method, respectively). Replace the partial derivatives by finite differences. Which approach do you expect to work best? Turn it into a programming exercise and write a simple finite-difference code and play around with different schemes (centred vs. non-centred finite differences). What do you observe?

(4.19) A seismometer consists of a spring with damping parameter ϵ, and eigenfrequency ω_0. The seismometer is excited by the (given) ground motion $\ddot{u}(t)$. The relative motion of the seismometer mass $x(t)$ is governed by the following equation

$$\ddot{x} + 2\epsilon\dot{x} + \omega_0^2 x = \ddot{u}.$$

Replace the derivatives on the left-hand side with finite differences. Solve for $x(t + dt)$. Note: a good strategy in this example is to centre the differences at the same point in time. The dots denote time derivative.

(4.20) Certain isotopes (e.g. $_9$Be) are washed into the sea by rivers and then mixed by advection through ocean currents and diffusion. In addition, the isotopes are removed from the system through biomechanical processes (e.g. death). These processes can be described by the diffusion–advection–reaction equation (concentration $C(x, t)$, diffusivity k (const), reactivity $R(x)$, source $p(x)$, advection velocity v). Substitute in the 1D equation below the partial differentials with finite differences and extrapolate to $C(t + dt)$:

$$\partial_t C = k\partial_x^2 C + v\partial_x C - RC + p.$$

How could a ring current be simulated with this 1D equation mimicking an oceanic gyre? What do you think is the best choice for the finite-difference formulation and why?

(4.21) You want to simulate 2D acoustic wave propagation in a medium with size 1,000 km × 1,000 km. You want to model wave propagation up to a period of 10 s. The maximum velocity c is 8 km/s, the minimum velocity is 4 km/s. Your numerical algorithm requires 20 grid points per wavelength to be accurate for the propagation distances of interest. What space increment dx do you need for the simulation? The stability criterion says that maximum velocity c, space increment dx and time increment dt are related by $\epsilon = cdt/dx$. You want a seismogram length of 500 s. How many time steps do you have to simulate, when $\epsilon = 0.5$?

(4.22) Show that when setting the Courant criterion to $\epsilon = 1$ for the homogeneous acoustic problem with constant dt and dx (in other words the physical velocity $c = dx/dt$) there is no numerical dispersion. Hint: Make use of equation Eq. 4.45. What is the relevance for practical applications?

(4.23) Choose an appropriately tight formulation for the discretized fields (see examples in the text) and write down the finite-difference extrapolation scheme for the 3D acoustic wave equation (Δ is the Laplace operator)

$$\partial_t^2 p(x, y, z, t) = c(x, y, z)^2 \Delta p(x, y, z) + s(x, y, z, t).$$

(4.24) Show that the Nyquist wavenumber corresponds to 2 grid increments per wavelength.

(4.25) Following the von Neumann analysis based on plane waves in the text calculate the stability limit (i.e. CFL criterion) for the 3D acoustic wave equation (see previous exercise).

(4.26) Following the developments in the section on staggered grids, write down the second-order 3D elastic wave equation in the displacement formulation, as well as the stress–strain relation, and the strain–displacement relation. Find an appropriate 3D staggered-finite-difference cell where the derivatives are calculated in between the functional values (e.g. strain components as the derivatives of displacement components).

(4.27) You want to simulate global wave propagation. The highest frequencies that we observe for global wave fields is 1 Hz. Let us for simplicity assume a homogeneous Earth. The P velocity $v_p = 10$ km/s and the v_p/v_s ratio is $\sqrt{3}$. Let us assume 20 grid points per wavelength. How many grid cells would you need (assume cubic cells)? What would their size be? Now let us be more realistic. The maximum P-velocity in the Earth is 14 km/s and the smallest P-velocity is 1.5 km/s in the oceans, or 5 km/s in the crust. Assume that you can only have one grid size for the whole Earth. Estimate the number of cells, their size, and the required time step. The CFL criterion $\epsilon = 0.5$.

(4.28) The strain–displacement relation is given by

$$\epsilon_{ij} = \frac{1}{2}(\partial_i u_j + \partial_j u_i).$$

Write down this relation in 2D. Allocate strain and displacement components to the four symbols such that there is a consistent scheme for a finite-difference method with a two-point operator for the first derivative. The central square corresponds to elements ij (e.g. $x \to i;\, y \to j$). Is the mapping unique?

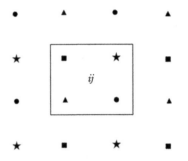

Programming exercises

For the following exercises, please also refer to the supplementary electronic material.

(4.29) Write a computer program for the 1D (2D) acoustic wave equation following the equations presented in this chapter. Implement the analytical solution (see Chapter 2) and try to match it with appropriate parameters.

(4.30) Determine numerically the stability limit of 1D and 2D implementations of the acoustic wave equation as accurately as possible by varying the stability criterion.

(4.31) Increase the dominant frequency of the wavefield in 2D. Investigate the behaviour of the wavefield as a function of azimuth. Why does the wavefield look anisotropic? Which direction is the most accurate and why?

(4.32) Extend the (1D and/or 2D) codes by adding the option to use a five-point operator for the second derivative. Compare simulations with the three-point and five-point operators. Is the stability limit still the same? Make it an option to change between three-point and five-point operators. Estimate the number of points per wavelength you are using and investigate the accuracy of the simulation by looking for signs of numerical dispersion in the resulting seismograms. The five-point weights are: $[-1/12, 4/3, -5/2, 4/3, -1/12]/dx^2$.

(4.33) Modify the program such that at the end of the calculation you can visualize and output synthetic seismograms.

(4.34) Modify the 2D velocity model c in *ac2d* and observe and discuss the resulting wavefield. (1) Add a low- (high-) velocity layer near the surface. Inject the source in this layer. (2) Add a vertical low-velocity zone (fault-zone) of a certain width (e.g. 10 grid points), and discuss the resulting wavefield (fault-zone trapped waves). (3) Simulate topography by setting the pressure to 0 above the surface. Use a Gaussian hill shape or a random topography.

(4.35) Use a spike source time function and look at the resulting seismogram. Examine the spectrum of this Green's function. Do you spot the numerical noise? Convolve the resulting seismograms with an appropriate source time function (e.g. a Gaussian of appropriate length). What happens with the numerical noise?

(4.36) Source–receiver reciprocity. Initialize a strongly heterogeneous 2D velocity model of your choice and simulate waves propagating from an internal source point (x_s, z_s) to an internal receiver (x_r, z_r). Show that by reversing source and receiver you obtain the same seismograms.

(4.37) Time reversal. Define a source at the centre of the domain in an arbitrary 2D velocity model and a receiver circle at an appropriate distance around the source. Simulate a wavefield, record it at the receiver ring, and store the results. Reverse the synthetic seismograms and inject them as sources at the receiver points. What happens? Can you explain the results?

The Pseudospectral Method

[1] Why *pseudo*? Well, pseudo means something like *sort of but not really* and in the context of spectral and pseudospectral has a specific significance. With spectral methods the equations are expressed and solved in the spectral domain, but with the pseudospectral method the spectral domain is used merely for the calculation of spatial derivatives—the equations are space–time formulations.

In terms of chronological order the pseudospectral[1] method was the first method that followed finite differences and was used extensively in several seismological research problems. We have seen in the previous chapter that the approximate spatial derivatives lead to quite dramatic problems when waves propagate over long distances. In the light of this, the desire was to find more accurate operators for the space derivatives. We have seen that extending the finite-difference operators leads to more accurate derivatives. In a sense, the pseudospectral method can be considered the most extreme case in which the length of the derivative operator is equivalent to the number of points along one of the space dimensions.

The attractive property of pseudospectral methods is that the space derivatives can be calculated *exactly*; that is, to at least machine precision. Of course this accuracy comes at a price. The price is that per calculation of space derivative many more floating-point operations have to be carried out. The two approaches we will introduce in what follows—the Fourier and the Chebyshev methods—can make use of the fast Fourier transform (FFT) to calculate derivatives. However, it is important to note that the pseudospectral method requires global communication; in other words, the future of a certain point in the grid depends on the current state of all other points.

The pseudospectral method was developed at a time when hardware architecture still primarily favoured serial processing. Global communication schemes are suboptimal for massively parallel computer architectures that favour minimal and preferably near-neighbour communications. This is the main reason why the pseudospectral method in its simplest form disappeared from the simulation market, when parallelization became a standard (some hybrid approaches are still used, and will be discussed here). However, it is important to note that because of the high accuracy of the space differentiation—requiring fewer grid points per wavelength compared to other methods—the pseudospectral method is very memory efficient.

On the other hand, of all the (regular grid-type) methods presented in this volume, the pseudospectral method is my favourite simply for its mathematical elegance and the fact that it does not require grid staggering. This latter property made it attractive in particular for the simulation of wave propagation in 3D anisotropic media or spherical sections (see Fig. 5.1), for which staggered finite-difference schemes suffer from additional errors.

In terms of the evolution of computational seismology, the mathematical concepts introduced in connection with the pseudospectral methods (series

Computational Seismology. First Edition. Heiner Igel.
© Heiner Igel 2017. Published in 2017 by Oxford University Press.

expansions, cardinal functions, exact interpolation, etc.) are elementary ingredients of methods that are state-of-the art today, such as the spectral-element method or the discontinuous Galerkin method. After a brief introduction to the history of the pseudospectral method, we will present the fundamentals of function interpolation using Fourier series and Chebyshev polynomials and then apply these concepts to the numerical solution of the acoustic and elastic wave equations.

5.1   History

Pseudospectral methods entered the arena in the early eighties as *transform methods* because their implementation was based on the Fourier transform (Gazdag, 1981; Kossloff and Bayssal, 1982). Later the term Fourier method was also used. Initial applications to the acoustic wave equation were extended to the elastic case (Kossloff et al., 1984), and to 3D (Reshef et al., 1988). Efficient time-integration schemes were developed (Tal-Ezer et al., 1987) that allowed large time steps to be used in the extrapolation procedure.

The biggest attraction of the pseudospectral method based on Fourier transforms was the fact that, compared to finite-difference schemes, substantially less memory was required, in particular for 3D calculations. This was possible because a smaller number of grid points was required due to the high accuracy of the derivative calculations.[2] The drawback of using Fourier transforms is the implicit assumption of periodicity along the spatial dimensions. This implies that boundary conditions such as the free surface condition are difficult to implement efficiently.

An elegant fix to this problem was achieved by replacing harmonic functions as bases for the function interpolation with Chebyshev polynomials (Kosloff et al., 1990). The originally infinite area calculations (because of periodicity) were converted to limited area calculations as the Chebyshev polynomials are defined in the interval $[-1, 1]$ (easily scaled to arbitrary length). This formulation allowed efficient implementation of free surface or absorbing boundaries by means of characteristic variables (Carcione and Wang, 1993).

However, as we all know, there is no free lunch, and a disadvantage also accompanies the use of Chebyshev polynomials. The collocation points at which the functions are exactly interpolated are irregular and densify towards the boundaries. The difference between shortest and largest grid point distance increases with the overall number of points along one dimension. In principle this can be compensated for by re-stretching the grid towards more regular grid distances (an approach presented in Carcione and Wang, 1993).

To improve the accurate modelling of curved internal interfaces and surface topography, grid stretching (see Fig. 5.2) as coordinate a transforms was introduced and applied to acoustic and elastic wave-propagation problems (Tessmer et al., 1992; Komatitsch et al., 1996). A further advantage of the pseudospectral

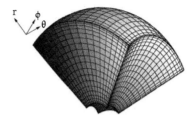

Fig. 5.1 *Computational grid for a pseudospectral simulation in spherical coordinates based on Chebyshev polynomials. Note the decreasing grid point distance towards the boundaries of the physical domain. In combination with the Chebyshev formulation this allows elegant implementation of free-surface or absorbing boundaries. From Igel (1999).*

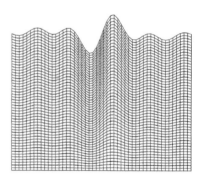

Fig. 5.2 *Illustration of grid stretching for a physical domain with curved boundaries. The grid lines are analytically generated such that they follow the desired boundary. The Jacobian maps the original Cartesian to the curvilinear grid (e.g. Carcione and Wang 1993).*

---

[2] The reduction of the number of grid points per wavelength by a factor of 2 leads to a memory reduction by a factor of 8 in 3D. In addition, because of the stability criterion, a time step twice as large reduces the overall computation time.

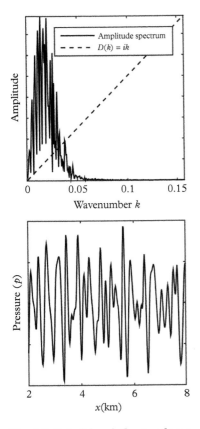

Fig. 5.3 *Principle of the pseudospectral method illustrated with a simulation of acoustic waves in 1D through a medium with random velocity perturbations.* **Bottom:** *Snapshot in space of the pressure field p.* **Top:** *Amplitude spectrum of the wavefield spectrum given below. From the sampling theorem the wavenumber axis is limited by the Nyquist wavenumber $k_N = \pi/dx$. The first derivative is calculated by multiplying the spectrum by ik followed by an inverse Fourier transform resulting in an exact (to machine precision) derivative $\partial_x p(x)$ at the regular grid points.*

approach is the fact that all fields (displacement, elastic parameters) are defined at the same grid locations (which is not the case in staggered-grid finite-difference schemes). This implies that in particular anisotropic media (Tessmer, 1995) can be efficiently implemented without the need for further interpolations. The same holds for the application of the pseudospectral method to the wave equation in other coordinate systems such as spherical coordinates (e.g. Igel, 1999).

Despite the difficulties of using the pseudospectral method on parallel hardware, the method has been used for interesting seismological problems (Furumura et al., 1998a; Furumura et al., 1998b; Furumura et al., 1999; Furumura and Kennett, 2005) partly by mixing finite-difference operators and pseudospectral operators in the different spatial directions (Furumura et al., 2002).

5.2   The pseudospectral method in a nutshell

In the introductory chapter we categorized the numerical methods presented in this volume roughly into grid point methods and series expansion methods. The pseudospectral method is both! One one hand, the spatial displacement field is expanded in (Fourier or Chebyshev) series; on the other hand, because of the specific interpolation properties of the basis functions, the actual values at the corresponding grid points are (exactly) the coefficients of the basis functions. This will become clear when we discuss the details. As stated before, the time derivatives in the wave equation will be replaced by finite differences, leaving us with

$$\frac{p(x, t + dt) - 2p(x, t) + p(x, t - dt)}{dt^2} = c(x)^2 \partial_x^2 p(x, t) + s(x, t) \qquad (5.1)$$

for the acoustic wave equation. The remaining task is to calculate the space derivative on the right-hand side. In general, the nth derivative of a space-dependent function can be expressed as

$$\partial_x^{(n)} p(x, t) = \mathscr{F}^{-1}[(ik)^n P(k, t)], \qquad (5.2)$$

where i is the imaginary unit, $\mathscr{F}^{-1}$ is the inverse Fourier transform, and $P(k, t)$ is the spatial Fourier transform of the pressure field $p(x, t)$, k being the wavenumber. When using the discrete Fourier transform of functions defined on a regular grid (which is the case in our applications), we obtain exact (machine precision) derivatives up to the Nyquist wavenumber $k_N = \pi/dx$ (two points per wavelength). The price to pay is the forward and inverse Fourier transform, which, depending on the number of grid points along the physical dimension, requires substantially more floating point operations than the finite-difference approach.

The principle of the pseudospectral method based on the Fourier series is illustrated in Fig. 5.3. The use of sine and cosine functions for the expansions implies periodicity at the boundaries of the physical domain.

This is not the case in most geophysical applications. In addition, the common boundary conditions (free surface, absorbing) are basically impossible to implement with similar (almost perfect) accuracy compared with the derivatives inside the medium. To some extent this can be improved by using other basis functions with similar interpolation behaviour such as Chebyshev polynomials. As they are defined in the interval [−1, 1] they are easily adapted to limited-area calculations, and an efficient implementation of boundary conditions is possible.

The pseudospectral method uses a mathematical principle (exact interpolation at grid points) that was later used extensively in the spectral-element method (in combination with the corresponding numerical integration scheme); therefore it deserves a prominent place in the history of computational seismology. In the end—despite its accuracy, the high memory efficiency, and its elegance—it did not replace the finite-difference method for geo-scientific applications. The reason is the communication-intensive algorithm that is difficult to implement efficiently on many-core systems.

5.3 Ingredients

When it was introduced to seismic wave propagation, the pseudospectral method was considered *more complicated than finite differences* (Kossloff and Bayssal, 1982), even though, by comparison with spectral elements or the discontinuous Galerkin method, the actual implementation using the often intrinsic fast Fourier transforms (*FFT*) is short and elegant. What is important to note is that the mathematical approach is indeed very different, and digs deeper into the world of numerical mathematics. Readers familiar with the concepts of function interpolation may jump directly to the actual solution of the wave equation in Section 5.4. However, the concepts of exact interpolation on specific spatial grids, cardinal functions, etc. play such an important role in the other methods, that we continue with a brief introduction.

5.3.1 Orthogonal functions, interpolation, derivative

In many situations, not only in natural sciences, we either (1) seek to approximate a known analytic function by an approximation, or (2) know a function only at a discrete set of points and would like to interpolate in between those points so that we have a representation everywhere. Let us start with the first problem and pose it such that we seek to approximate our known function by a finite sum over some N basis functions Φ_i:

$$f(x) \approx g_N(x) = \sum_{i=0}^{N} a_i \Phi_i(x) \tag{5.3}$$

and assume that the basis functions form an orthogonal set.[3]

Why would one want to replace a known function with something else? In many fields of science dynamic phenomena are expressed by partial differential

[3] What are orthogonal functions? Functions $f(x)$ and $g(x)$ are orthogonal in the interval $[a, b]$ if $\int_a^b f(x)g(x)dx = 0$. The concept of orthogonality is more easily grasped with vectors. Thinking of a discrete representation $f(x)$ as f_i and $g(x)$ as and g_i as vectors of length n the integral can be interpreted as an n-dimensional scalar vector product with the number of elements going to infinity.

equations. This implies that in order to find solutions we must be able to deal with derivatives. Yet, in many cases either nature does not do us the favour of being smooth and differentiable (e.g. layered Earth models with sudden changes of physical properties, sudden bursts of energy, such as explosions) or there are mathematical reasons to deal with functions that are non-differentiable (e.g. turn-on–turn-off phenomena in circuits, saw-tooth-like behaviour (see Fig. 5.4)).

Once we find approximations to the original function with sufficient accuracy (the criteria will be defined below), we are in good shape. With the right choice of differentiable basis functions Φ_i the calculation of the (approximate) derivative becomes trivial as

$$\partial_x f(x) \approx \partial_x g_N(x) = \sum_{i=0}^{N} a_i \partial_x \Phi_i(x). \tag{5.4}$$

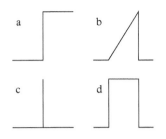

Fig. 5.4 *Examples of non-differentiable functions (in the classic sense). a: Heaviside function; b: ramp; and their derivatives c: spike; d: boxcar function. All the functions illustrated play a role in seismology as possible source time functions for earthquakes or explosions.*

There are many possible choices for basis functions and we will encounter a variety of them (e.g. Chebyshev polynomials, Lagrange polynomials, radial basis functions) in the course of this volume. Let us start with the most commonly used trigonometric basis functions (at least in spectral analysis). Consider the set of basis functions

$$\begin{aligned} \cos(nx) & \qquad n = 0, 1, \dots, \infty \\ \sin(nx) & \qquad n = 0, 1, \dots, \infty, \end{aligned} \tag{5.5}$$

with

$$\begin{aligned} & 1, \cos(x), \cos(2x), \cos(3x), \dots \\ & 0, \sin(x), \sin(2x), \sin(3x), \dots \end{aligned} \tag{5.6}$$

in the interval $[-\pi, \pi]$. We can proceed by checking whether these functions are orthogonal by evaluating integrals with all possible combinations

$$\int_{-\pi}^{\pi} \cos(jx) \cos(kx) \, dx = \begin{cases} 0 & \text{for } j \neq k \\ 2\pi & \text{for } j = k = 0 \\ \pi & \text{for } j = k > 0 \end{cases}$$

$$\int_{-\pi}^{\pi} \sin(jx) \sin(kx) \, dx = \begin{cases} 0 & \text{for } j \neq k; \ j, k > 0 \\ \pi & \text{for } j = k > 0 \end{cases} \tag{5.7}$$

$$\int_{-\pi}^{\pi} \cos(jx) \sin(kx) \, dx = 0 \text{ for } j \geqslant 0, \ k > 0$$

to find that these trigonometric functions form indeed an orthogonal set (see Fig. 5.5 for an illustration). Our problem can consequently be stated as finding an approximate function $g_N(x)$ so that

$$f(x) \approx g_N(x) = \sum_{k=0}^{N} a_k \cos(kx) + b_k \sin(kx). \tag{5.8}$$

How can we find the coefficients a_k, b_k given function $f(x)$? An obvious way of posing this problem mathematically is to seek coefficients that minimize the difference between approximation $g_N(x)$ and the original function $f(x)$. However, there are many ways of defining the difference (i.e. distance, norm) between two functions. The preferred choice is the so-called l_2-norm[4] basically quantifying the misfit-energy between the two functions:

$$\|f(x) - g_N(x)\|_{l_2} = \left[\int_a^b \{f(x) - g_N(x)\}^2 \, dx \right]^{\frac{1}{2}} = \text{Min.} \qquad (5.9)$$

This equivalence is independent of the choice of basis functions. In the case of trigonometric basis functions this requirement leads to the well-known formulations of Fourier series and the Fourier transform that will be discussed in the next section.

5.3.2 Fourier series and transforms

The concepts of Fourier series and transforms are so central to seismology (data processing, spectral analysis, instrument correction, filtering, etc.) that, even though here they are used just to calculate space derivatives efficiently, we will present the fundamental equations. The general concepts introduced above—approximating a function in a certain interval by summing over weighted basis functions—already implies a discretization, as we assume that the sum is finite. The most important concept of this section will consist of the properties of Fourier series on regular grids. It is important to note that we really only scratch the surface of Fourier analysis here, and the interested reader is referred to the literature suggested at the end of this chapter.

Let us start by presenting the result of the minimization problem presented in Eq. 5.9. The requirement that we approximate a 2π-periodic arbitrary function in the interval $[-\pi, \pi]$ (this can be relaxed) by a sum over sine and cosine functions of the form

$$g_N(x) = \frac{1}{2}a_0 + \sum_{k=1}^{n} a_k \cos(kx) + b_k \sin(kx) \qquad (5.10)$$

leads to the coefficients

$$a_k = \frac{1}{\pi} \int_{-\pi}^{\pi} f(x) \cos(kx) \, dx \qquad k = 0, 1, \ldots, N$$

$$\qquad\qquad\qquad\qquad\qquad\qquad\qquad (5.11)$$

$$b_k = \frac{1}{\pi} \int_{-\pi}^{\pi} f(x) \sin(kx) \, dx \qquad k = 0, 1, \ldots, N.$$

Note that k is the wavenumber $2\pi/\lambda$. In light of this, the coefficients can be interpreted as the amplitude of the harmonic waves that make up the function $f(x)$ and thus describe its spectral content.

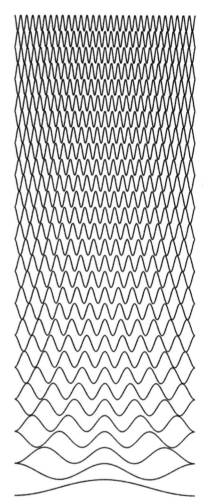

Fig. 5.5 *Illustration of orthogonal functions using cosine functions. $f(x) = \cos(nx)$ is shown for $n = 1 \ldots 30$ for increasing wavenumber n.*

[4] This misfit criterion is also commonly used for solving inverse problems in seismology. In this case the problem is formulated as minimizing the difference between theoretical calculations (e.g. travel times, polarities, waveforms) and observations. The coefficients that are sought are physical parameters of the Earth's interior, or seismic sources.

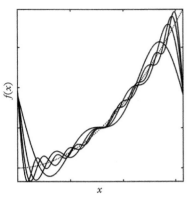

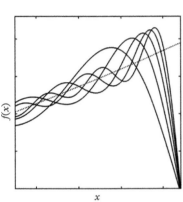

Fig. 5.6 *Illustration of Fourier series.* **Top:** *The function $f(x) = x^2$ defined in the interval $[0, 2\pi]$ (dotted line) is approximated by Fourier series of increasing order $N = 2 - 8$ (solid lines). The approximation $g_N(x)$ oscillates around the exact function.* **Bottom:** *Detail of the approximation behaviour towards the upper boundary for $N = 8 - 24$. Note the unavoidable overshoot of the approximation at the boundary caused by the periodicity requirement (the Gibbs phenomenon).*

This formulation of Fourier series can be written in an elegant way using Euler's formula to obtain

$$g_N(x) = \sum_{k=-N}^{k=N} c_k e^{ikx}, \tag{5.12}$$

with complex coefficients c_k given by

$$\begin{aligned} c_k &= \tfrac{1}{2}(a_k - ib_k) \\ c_{-k} &= \tfrac{1}{2}(a_k + ib_k) \quad k > 0, \\ c_0 &= \tfrac{1}{2}a_0. \end{aligned} \tag{5.13}$$

while a_k and b_k are obtained through Eq. 5.11.

Let us see how the approximation in the continuous case works by finding the interpolating trigonometric polynomial for the 2π-periodic function

$$f(x + 2\pi n) = f(x) = x^2, \qquad x \in [0, 2\pi], \; n \in \mathbb{N}. \tag{5.14}$$

Applying Eq. 5.11 we obtain (see exercises) the approximation $g_N(x)$

$$g_N(x) = \frac{4\pi^2}{3} + \sum_{k=1}^{N}\left[\frac{4}{k^2}\cos(kx) - \frac{4\pi}{k}\sin(kx)\right], \tag{5.15}$$

where N is the number of terms used for the approximation. The approximation behaviour is shown in Fig. 5.6. While in general the approximation wraps around the original function and seems to converge, the overshoot at the boundary raises some questions about the actual convergence behaviour of Fourier series for increasing N. In fact it is important to note that while Fourier postulated that arbitrary (i.e. even non-differentiable) functions could be represented by a single analytical expression (the Fourier series) in the late eighteenth century, the idea found substantial resistance amongst mathematicians. Today Fourier series and all related concepts such as the Fourier transform are ubiquitous in all fields of science. However the question of convergence remained open until the twentieth century!

To make use of these concepts for the numerical solution of partial differential equations let us move to the discrete world. We assume that we know our function $f(x)$ at a discrete set of points x_i given by

$$x_k = \frac{2\pi}{N}k, \quad k = 0, \ldots, N. \tag{5.16}$$

Using the 'trapezoidal rule' for the integration of a definite integral we obtain for the Fourier coefficients

$$a_k^* = \frac{2}{N}\sum_{j=1}^{N} f(x_j)\cos(kx_j) \qquad\qquad k = 0, 1, \ldots \tag{5.17}$$

$$b_k^* = \frac{2}{N}\sum_{j=1}^{N} f(x_j)\sin(kx_j) \qquad\qquad k = 0, 1, \ldots, \tag{5.18}$$

where the upper asterisk denotes the discrete case. Note that the integrals have been replaced by sums over (weighted) values of function f at grid points x_j. We thus obtain the specific Fourier polynomial with $N = 2n$

$$g_N^* =: \frac{1}{2}a_0^* + \sum_{k=1}^{N-1}\{a_k^* \cos(kx) - b_k^* \sin(kx)\} + \frac{1}{2}a_N^* \cos(Nx), \tag{5.19}$$

with the tremendously important property that

$$g_N^*(x_i) = f(x_i). \tag{5.20}$$

This behaviour is illustrated in Fig. 5.7. At the discrete points x_i (in fact the integration points for the calculation of the Fourier coefficients), the approximating function *exactly* (that means to machine precision including possible rounding errors) interpolates the original function $f(x)$. Even though you might argue that the function is not well represented in between the grid points (see Fig. 5.7), in the context of our desire to solve a partial differential equation on a discrete grid we do not really care. We are only interested in the function itself and its derivatives at the grid point locations. In formulations that require the evaluation of integrals (e.g. finite-element-type methods) this is a different story!

At this point it is instructive to introduce the concept of *cardinal functions*, which will play an important role in some of the other methods that we will encounter. The fact that we exactly recover the original discrete function at the grid points is thanks to the specific form of the cardinal function when interpolating with trigonometric basis functions. It is the result of approximating a spike function (equivalent to a delta function in the continuous world) with Fourier series. The concept is illustrated in Fig. 5.8. Discrete interpolation and derivative operations can also be formulated in terms of convolutions (this should become clear after we introduce the Fourier transform and its properties). Cardinal functions are used in the convolutional formulation for the interpolation problem.

What is missing in our discussion on orthogonal functions and function approximations is the calculation of derivatives needed to solve the elastic wave equation. To introduce the powerful concept of spectral derivatives, we go back to the continuous world for a moment and introduce the continuous Fourier transform, which generalizes the concepts of Fourier series. One possible definition for the Fourier transform of function $f(x)$ is

$$F(k) = \mathcal{F}[f(x)] = \frac{1}{\sqrt{2\pi}} \int_{-\infty}^{\infty} f(x)e^{-ikx}dx, \tag{5.21}$$

where $F(k)$ is the spectrum of the original space-dependent function $f(x)$ in the wavenumber domain k.[5] In terms of spectral content the absolute values of the complex Fourier transform $|F(k)|$ correspond to the spectral amplitudes. This is called the *forward* transform in the sense that we transform from the common

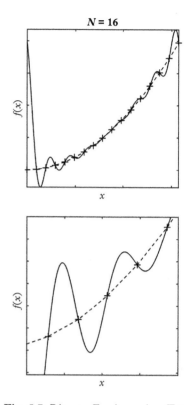

Fig. 5.7 *Discrete Fourier series.* **Top:** *Approximation of the function $f(x) = x^2$ known at a discrete set of $N = 16$ points indicated by '+'.* **Bottom:** *Detail with an illustration of the exact interpolation property at the so-called collocation points.*

[5] The various definitions differ in particular concerning the sign of the exponent and the factors in front of the integral. When working with intrinsic numerical implementations of the Fourier transform it is always advisable to carefully check which formulation is being used!

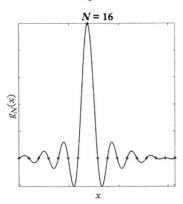

Fig. 5.8 *Illustration of cardinal functions. The function shown in this graph is the interpolating (cardinal) function for grid point* $f_i = 1$, $i = N/2$. *Note that it is unity at grid point* x_i *and zero at all other points on the discrete grid denoted by '+'. The cardinal function has the form of a* sinc *function.*

physical domain (space x or time t) to the spectral domain (spatial wavenumber k or temporal frequency f). From now on we will denote the forward transform of $f(x)$ (or its discrete representation) by $\mathscr{F}[f(x)]$. Note that we attempt to use a consistent notation for values defined in continuous or discrete physical space with small letters and values defined in the spectral domain with capital letters, respectively. To get back from the spectral domain to the physical space we apply the inverse transform, which we denote as $\mathscr{F}^{-1}[F(k)]$

$$f(x) = \mathscr{F}^{-1}[F(k)] = \frac{1}{\sqrt{2\pi}} \int_{-\infty}^{\infty} F(k) e^{ikx} dk, \qquad (5.22)$$

with $F(k)$ being the complex spectrum. Note the infinite integral boundaries that, because of the convergence behaviour, imply the equivalence of the representation of function $f(x)$ in the physical space and the spectral domain.

Taking the formulation of the inverse transform (Eq. 5.22) it is straightforward (see exercise) to obtain the derivative of function $f(x)$ with respect to the spatial coordinate

$$\begin{aligned}
\frac{d}{dx} f(x) &= \frac{d}{dx} \frac{1}{\sqrt{2\pi}} \int_{-\infty}^{\infty} F(k) e^{ikx} dk \\
&= \frac{1}{\sqrt{2\pi}} \int_{-\infty}^{\infty} ik\, F(k) e^{ikx} dk \\
&= \frac{1}{\sqrt{2\pi}} \int_{-\infty}^{\infty} D(k)\, F(k) e^{ikx} dk,
\end{aligned} \qquad (5.23)$$

with $D(k) = ik$ corresponding to the derivative operator in the spectral domain. Intuitively, in the continuous world—with infinite integration domain—the derivative should be exact (in the sense of convergence behaviour). Equivalently we can extend this formulation to the calculation of the nth derivative of $f(x)$, to obtain

$$F^{(n)}(k) = D(k)^n\, F(k) = (ik)^n\, F(k), \qquad (5.24)$$

followed by an inverse Fourier transform to return to physical space. Thus using the Fourier transform operator $\mathscr{F}$ we can obtain an exact *nth* derivative using

$$\begin{aligned}
f^{(n)}(x) &= \mathscr{F}^{-1}[(ik)^n\, F(k)] \\
&= \mathscr{F}^{-1}[(ik)^n\, \mathscr{F}[f(x)]].
\end{aligned} \qquad (5.25)$$

How does this work in the discrete world? With what accuracy can we obtain a derivative for a function defined on a discrete set of points as introduced in connection with Fourier series? To answer these questions let us write down the discrete Fourier transform that is so widely used for data analysis, filtering, etc. in all areas of science. Again there are several possibilities concerning the – signs. Adopting the complex notation of the forward transform we obtain

$$F_k = \sum_{j=0}^{N-1} f_j \, e^{-i \, 2\pi jk/N}, \quad k = 0, \ldots, N-1 \tag{5.26}$$

and the inverse transform

$$f_j = \frac{1}{N} \sum_{k=0}^{N-1} F_k \, e^{i \, 2\pi jk/N}, \, j = 0, \ldots, N-1, \tag{5.27}$$

essentially the complex formulation of Eq. 5.19 with the same interpolation properties. Here f_j is the vector describing the space-dependent function (e.g. the seismic wave field) and F_k is its complex wavenumber spectrum.

In this formulation the number of calculations to be carried out for a Fourier transform of a vector with length N is proportional to N^2. In terms of overall number of operations in connection with long vectors of multi-dimensional transforms this is serious. An ingenious exploitation of symmetry properties introduced by Cooley and Tukey (1965)[6] reduces the proportionality to $N \log N$. It is worth exploring the consequence of this improvement for realistic 3D calculations (see exercises).

By analogy with the continuous formulation of the Fourier transformation, noting that we exactly interpolate at the collocation points, we are able to obtain exact (to machine precision) nth derivatives on our regular grid by performing the following operations (here the $\mathscr{F}$ operator stands for the discrete Fourier transform often realized by the fast Fourier transform) on vector f_j defined at grid points x_j:

$$\partial_x^{(n)} f_j = \mathscr{F}^{-1} [(ik)^n \, F_k], \tag{5.28}$$

where

$$F_k = \mathscr{F}[f_j], \tag{5.29}$$

and we used the partial derivative symbol since the discrete space-dependent function might also depend on time (as is the case for the displacement field in the wave equation).

Let us see how this works in practice and take an example. We initialize a 2π-periodic Gauss function in the interval $x \in [0, 2\pi]$ as

$$f(x) = e^{-1/\sigma^2 \, (x-x_0)^2} \tag{5.30}$$

with $x_0 = \pi$ and the derivative

$$f'(x) = -2\frac{(x-x_0)}{\sigma^2} \, e^{-1/\sigma^2 \, (x-x_0)^2} \tag{5.31}$$

allowing us an easy evaluation of the numerical accuracy of the Fourier-based derivative. The vector with values f_j is required to have an even number of

[6] You don't always have to find a new particle, or discover plate tectonics or relativity, to leave your mark! This four-page FFT paper probably beats them all in terms of citations!

uniformly sampled elements. In our example this is realized with a grid spacing of $dx = 2\pi/N$ with $N = 128$ and $x_j = j\,2\pi/N$, $j = 1, \ldots, N$, $\sigma = 0.5$, and $x_0 = \pi$.

The results are shown in Fig. 5.9. They were obtained using the intrinsic fast Fourier transform (*numpy* library) with the following Python commands:

```
# [...]
# Basic parameters
nx = 128
x0 = pi
def fourier_derivative(f, dx):
    # Length of vector f
    nx = f.size
    # Initialize k vector up to Nyquist wavenumber
    kmax = pi/dx
    dk = kmax/(nx/2)
    k = arange(float(nx))
    k[:nx/2] = k[:nx/2] * dk
    k[nx/2:] = k[:nx/2] - kmax
    # Fourier derivative
    ff = 1j*k*fft(f)
    df = ifft(ff).real
    return df
# [...]
# Main program
# Initialize space and Gauss function (also return dx)
x = linspace(2*pi/nx, 2*pi, nx)
dx = x[1]-x[0]
sigma = 0.5
f = exp(-1/sigma**2 * (x - x0)**2)
# Calculate derivative of vector f
df = fourier_derivative(f, dx)
# [...]
```

Note that $1j$ represents the imaginary unit. The reader is referred to Python tutorials or the supplementary electronic material.

This calculation of the derivative based on the Fourier transform is very elegant. What needs some care is the correct initialization of the wavenumber axis k, that is, the imaginary part of the derivative operator. The specific form of this vector is defined by the requirements of the fast fourier algorithm and the way the frequencies are arranged. The reader is referred to the documentation of the specific Fourier transform routines.

From Fig. 5.9 we infer that the numerical approximation of the derivative is indeed obtained with errors only coming from round-off problems caused by the bit depth of the floating-point definition (here double precision). This is as good

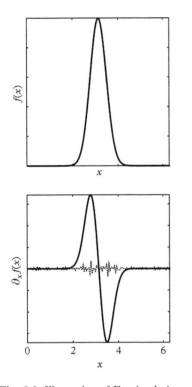

Fig. 5.9 *Illustration of Fourier derivatives.* **Top:** *Periodic Gauss function* f(x) *defined in the interval* $[0, 2\pi]$. **Bottom:** *Superposition of the analytical derivative (solid line) and the numerical derivative (dot-dashed line) based on the fast Fourier transform. The differences are indistinguishable, therefore they are illustrated by the dotted line, multiplied by a factor 10^{13}! This is the expected level of inaccuracy due to rounding errors at double precision (8 bytes per number).*

as it gets, at least for the conditions under which these numerical calculations were carried out (regular spacing, periodicity). We will proceed with assembling this powerful tool to obtain numerical solutions to the wave equation.

5.4   The Fourier pseudospectral method

With the simple recipe to calculate *exact* nth derivatives on a regular-spaced grid (assuming periodicity) we now have all it takes to solve a wave equation-type problem. For the moment we disregard the implementation of spatial boundary conditions. Even though our main focus in this volume is to illustrate the concepts in 1D, we proceed in a similiar way as in the previous chapter presenting first the acoustic 1D and 2D cases. The reason is that the numerical solutions can still be treated analytically and important results on numerical dispersion emerge. This will be followed by the elastic 1D case.

5.4.1   Acoustic waves in 1D

Assuming implicitly the space–time dependence, the constant-density acoustic wave equation in 1D is given as

$$\partial_t^2 p = c^2 \partial_x^2 p + s \tag{5.32}$$

and we seek solutions for pressure field $p(x, t)$ for a velocity model $c(x)$ with (in principle) arbitrary heterogeneous variations. The source is injected through $s(x, t)$. The time-dependent part is solved using a standard three-point finite-difference operator, leading to

$$\frac{p_j^{n+1} - 2p_j^n + p_j^{n-1}}{dt^2} = c_j^2 \partial_x^2 p_j^n + s_j^n, \tag{5.33}$$

where upper indices represent time and lower indices space. The remaining task is to calculate the second derivatives on the right-hand side.

Based on the developments in the previous sections we proceed by calculating the second derivatives using the Fourier transform (in practice this is usually realized by applying the discrete fast Fourier transform):

$$\partial_x^2 p_j^n = \mathscr{F}^{-1}[(ik)^2 P_\nu^n] = \mathscr{F}^{-1}[-k^2 P_\nu^n], \tag{5.34}$$

leading (within some limits) to an exact derivative with only numerical rounding errors. Here, P_ν^n is the discrete complex wavenumber spectrum at time n. As a consequence the main overall error of the numerical solutions comes from the time integration scheme. The following Python code snippet illustrates the compact algorithm that results when this operation is carried out with the help of a function (or subroutine) calculating the nth derivative based on Fourier transforms.[7]

[7] In the code below space-dependent fields (pressure p, function f, complex spectrum ff, spatial source function sg, numerical derivative df, second derivative $d2p$, pressure time levels *pold* and *pnew*) are vectors of length [*nx*] where *nx* is the number of grid points. j is the imaginary unit.

```
# [...]
# Fourier derivative
def fourier_derivative_2nd(f, dx):
    # [...]
    # Fourier derivative
    ff = (1j * k)**2 * fft(f)
    df = ifft(ff).real
    return df
# [...]
# Time extrapolation
for it in range(nt):
    # 2nd space derivative
    d2p = fourier_derivative_2nd(p, dx)
    # Extrapolation
    pnew = 2 * p - pold + c**2 * dt**2 * d2p
    # Add sources
    pnew = pnew + sg * src[it] * dt**2
    # Remap pressure field
    pold, p = p, pnew
# [...]
```

The function *fourier_derivative_2nd* returns the second derivative of function f discretized with grid increment dx. The source is injected via a smooth space-dependent field sg of Gaussian shape to avoid the Gibbs phenomenon. Otherwise the code differs from the finite-difference solution only in the calculation of the space derivatives. Let us take an example and compare the result with the finite-difference method. The parameters for the simulation are given in Table 5.1.

Before showing the results we need to discuss a specific feature of source injection when series-based methods are used. In the case of finite differences it was possible and straightforward to initiate a point-like source at one grid point. This is no longer the case for pseudospectral methods. The Fourier transform of a spike-like function creates oscillations that damage the accuracy of the solution. To avoid this it is common practice to define a space-dependent part of the source using a Gaussian function. In the example shown in the following a Gaussian function $e^{-1/\sigma^2 (x-x_0)^2}$ was used for pseudospectral and finite-difference algorithms with $\sigma = 2dx$, dx being the grid interval and x_0 the source location. To match with the analytical solution the integral over this function should be scaled to $1/dx$ (see previous chapter).

The results of a simulation in a homogeneous medium are shown in Fig. 5.10 for various propagation distances given in terms of dominant wavelengths $n\lambda$. It is instructive to compare the pseudospectral results with those obtained with the finite-difference approximation of the space derivatives using exactly the same set-up. The results indicate that while the pseudospectral solutions show very

Table 5.1 *Simulation parameters for 1D acoustic simulation with the Fourier method*

Parameter	Value
x_{max}	1,250 m
nx	2048
c	343 m/s
dt	0.00036 s
dx	0.62 m
f_0	60 Hz
ϵ	0.2

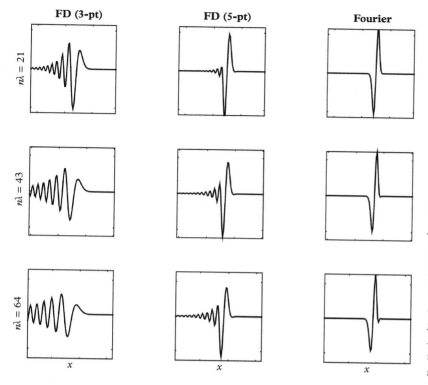

FD (3-pt)   **FD (5-pt)**   **Fourier**

$n\lambda = 21$

$n\lambda = 43$

$n\lambda = 64$

x x x

Fig. 5.10 *Application of the Fourier method to the acoustic wave equation: comparison with finite differences. Simulation results after propagation distances given in terms of wavelengths.* **Left column:** *finite differences with three-point operator.* **Middle column:** *finite differences with five-point operator.* **Right column:** *Fourier method. The correct waveform corresponds to the shape in the top-right graph. All windows are scaled to the same space and amplitude interval. Note the superior stability of the waveform for the Fourier method and the strong numerical dispersion for both finite-difference implementations.*

small increasing effects of numerical dispersion with distance, the finite-difference solutions are too inaccurate to be usable. However, note the considerable improvement of five-point over the three-point operator already discussed in the previous chapter. Another observation is that while the dispersive waves in the finite-difference method are slower (arrive later), the small dispersion visible in the pseudospectral results arrive earlier (are faster).

Results like the ones shown in Fig. 5.10 seem to suggest that the pseudospectral method is far superior to the finite-difference method, as was often claimed in the first papers.[8] In fact, so far we have only looked at part of the story and the comparison is not entirely fair. First, we used the same number of grid points in all cases. That means that the number of floating point operations and thus run time is much higher in the case of the Fourier method. Second, we used the same stability criterion for all simulations, which creates a disadvantage for the finite-difference method. We will dwell more on comparing solutions from different methods when we introduce the elastic case.

To understand what is happening, let us proceed in an analoguous way to the finite-difference method and seek analytical solutions for the numerical algorithm.

[8] In fact, several methods were introduced with the notion that they were superior to others. In some cases, when more issues were taken into account, these claims had to be substantially revised. The lesson is that you have to be very careful when comparative statements between methods are made (including in this volume).

5.4.2   Stability, convergence, dispersion

An effective way of understanding the behaviour of numerical approximations is to seek solutions of the algorithms using discrete plane waves of the form

$$
\begin{aligned}
p_j^n &= e^{i(kjdx - \omega n dt)} \\
\partial_x^2 p_j^n &= -k^2 e^{i(kjdx - \omega n dt)},
\end{aligned}
\tag{5.35}
$$

where the second space derivatives are given in the way they are calculated using Fourier transforms. Following the developments in the chapter on finite differences, the time-dependent part can be expressed as

$$
\partial_t^2 p_j^n = -\frac{4}{dt^2}\, \sin^2\left(\frac{\omega dt}{2}\right) e^{i(kjdx - \omega n dt)},
\tag{5.36}
$$

where we made use of the Euler formula and the fact that

$$
2\sin^2 x = \frac{1}{2}(1 - \cos 2x).
\tag{5.37}
$$

This is the so-called von Neumann analysis and we proceed in the same way as before and replace the partial derivatives in the wave equation (Eq. 5.32) with these results, and extract the frequency to obtain a formula for the phase velocity $c(k)$ after division by wavenumber k:

$$
c(k) = \frac{\omega}{k} = \frac{2}{kdt}\, \sin^{-1}\left(\frac{kcdt}{2}\right).
\tag{5.38}
$$

This result has some important consequences. First, when dt becomes small, $\sin^{-1}(kcdt/2) \approx kcdt/2$ and we recover $c = \omega/k$ as the analytical phase velocity. This demonstrates the convergence of the scheme and an important aspect is that the space increment dx does not appear in this equation (as was the case in the finite-difference method), with the result that making the time step smaller always decreases the error of the overall solution. Second, as the argument of the inverse sine must be smaller than one, the stability limit requires $k_{max}(cdt/2) \leq 1$. As $k_{max} = \pi/dx$ (sampling theorem) the stability criterion for the 1D case is $\epsilon = cdt/dx = 2/\pi \approx 0.64$.

Eq. 5.38 allows us to calculate the numerical phase velocity as a function of the number of grid points used for various stability criteria. This dispersion curve illustrated in Fig. 5.11 shows the tremendous accuracy of the phase velocity in the case of $\epsilon = 0.2$ for more than ten grid points per wavelength. The results also indicate that the error due to time discretization leads to an increase in phase velocity with decreasing number of grid points per wavelength. This is the behaviour we saw in the numerical simulation presented in Fig. 5.10.

One of the features we discovered for finite-difference approximations to the wave equation in higher dimensions was the fact that the error of the wavefield depends on the propagation direction (numerical anisotropy). How does the Fourier method behave in this context? To answer this question we now turn to the 2D acoustic case.

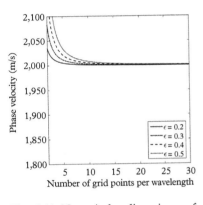

Fig. 5.11 *Numerical dispersion of Fourier method. The numerical phase velocity is shown as a function of number of grid points per wavelength for various CFL criteria (ϵ). The correct propagation velocity is 2,000 m/s. As the number of grid points per wavelength increases the correct velocity is recovered. Note that the error due to temporal discretization leads to higher phase velocities (as opposed to a decrease, in the case of finite differences).*

5.4.3   Acoustic waves in 2D

The acoustic wave equation in 2D reads

$$\partial_t^2 p = c^2(\partial_x^2 p + \partial_z^2 p) + s, \tag{5.39}$$

where the difference with the 1D case is simply the fact that all space-dependent fields are now defined in (x, z), stored in the computer as matrices, and structure and source are translationally invariant in the third dimension (here: y). As in the previous section we replace the time-dependent part by a standard three-point finite-difference approximation and extrapolate the pressure field accordingly:

$$\frac{p_{j,k}^{n+1} - 2p_{j,k}^n + p_{j,k}^{n-1}}{dt^2} = c_{j,k}^2(\partial_x^2 p + \partial_z^2 p)_{j,k} + s_{j,k}^n. \tag{5.40}$$

We are left with approximating the second partial derivatives with respect to x and z (denoted by indices j and k, respectively) using the Fourier approach. The sequence of operations to achieve this is

$$\partial_x^2 p + \partial_z^2 p = \mathscr{F}^{-1}[-k_x^2 \,\mathscr{F}[p]] + \mathscr{F}^{-1}[-k_z^2 \,\mathscr{F}[p]], \tag{5.41}$$

where the discrete Fourier transforms are usually calculated with the fast Fourier transform. In practice, as is the case in the finite-difference method, we have to loop through all grid points in x and z and independently calculate the space derivatives, keeping one of the coordinates constant, while calculating the derivative with respect to the other. This is illustrated in the following code snippet:

```
# [...]
# second space derivatives
for j in range(nz):
    d2px[:,j] = fourier_derivative_2nd(p[:,j].T, dx)
for i in range(nx):
    d2pz[i,:] = fourier_derivative_2nd(p[i,:], dx)
# Extrapolation
pnew = 2 * p - pold + c**2 * dt**2 * (d2px + d2pz)
# [...]
```

Table 5.2 *Simulation parameters for 2D acoustic simulation with the Fourier method*

Parameter	Value
x_{max}	200 m
nx	256
c	343 m/s
dt	0.00046 s
dx	0.78 m
f_0	200 Hz
ϵ	0.2

where we assume identical space increments dx in both directions. In order to investigate the wave field behaviour in various directions we perform a simulation with parameters as listed in Table 5.2 and compare with the finite-difference solution using a five-point operator for the space derivatives. The snapshots shown in Fig. 5.12 indicate that (1) there is strong anisotropic dispersion behaviour visible for the finite-difference solution with the most accurate direction at 45° to the co-ordinate axes, and (2) the Fourier solution shows weak signs of dispersion—but

Fig. 5.12 *Comparison of Fourier and finite-difference method in 2D:* **Left:** *Snapshot of wavefield simulated with the Fourier method. The wavefront is spherical and does not show dispersive behaviour.* **Right:** *Snapshot obtained with the finite-difference method with exactly the same set-up (which is not fair to finite differences). Strong anisotropic dispersion behaviour is visible. Note, however, that the Fourier method requires substantially more calculations.*

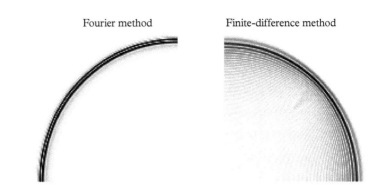

Fourier method   Finite-difference method

the most important observation is that there does not seem to be a directional dependence: the error is isotropic.

Looking at the simulation parameters, the wave field is actually sampled only with 2–3 grid points per wavelength. This is not entirely true as there is a Gaussian-shaped spatial source function that acts as a low-pass filter. Nevertheless, it is clear that with this set-up we are way beyond the healthy operating range for a low-order finite-difference approximation. Again, please note that the point here is not to sell the Fourier method as the better choice! The point is to illustrate an interesting property, namely an isotropic dispersion error for the Fourier method despite the fact that the cubic grid is not isotropic.[9]

This isotropic error behaviour is certainly an advantage when simulating *physical* anisotropy that is often weak. Then, the physical effects cannot be confused with numerical anisotropy caused by discretization. To bring this point home, let us look at the analytical result for the dispersion problem in 2D.

5.4.4   Numerical anisotropy

To investigate the dispersion behaviour of our numerical scheme based on the Fourier method we proceed by finding solutions to monochromatic plane waves propagating in the direction $\mathbf{k} = (k_x, k_z)$:

$$p_{j,k}^n = e^{i(k_x j dx + k_z k dx - \omega n dt)}, \tag{5.42}$$

assuming regular discretization in space and time with increments dx and dt, respectively. In the case of the Fourier method the derivatives can be calculated exactly, leading to the following formulation:

$$\begin{aligned} \partial_x p_{j,k}^n &= -k_x^2 \, e^{i(k_x j dx + k_z k dx - \omega n dt)} \\ \partial_z p_{j,k}^n &= -k_z^2 \, e^{i(k_x j dx + k_z k dx - \omega n dt)}. \end{aligned} \tag{5.43}$$

Combining this with the analytical form of the three-point operator for the time derivative (see previous sections)

$$\partial_t^2 p_{j,k}^n = -\frac{4}{dt^2} \, \sin^2\left(\frac{\omega dt}{2}\right) \, e^{i(k_x j dx + k_z k dx - \omega n dt)} \tag{5.44}$$

[9] There exists an isotropic grid in 2D that is hexagonal; in 3D there is no such thing as an isotropic grid.

we obtain the numerical dispersion relation in 2D for arbitrary wave number vectors (i.e. propagation directions) **k** as

$$c(\mathbf{k}) = \frac{\omega}{|\mathbf{k}|} = \frac{2}{|\mathbf{k}|\,dt}\,\sin^{-1}\!\left(\frac{c\,dt\sqrt{k_x^2 + k_z^2}}{2}\right). \tag{5.45}$$

The direction-dependent error of the phase velocity is illustrated in Fig. 5.13, confirming the observation of our 2D simulation example. The error is isotropic and in the case of 10 points per wavelength below 0.5%.

5.4.5 Elastic waves in 1D

Finally we want to solve the 1D elastic case

$$\rho(x)\partial_t^2 u(x,t) = \partial_x\left[\mu(x)\partial_x u(x,t)\right] + f(x,t), \tag{5.46}$$

which contains a sequence of first derivatives with respect to space of the displacement field u and the space-dependent shear modulus μ, and combinations thereof. Again, the finite-difference approximation of the extrapolation part leads to

$$\rho_i\frac{u_i^{j+1} - 2u_i^j + u_i^{j-1}}{dt^2} = (\partial_x\left[\mu(x)\partial_x u(x,t)\right])_i^j + f_i^j, \tag{5.47}$$

with space derivatives to be calculated using the Fourier method. The sequence of operations required to obtain the right-hand side of Eq. 5.46 (without sources) reads

$$\begin{aligned}u_i^j &\to \mathscr{F}[u_i^j] \to U_\nu^j \to -ikU_\nu^j \to \mathscr{F}^{-1}[-ikU_\nu^j] \to \partial_x u_i^j\\ \partial_x u_i^j &\to \mathscr{F}[\mu_i\partial_x u_i^j] \to \tilde U_\nu^j \to \mathscr{F}^{-1}[-ik\tilde U_\nu^j] \to \partial_x\left[\mu(x)\partial_x u(x,t)\right],\end{aligned} \tag{5.48}$$

where, as a reminder, capital letters denote fields in the spectral domain, lower indices with Greek letters indicate discrete frequencies, and $\tilde U_\nu^j = \mu_i\partial_x u_i^j$ was introduced as an intermediate result to facilitate notation. It is important to note that, as demonstrated in the preceding sections, this entire result is accurate to machine precision (provided the space-dependent fields obey the requirements for the discrete Fourier transform to be accurate). Note also that, in comparison with the acoustic case, two Fourier transform operations are necessary, basically doubling the computational effort to evaluate the space-dependent part of the wave equation.

Even though you might get bored with looking at solutions to the homogeneous problem, it is actually instructive to investigate the performance difference to the finite-difference method. Throughout this volume I will stress again and again that a fair comparison between methods is extremely difficult. However,

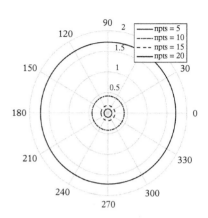

Fig. 5.13 *Numerical anisotropy of the Fourier method. The error of the phase velocity is shown as a function of propagation direction (in %) for varying numbers of grid points per wavelength. Note that there is no directional dependence of the phase velocity error. Thus, unlike with the finite difference method, there is no numerical anisotropy.*

Table 5.3 *Simulation parameters for 1D elastic simulation with pseudospectral Fourier method (PS), comparison with finite differences (FD)*

	FD	PS
nx	3,000	1,000
nt	2,699	3,211
c	3,000 m/s	3,000 m/s
dx	0.33 m	1.0 m
dt	5.5 e–5 s	4.7 e–5 s
f_0	260 Hz	260 Hz
ϵ	0.5	0.14
n/λ	34	11

FD (5-pt), run time: 2.50875 s

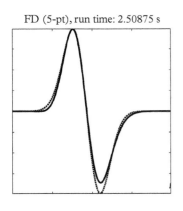

Fourier, run time: 3.518 s

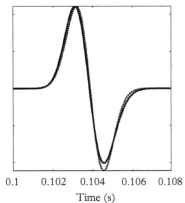

0.1   0.102   0.104   0.106   0.108

Time (s)

Fig. 5.14 *Elastic Fourier method in 1D: comparison with finite differences. An attempt is made to compare memory requirements and computation speed between the Fourier method (**bottom**) and a fourth-order finite-difference scheme (**top**), solving the same problem. In both cases the relative error compared to the analytical solution (misfit energy) is approximately 1%. The run time is comparable. The big difference is the number of grid points along the x dimension. The ratio is 3:1 (FD:Fourier).*

the comparison can serve to illustrate some specific aspects of one method vs. the other, which are not usually generalizable to arbitrary applications.

We proceed by finding a set-up for a classic staggered-grid finite-difference solution to the elastic 1D problem (see previous chapter) that leads to an energy misfit to the analytical solution u_a of 1% (which is commonly required for realistic simulations). The energy misfit is simply calculated by $(u_{FD} - u_a)^2/u_a^2$. Then we adjust the set-up for the Fourier solution such that we obtain roughly the same error (1%) and log the run time. The parameters for these set-ups are listed in Table 5.3 and the seismograms are shown in Fig. 5.14.

It is worth having a close look at the simulation set-up. The number of grid points (and thus the number per wavelength) is three times larger in the case of the finite-difference method. For the Fourier simulation the stability criterion ϵ was adjusted (leading to a change in the time step) until the error was in the same range.

It is interesting to note (but not necessarily representative) that we end up with a quite similar time increment (and thus overall number of simulation time steps). The results in Fig. 5.14 indicate that the overall run time (using a simple elapsed time routine) is also quite similar. Thus the main difference comes with the memory used to obtain this result. Obviously, the memory reduction by using the Fourier method is a factor of 3, while more floating point operations per iteration lead in the end to the same accuracy. For simulations of higher dimensions this effect is much more dramatic.

The actual numbers in this example should not be taken too seriously, but the effect on the memory economization was one of the key reasons for the great interest in this method at a time when (1) there were no parallel architectures other than vector processors, and (2) rapid-access memory was expensive (and small). As indicated in the introduction, the global communication property of pseudospectral methods and the associated difficulties with efficient parallelization led to a waning interest in those techniques soon after the emergence of parallel hardware.

5.5   Infinite order finite differences

You might be surprised to find a section on finite differences in this chapter. However, after having learned the connection between derivative calculations and the Fourier transform, we can take an alternative look at difference operators from a spectral point of view. We will find that any finite-difference operation can in fact be described as a convolution operation, which implies that the specific finite-difference operator has a spectral representation that can be compared with the now familiar exact *ik* operator.

We first look at this schematically. The least accurate approximation of a centred partial derivative is a two-point operator, as introduced in the previous chapter. The accuracy can be improved by increasing the length of the operator, for example using Taylor series. On the other hand, in this chapter we have learned that by using the Fourier concepts we can calculate basically an exact derivative (to machine precision, provided we have frequencies below Nyquist). Thus we can consider the two-point finite-difference scheme and the Fourier method (*n*-point) as two ends of an axis describing the length of the difference operator, *n* being the length of the vector describing the spatial domain. This is illustrated in Fig. 5.15. It is important to note that from a practical/computational point of view it makes sense to use *either* (1) a low-order spatial scheme, or (2) an infinite-order (Fourier) scheme—whereas approaches in between are less optimal (see exercises).

To demonstrate why the Fourier method can be interpreted as an infinite-order finite-difference scheme let us recall one of the most fundamental (and useful) mathematical results; the convolution theorem. Expressed in words, a multiplication in the spectral domain is a convolution in the space domain and vice versa. In mathematical terms, for two functions $d(x)$ and $f(x)$ with complex spectra $D(k)$ and $F(k)$, the convolution theorem says that if

$$D(k) = \mathscr{F}[d]$$
$$F(k) = \mathscr{F}[f]$$

(5.49)

then $d * f = \mathscr{F}^{-1}[D(k)F(k)]$,

where $\mathscr{F}$ represents the Fourier transform, and $*$ denotes convolution, defined in the continuous case as

$$(d * f)(x) := \int_{-\infty}^{\infty} d(x')f(x - x')dx'$$

(5.50)

and in the discrete case with vectors d_i, $i = 0, 1, \ldots, m$, and f_j, $j = 0, 1, \ldots, n$,

$$(d * f)_k = \sum_{i=0}^{m} d_k f_{k-i} \quad k = 0, 1, \ldots, m + n.$$

(5.51)

Because of the tremendous practial importance in seismology in particular of the discrete convolution in connection with filtering, instrument correction, Green's function analysis, and data processing, I strongly encourage the reader at some point to play around with the numerical implementation of these operations with software systems such as Python, Matlab, Mathematica, Maple, or others (see exercises). Note that the indexing in the previous equation is not unique. Other schemes lead to identical results.

Let us restate the previous result of the partial derivative as an inverse Fourier transform defined as

$$\partial_x f(x) = \frac{1}{\sqrt{2\pi}} \int_{\infty}^{\infty} \partial_x F(k) e^{ikx} dk$$
$$= \frac{1}{\sqrt{2\pi}} \int_{-\infty}^{\infty} ikF(k) e^{ikx} dk.$$

(5.52)

Fourier n-point

nothing
happening
here

FD 6-point
FD 4-point
FD 2-point

increasing difference operator length

Fig. 5.15 *Finite differences and Fourier method as two extreme cases. The two-point finite-difference operator is the least accurate. Increasing the number of points contributing to the derivative decreases the error. In practical applications rarely more than 4–8 points are used. Using all points n along a regularly spaced dimension allows the exact calculation of a derivative at the cost of many more calculations.*

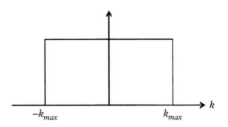

Fig. 5.16 *Band limitation of the derivative operator. The discretization with increment dx requires the wavenumber domain to be restricted to the interval $k \in [-k_{max}, k_{max}]$, where $k_{max} = \pi/dx$. Negative frequencies are of course unphysical but originate in the formulation of the Fourier transform using complex numbers.*

Defining the factors in front of the complex amplitude spectrum $F(k)$ of function $f(x)$ as

$$\partial_x f(x) = \int_{-\infty}^{\infty} D(k)F(k)e^{ikx}dk, \quad D(k) = ik \tag{5.53}$$

we can now interpret this result in connection with the convolution theorem. $D(k)$ in general is nothing else but a function defined in the spectral domain acting like a filter on the complex spectrum $F(k)$. The convolution theorem implies that

$$\partial_x f(x) = \int_{-\infty}^{\infty} d(x - x')f(x')dx', \tag{5.54}$$

where $d(x)$ is a real function, the spatial representation of spectrum $D(k)$, in other words

$$d(x) = \mathscr{F}^{-1}[D(k)]. \tag{5.55}$$

Are you curious to see, what $d(x)$ looks like? Luckily, this can be derived analytically in a straightforward way. However, it is not realistic to work with infinite integral boundaries. We thus limit the wavenumber domain to the Nyquist wavenumber $k_{max} = \pi/dx$, knowing that we are operating on regular spaced grids with sampling interval dx (see Fig. 5.16). Thus $D(k)$ becomes

$$D(k) = ik[H(k + k_{max}) - H(k_{kmax})], \tag{5.56}$$

where $H()$ denotes the Heaviside function, and to obtain $d(x)$ we simply have to inverse transform

$$d(x) = \mathscr{F}^{-1}[ik[H(k + k_{max}) - H(k - k_{kmax})]], \tag{5.57}$$

leading to (see exercises)

$$d(x) = \frac{1}{\pi x^2}[sin(k_{max}x) - k_{max}x \cos(k_{max}x)] \tag{5.58}$$

in the space time domain.

It is again instructive to use the convolution theorem to understand this result. The right-hand side of Eq. 5.57 inside the Fourier integral is again a multiplication of two spectra; one is the derivative operator ik, and the other is a boxcar function expressed in terms of Heaviside functions. The inverse transform of a boxcar is a well-known result frequently encountered in filter analysis. Without formal demonstration we note that it is a *sinc* function of the form $\sin(x)/x$. As a consequence, the result of Eq. 5.57 is the derivative of a *sinc* function, and is illustrated in Fig. 5.17.

Up to this point this has applied to the continuous world; so let us see what happens if space is discretized according to

$$x_n = n\,dx, \quad n = -N, \ldots, 0, \ldots, N. \tag{5.59}$$

In this case the convolution integral becomes a convolution sum

$$\partial_x f(x) \approx \sum_{n=-N}^{n=N} d_n f(x - ndx), \tag{5.60}$$

where d_n is the difference operator, in other words, the weight with which the function value $f(x - ndx)$ has to be multiplied before summing up to obtain the derivative approximation. Note that, even though we derived this result using Fourier transformations, this result is general and applies to any operator half-length $N \geq 1$, including the case of first-derivative staggered-grid calculations of second order ($N = 1$) in which case the derivative would be defined halfway between grid points.

If we insert the discretization into Eq. 5.58 we obtain analytically the discrete difference operator based on the Fourier transform, now expressed in the discretized space, a beautifully simple result:

$$d_n = \begin{cases} 0 & \text{for } n = 0 \\ \frac{(-1)^n}{ndx} & \text{for } n \neq 0. \end{cases} \tag{5.61}$$

Let us have a closer look at the illustration of this operator in Fig. 5.17. The figure shows the analytical solution of the operator $d(x)$ (solid line, the derivative of a sinc function) as a reference using a nominal space increment $dx = 1$. The discrete operator is indicated by + marks.

First, note the anti-symmetry of the analytical (and discrete) result, with the consequence that the information at the central point, where the derivative is defined, is not influencing the derivative (one of the arguments for the use of a staggered-grid approach). Second, note that the standard two-point finite-difference operator (in this case $[-1, 1]$) is included (black dots) in the Fourier operator (that, however, extends to $2N + 1$ points in total). Third, the Fourier derivative operator seems to decay slowly with distance from the central point (in fact the decay is proportional to n^{-1}, indicating that a long operator is required to

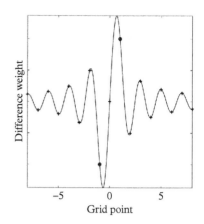

Fig. 5.17 *Exact convolutional derivative operator. The band-limited discrete central first derivative operator (+) is compared with the analytical representation (solid line). The discrete operator $d(x)$ is obtained by inverse transforming the complex representation of the derivative operator $D(k) = ik$ in the Fourier domain, thus $d(x) = \mathscr{F}^{-1} D(k)$. The band limitation is a consequence of the sampling theorem with a maximum wavenumber of $k_{max} = \pi/dx$.*

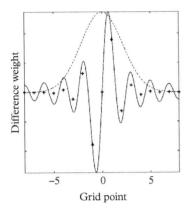

Fig. 5.18 *Truncated Fourier operator. The original exact, discrete Fourier operator d_n is mutiplied by a Gaussian and limited to N = 6 points on one side. Original exact operator (solid line), truncated operator (+).*

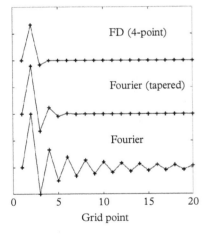

Fig. 5.19 *Convolutional difference operators. A standard (Taylor-series-based) four-point finite-difference operator (top) is compared with a truncated Fourier operator obtained after multiplication with a Gauss function and limitation to N = 5 points (middle), and the exact discrete Fourier operator (bottom).*

obtain high accuracy (see Fichtner, 2010, for a more detailed discussion of such operators also in the case of staggered grids and comparison with finite-difference operators).

This raises the question of whether it would be possible to define short-difference operators by cutting off the Fourier operator at some appropriate distance from the central point. A sensible strategy to do this seems to be to multiply this operator with a Gaussian centred at $n = 0$. Such a procedure is illustrated in Fig. 5.18. This approach was pioneered for elastic wave problems (to my knowledge) in the Stanford group (Mora, 1986) and later extensively used for derivatives and interpolations (Igel et al., 1995) for anisotropic wave propagation and waveform inversion problems. Several convolutional difference operators are compared in Fig. 5.19.

How can we conveniently compare the accuracy of such operators? To do this we turn around the procedure given above that provided us with the space representation of the exact difference operator $D(k) = ik$ in the wavenumber domain. Nothing keeps us from simply Fourier transforming any difference operator into the wavenumber domain, which will allow us to compare the result with the exact $D(k)$. Thus, for a finite-difference operator d_n^{FD} we will obtain

$$D^{FD}(k) = i\,\tilde{k}_v(k) = \mathscr{F}[d_n^{FD}] \tag{5.62}$$

which—at least intuitively—we expect to be close to the exact solution for low wavenumbers (meaning large numbers of grid points per wavelength). Let us transform the operators illustrated in Fig. 5.19 to the wavenumber domain and compare the imaginary part of the approximate operator spectra with the exact solution (i.e, $\mathrm{Imag}[D(k)] = k$). The results are shown in Fig. 5.20.

We can see that the short-finite-difference operator is accurate for large numbers of grid points per wavelength, but the truncated Fourier operator already performs substantially better.

I find this alternative description in the spectral domain a very powerful concept to classify explicit numerical approximations to derivatives with the added value of providing a means to generate very accurate short-difference operators. The curse of the Fourier method is its requirement of periodicity, which is very rarely what we need in Earth sciences. How this can be fixed is the topic of the next few sections.

5.6   The Chebyshev pseudospectral method

Historically, the substantial accuracy improvement of the space derivative calculations using the Fourier transform was a major step forward at a time when serial (vectorized) hardware architecture was the standard. The major drawback was the requirement that the space-dependent functions have periodic behaviour. In most cases in Earth sciences we are dealing with *limited-area* calculations (e.g. a reservoir, a volcano, a piece of the Earth's crust), which requires the accurate

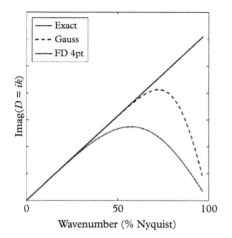

Fig. 5.20 *Difference operators compared in the wavenumber domain. The spectra of the truncated difference operators are compared with the exact solution. In principle the vertical axis corresponds to the numerical wavenumber $\tilde{k}$ as a function of the exact wavenumber k.*

implementation of free-surface or absorbing boundary conditions that are very hard to achieve when using trigonometric basis functions.

For those reasons computational seismologists looked for alternative function approximations with similar convergence properties to Fourier series. Chebyshev[10] polynomials were the natural choice as the preferred solution for non-periodic problems with spectral convergence. In the first applications, Chebyshev polynomials were used to describe the space-dependent functions in the whole physical domain. The required irregular spatial grid creates other problems, however, reducing the maximum time step allowed for large models. However, a major advantage is the possibility of accurate implementation of boundary conditions.

There is another reason why Chebyshev polynomials are important for the further evolution of computational seismology. Their superior interpolation properties in combination with an accurate integration scheme led to the first Chebyshev-based *spectral* finite-element implementation, albeit with non-diagonal mass matrix formulation. This is why the concept of limited-area function interpolation is a key ingredient to some of the methods we use today.

5.6.1 Chebyshev polynomials

To introduce Chebyshev polynomials, let us start with the trigonometric relation.

$$\cos\left[(n+1)\phi\right] + \cos\left[(n-1)\phi\right] = 2\cos(\phi)\cos(n\phi), \qquad n \in \mathbb{N}. \quad (5.63)$$

Inserting $n = 0$ leads to a trivial statement. However, for $n \geq 1$ we obtain statements like

$$\begin{aligned}
\cos(2\phi) &= 2\cos^2(\phi) - 1 \\
\cos(3\phi) &= 4\cos^3(\phi) - 3\cos(\phi) \\
\cos(4\phi) &= 8\cos^4(\phi) - 8\cos^2(\phi) + 1 \\
&\vdots
\end{aligned} \quad (5.64)$$

[10] Born as Pafnuti Lwowitsch Tschebyschow (1821–1894)—his surname would be spelled in many different ways—he is considered one of the greatest mathematicians of the nineteenth century. He worked in St. Petersburg. His pupils included Marcov, Voronoi, and Ljapunov.

indicating that we can express $\cos(n\phi)$ in terms of polynomials in $\cos(\phi)$. This in fact leads to the definition of Chebychev polynomials with

$$\cos(n\phi) := T_n(\cos(\phi)) = T_n(x) \tag{5.65}$$

with

$$x = \cos(\phi) \qquad x \in [-1, 1], \qquad n \in \mathbb{N}_0, \tag{5.66}$$

T_n being the nth order Chebyshev polynomial. The important step here is the mapping of $x = \cos(\phi)$ which limits the definition of these polynomials to the interval $[-1, 1]$. Furthermore

$$|T_n(x)| \leqslant 1 \text{ for } [-1, 1], \qquad n \in \mathbb{N}_0. \tag{5.67}$$

Finally, we can write down the first polynomials in $x \in [-1, 1]$:

$$\begin{aligned}
T_0(x) &= 1 \\
T_1(x) &= x \\
T_2(x) &= 2x^2 - 1 \\
T_3(x) &= 4x^3 - 3x \\
T_4(x) &= 8x^4 - 8x^2 - 1
\end{aligned} \tag{5.68}$$

$$\vdots$$

and an illustration of some polynomials is given in Fig. 5.21. There is a recursive relation that can be conveniently used to calculate the Chebyshev polynomials of any order n:

$$T_{n+1}(x) = 2xT_n(x) - T_{n-1}(x), \quad n \geq 1. \tag{5.69}$$

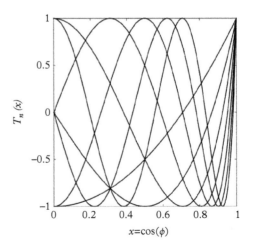

Fig. 5.21 *Chebyshev polynomials $T_n(x)$ in the interval $[0, 1]$ for $n = 2, \ldots, 8$.*

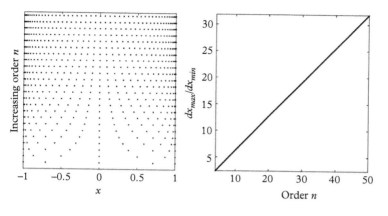

Fig. 5.22 *Extrema of the (even-order) Chebyshev polynomials (collocation points) for N = 4,...,50.* **Left:** *Location of the extrema with increasing order. Note the densification of points near the boundaries [−1,1] and the increasing difference between largest and smallest separation.* **Right:** *Ratio between largest and smallest grid point distance as a function of order.*

The extremal values $x_k^{(e)}$ of these polynomials have a very simple form

$$x_k^{(e)} = \cos\left(\frac{k\pi}{n}\right) \qquad k = 0, 1, 2, \ldots, n \tag{5.70}$$

and they are shown in Fig. 5.22 for varying n. In fact, these points, and in particular this form of irregular grid with densification at the edges, will play an important role in the story that unfolds. We anticipate that these extremal points will be the Chebyshev collocation points at which the polynomials exactly interpolate an arbitrary discrete function.

With this in mind—assuming that we have a spatial domain discretized with these grid points—the linear function shown in Fig. 5.22 (right) has a strong message. The fact that this ratio increases with order implies that, according to the Courant criterion that we can not escape from, the required time step will become very small for increasing order while any spatial field is sampled coarsely in the centre of the domain. This behaviour will also be relevant later for spectral element methods.

It can be shown that the Chebyshev polynomials form an orthogonal set with respect to the weighting function $w(x) = 1/\sqrt{1-x^2}$. This implies that we can use them as a basis for function interpolation. By analogy with the discussion of Fourier series, we pose the problem of finding an approximation $g_n(x)$ to an arbitray function $f(x)$ defined in the interval $[-1, 1]$ (a condition that can be easily relaxed). In mathematical terms

$$f(x) \approx g_n(x) = \frac{1}{2}c_0 T_0(x) + \sum_{k=1}^{n} c_k T_k(x), \tag{5.71}$$

where $T_n(x)$ are the Chebyshev polynomials and c_k are real coefficients. The coefficients c_k can be found in the same way as the Fourier coefficients by minimizing the least-squares misfit between original function $f(x)$ and approximation $g_n(x)$ to obtain

$$c_k = \frac{2}{\pi} \int_{-1}^{1} f(x) T_k(x) \frac{dx}{\sqrt{1-x^2}} \qquad k = 0, 1, \ldots, n \tag{5.72}$$

which—after substituting $x = \cos(\phi)$—can be written as

$$c_k = \frac{1}{\pi} \int_{-\pi}^{\pi} f(\cos(\phi)) \cos(k\phi) d\phi \qquad k = 0, 1, \ldots, n. \tag{5.73}$$

These coefficients turn out to be the Fourier coefficients for the even 2π-periodic function $f(\cos(\phi))$ with $x = \cos(\phi)$.[11]

Is there a set of points on which Chebyshev polynomials interpolate exactly as was the case with the discrete Fourier transform? Well, as stated before, the answer is yes, and we thus can write down a corresponding discrete Chebyshev transform with similar properties to the Fourier transform but without the requirement of periodicity. The points we need are the extrema of the Chebyshev polynomials the (Chebyshev-) Gauss-Lobatto points defined as[12]

$$x_i = \cos\left(\frac{\pi}{N}i\right) \qquad i = 0, 1, \ldots, n. \tag{5.74}$$

With these unevenly distributed grid points we can define the discrete Chebyshev transform as follows. The approximating function is

$$g_n^*(x) = \frac{1}{2} c_0^* T_0 + \sum_{k=1}^{n-1} c_k^* T_k(x) + \frac{1}{2} c_n^* T_n \tag{5.75}$$

with the coefficients defined as

$$c_k^* = \frac{2}{m} \left[\frac{1}{2}(f(1) + (-1)^k f(-1)) + \sum_{j=1}^{m-1} f_j \cos\left(\frac{kj\pi}{m}\right) \right] \tag{5.76}$$

$$k = 0, 1, \ldots, n, \qquad n = m.$$

Here, $f(1)$ and $f(-1)$ are the function values at the interval boundaries and f_j are the values at the collocation points $f(x = \cos(j\pi/m))$. With these definitions we recover the fundamental property

$$g_m^*(x_i) = f(x_i), \tag{5.77}$$

where x_i are the collocation points with the implication that we exactly (to machine precision) recover the original function. These equations should be compared with those of the discrete Fourier transform (Eq. 5.16–Eq. 5.20) as an equivalent formulation for limited functions.

[11] This has practical consequences. It implies that the fast Fourier transform can be used to calculate the coefficients (and derivatives). However, for the purpose of simplicity we stick to the approach of using derivative matrices for our numerical examples.

[12] Careful! In this notation the x coordinate starts at 1 and ends at −1. Sometimes this definition appears with a -sign but this affects the definitions that follow.

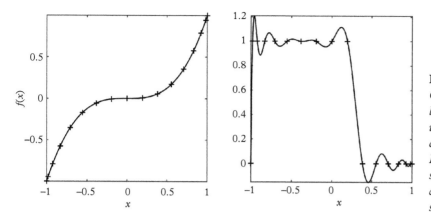

Fig. 5.23 *Discrete interpolation with Chebyshev polynomials.* **Left:** *Interpolation of the function $f(x) = x^3$ in the interval $[-1, 1]$ on the Chebyshev collocation points.* **Right:** *Interpolation of a Heaviside function with discrete Chebyshev polynomials. Note the occurence of an overshoot as we observed with Fourier series (the Gibbs phenomenon).*

Let us again see how this works in practice and approximate a simple function like $f(x) = x^3$ in the interval $[-1, 1]$ using the Chebyshev transform. The results are shown in Fig. 5.23 (left), indicating that a non-periodic function like $f(x)$: (1) can be exactly interpolated at the collocation points, and (2) converges very rapidly, with just a few polynomials. This raises the question of how the Chebyshev transform deals with discontinuous behaviour inside the domain. This is illustrated in the second example of Fig. 5.23 (right). We note that, despite the exact interpolation, convergence of the interpolating function is slow and an overshoot occurs, just as with the Fourier method. The consequence is that similar smoothness constraints have to be imposed when using these techniques for the solution of wave equations.

Finally, it is instructive again to demonstrate the interpolation behaviour showing the corresponding cardinal functions of Chebyshev polynomials. They are obtained by finding the interpolating function for spikes of the form $f_j = (0, 1, \ldots, 0)$ at any point on the grid. Two examples are shown in Fig. 5.24. In

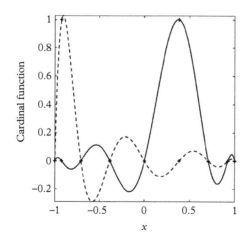

Fig. 5.24 *Cardinal function with Chebyshev polynomials. Two examples of cardinal functions for grid points $i = 2$ (dashed line) and $i = 6$ (solid line) are shown for $n = 8$.*

the case of the Fourier method they were of identical form and only translated for each grid point. Here they are different as a result of the uneven spacing.[13]

For the numerical solution of the wave equation we need derivatives. How to obtain derivatives with the Chebyshev transform is discussed in the next section.

5.6.2   Chebyshev derivatives, differentiation matrices

It is well known that a convolution operation as introduced in the section on Fourier derivatives can be formulated as a matrix–vector product involving so-called Toeplitz matrices. An elegant (but inefficient) way of performing a derivative operation on a space-dependent function described on the Chebyshev collocation points is by defining a derivative matrix D_{ij}

$$D_{ij} = \begin{cases} -\frac{2N^2+1}{6} & \text{for } i = j = N \\ -\frac{1}{2}\frac{x_i}{1-x_i^2} & \text{for } i = j = 1,2,...,N\text{-}1 \\ \frac{c_i}{c_j}\frac{(-1)^{i+j}}{x_i-x_j} & \text{for } i \neq j = 0,1,...,N \end{cases} \tag{5.78}$$

where $N + 1$ is the number of Chebyshev collocation points $x_i = \cos(i\pi/N)$, $i = 0,\ldots,N$, $D_{00} = -D_{NN}$, and the c_i are given as

$$c_i = \begin{cases} 2 & \text{for i=0 or N} \\ 1 & \text{otherwise.} \end{cases} \tag{5.79}$$

This differentiation matrix allows us to write the derivative of function $u_i = u(x_i)$ (possibly depending on time) simply as

$$\partial_x u_i = D_{ij}\, u_j, \tag{5.80}$$

where the right-hand side is a matrix–vector product, and the Einstein summation convention applies. Following the lengthy foreword about Chebyshev polynomials, we finally arrive at an exact (polynomial-based) derivative at the Chebyshev collocation points.

Before we proceed to demonstrate its behaviour, let us have a closer look at the concept of differentiation matrices. In the preceding sections we demonstrated that any finite-difference-type calculation can be expressed as a discrete convolution. In that respect, any convolution of two vectors can be expressed as a matrix–vector product. This is well known and frequently used in filter theory. Finding these matrices involves the concept of Toeplitz matrices. Thus, if we can write down the convolutional operator for a differentiation we can also determine a differentiation matrix. In practice this is rarely used in computer programs as the operation scales with n^2, with n being the length of the original vector. However,

[13] It is worth noting that the cardinal functions may exceed 1 in this case. Later we will use Lagrange polynomials (spectral-element method) that do not exceed 1 in the corresponding interval. This has advantages (see Fichtner, 2010 for an in-depth discussion).

Fourier   **Fourier tapered**

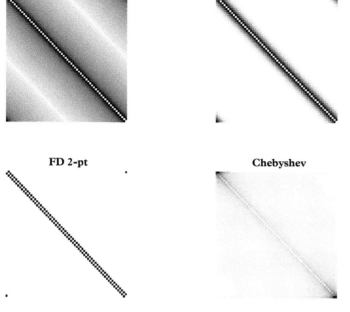

FD 2-pt   **Chebyshev**

Fig. 5.25 *Illustration of differentiation matrices (n = 64).* **Top left:** *Exact Fourier differentiation matrix for regular grid (full).* **Top right:** *Tapered Fourier operator (12-point). Matrix is banded. For illustration purposes the square root of the absolute values is shown.* **Bottom Left:** *Standard 2-point finite difference operator (banded).* **Bottom Right:** *Exact Chebyshev differentiation matrix for Chebyshev collocation points. Note that the matrix is full. Increasing weights at the corners dwarf interior values.*

again this is a powerful concept for comparing the structure and properties of the various approaches.

In Figure 5.25 a number of differentiation matrices are shown, including the Chebyshev case. Both *exact* cases (Chebyshev and Fourier) consist of full matrices. The finite-difference operators are banded matrices with the band-width equal to the length of the operator. It is noteworthy that, for the same number of points, the Chebyshev weights increase drastically towards the edges. This is related to the decreasing grid point distances at the interval boundaries. Another interesting point is that the Chebyshev differentiation matrix consists of the standard two-point operator for N = 1.

Let us proceed by testing the differentiation using Chebyshev polynomials. We define a function akin to seismic wavefield calculations as

$$f(x_i) = \sin(2x_i) - \sin(3x_i) + \sin(4x_i) - \sin(10x_i) \qquad (5.81)$$

in the interval $x_i \in [-1, 1]$, where the discrete points are the Chebyshev collocation points $x_i = \cos(\pi i/n)$, $i = 0, \ldots, n$. Following the approach shown earlier the derivative of the function $f(x_i)$ is simply obtained by performing the matrix–vector product presented in Eq. 5.80 to the vector containing the function values. The results for $n = 63$ (which implies $n + 1$ points including the boundaries) are shown in Fig. 5.26. We expect a (close-to) exact derivative, and this is what we obtain if we look at the error of the derivative (compared to the analytical solution). Numerically the error behaves slightly different from the Fourier method

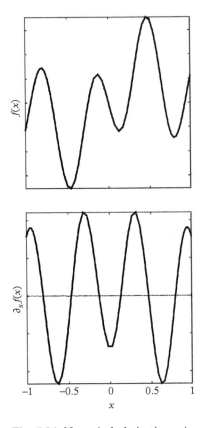

Fig. 5.26 *Numerical derivative using the Chebyshev differentiation matrix.* **Top:** *Discrete analytical function (see text) defined on the Chebyshev collocation points ($n = 63$).* **Bottom:** *Superposition of exact derivative and numerical derivative based on matrix–vector multiplication (indistinguishable solid lines). The difference between numerical derivative and analytical solution is given by the dotted line after multiplication by 10^{11}.*

related to the irregular sampling intervals. The round-off error increases at the boundaries and also depends on the overall number of grid points used in the entire interval.

We can now assemble these tools to come up with another solution to our 1D elastic wave equation, using Chebyshev polynomials.

5.6.3 Elastic waves in 1D

Let us state the elastic 1D wave equation for the unknown discrete displacement field u_i^j right away as an extrapolation problem of the form

$$\rho_i \frac{u_i^{j+1} - 2u_i^j + u_i^{j-1}}{dt^2} = (\partial_x [\mu(x)\partial_x u(x,t)])_i^j + f_i^j \tag{5.82}$$

using the standard three-point operator for the time-dependent part. Again the lower index i corresponds to the spatial discretization and the upper index j to the discrete time levels. The main difference with any previous methods is that the displacement field, and the geophysical parameters like density ρ_i and shear modulus μ_i, are defined on the irregular Chebyshev collocation points. The parameters used in our example simulation are given in Table 5.4.

The fact that we use the Chebyshev collocation points with densification near the interval boundaries has (dramatic) consequences for the simulation set-up. As we can see from the parameters, the distance between the grid points is 80 times smaller at the boundaries compared to the centre of the physical domain. The time step for a stable simulation, according to the Courant criterion, requires $c\,dt/dx \leq \epsilon$, where c is the maximum velocity, dt is the time step, dx is the minimum(!) grid interval, and ϵ is some value close to 1. That means that the grid distance near the boundary is responsible for the global simulation time step.

In our example we stick to the interval boundaries $x_i \in [-1, 1]$ and thus the simulation would correspond to a 2 m block with wave propagation in the kHz range. The results of the simulation are shown in Fig. 5.27. The figure shows two snapshots of the propagating phase (the first derivative of a Gaussian) near the centre of the domain (sampled with ≈ 10 points) and near the edge (sampled with ≈ 35 points). In the graphical representation the discrete displacement values are interpolated using the Chebyshev transform for illustration purposes.

The following code snippet shows the implementation of the Chebyshev method with Python. We assume that the differentiation matrix D has been initialized and current values of the displacement field are stored in vector u. The derivatives are obtained by the implicit matrix–vector multiplication requiring the vectors to be transposed (e.g. $u.T$). The boundary points at the edges [-1,1] are initialized to 0 at each time step.

```
# Time extrapolation
for it in range(nt):
```

```
# Space derivatives
du = D @ u.T
du = mu/rho * du
du = D @ du
# Extrapolation
unew = 2 * u - uold + du.T * dt**2
# Source injection
unew = unew + gauss/rho * src[it] * dt**2
# Remap displacements
uold, u = u, unew
```

Table 5.4 *Simulation parameters for 1D elastic simulation with the Chebyshev method*

Parameter	Value
nx	200
c	3,000 m/s
ρ	2,500 kg/m^3
dt	6×10^{-8} s
dx_{min}	1.2×10^{-4} m
dx_{max}	0.015 m
f_0	100 kHz
ϵ	1.4

Looking at our simulation example we can identify a major disadvantage of the Chebyshev method. While we benefit from the fact that we are working in a limited domain, the price is high. To obtain a stable solution we need a very small time step, which is basically only needed at the boundaries. In fact, mathematically the time step scales with $O(N^{-2})$. This implies that inside the domain we hugely oversample the wavefield in time.

However, there is (at least to some extent) a fix to that problem. In principle we can (re-) stretch the spatial grid such that the grid points close to the boundaries are further apart while the grid point distances at the centre remain basically unchanged. If that stretching function is $\xi(x)$ then the derivative of a function $f(x)$ on the stretched grid is defined as

$$\partial_x f(x) = \frac{\partial f}{\partial \xi} \frac{d\xi}{dx}. \tag{5.83}$$

This is a trivial additional operation if the coordinate change is an analytical function, which is the case. The procedure is explained for example in Carcione and Wang (1993) and is an essential ingredient in all Chebyshev methods that are used for scientific applications.

This concludes the presentation of the pseudospectral method based on limited-area calculations using Chebyshev polynomials. Spectral interpolation properties in limited areas—possible also with Lagrange polynomials—will reappear when we discuss the spectral element method.

5.7 The road to 3D

The extension of pseudospectral methods to 3D is straightforward. Consider the acoustic equation using the Fourier method with

$$\partial_t^2 p = c^2 \left(\partial_x^2 p + \partial_y^2 p + \partial_z^2 p \right) + s \tag{5.84}$$

and

$$\partial_x^2 p_{ijk}^n = \mathscr{F}^{-1} [-k^2 \, P_v^n]. \tag{5.85}$$

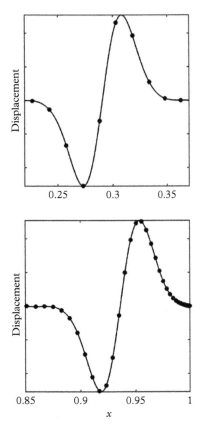

Fig. 5.27 *Simulation of the 1D elastic wave equation with the Chebyshev method: snapshots of the propagating phase. The discrete values of the numerical simulation are indicated by dots. The solid line is the (exact) interpolation function obtained by the Chebyshev transform.* **Top:** *Wavefield near the centre of the domain.* **Bottom:** *Wavefield near the boundary. Note the drastic difference of the wavefield sampling in both cases.*

Here, P_v^n corresponds to the Fourier transform of p with respect to x. Derivatives with respect to other coordinates can be calculated accordingly. Furthermore, ijk denote space increments and n the time level. With $p_i \rightarrow p_{ijk}$ we simply add two more dimension to the space-dependent fields and calculate the derivatives using the same routines as in 1D. The time extrapolation remains unchanged.

The same applies to the Chebyshev method in matrix form, where Eq. 5.85 would be replaced by

$$\partial_x^2 p_{ijk}^n = D_{im}^2 p_{mjk}^n, \tag{5.86}$$

where D_{im}^2 is a differentiation matrix for the second derivative and the summation convention applies.

However, note that *pure* 3D pseudospectral schemes only make sense in a serial computing environment, because of the global communication schemes. Furumura et al. (1998b) presented a parallel Fourier scheme for 3D using a clever partitioning. Attempts have been made to combine pseudospectral schemes with finite differences (e.g. using pseudospectral derivatives in horizontal directions and finite differences in the vertical direction). Furumura et al. (2002) presents examples for ground motion modelling using such a hybrid scheme. Furumura and collaborators also applied the pseudospectral method to global wave propagation in the 2.5D axi-symmetric approximation (Furumura et al., 1998a; Wang et al., 2001) with several interesting applications modelling heterogeneous structures (e.g. Furumura et al., 1999; Furumura and Kennett, 2005).

An excellent description of a 3D Chebyshev implementation of anisotropic wave propagation with absorbing and free-surface boundary conditions is given in Tessmer (1995). Igel (1999) applied this scheme to regional-scale wave propagation in spherical coordinates. Recent developments include applications to strongly heterogeneous media using a poly-grid Chebyshev method (Seriani and Su, 2012).

Chapter summary

- Pseudospectral methods are based on discrete function approximations that allow exact interpolation at so-called collocation points. The most prominent examples are the Fourier method based on trigonometric basis functions and the Chebyshev method based on Chebyshev polynomials.

- The Fourier method implicitly assumes periodic behaviour. Boundary conditions like the free surface or absorbing behaviour are difficult to implement.

- The Chebyshev method is based on the description of spatial fields using Chebyshev polynomials defined in the interval $[-1, 1]$ (easily generalized to arbitrary domain sizes). Exact interpolation (and derivatives) are possible when the discrete fields are defined at the Chebyshev collocation points given by $x_i = \cos(\pi i/n)$, $i = 0, \ldots, n$.

- Because of the grid densification at the boundaries when using Chebyshev collocation points, very small time steps are required for stable simulations for increasing number of grid points. This can be avoided by stretching the grids near the boundaries by a coordinate transformation.

- A major advantage of the Chebyshev method is an elegant formulation of boundary conditions (free-surface or absorbing) through the definition of so-called characteristic variables.

- Pseudospectral methods have isotropic errors. Therefore they lend themselves to the study of physical anisotropy.

- The derivative operations of pseudospectral methods are of a global nature. That means every point on a spatial grid contributes to the gradient. While this is the basis for the high precision, it creates problems when implementing pseudospectral algorithms on parallel computers with distributed memory architectures. As communication is usually the bottleneck, efficient and scalable parallelization of pseudospectral methods is difficult.

....

FURTHER READING

- Fornberg (1996) is an excellent book illustrating the properties of the pseudospectral method in comparison with finite-difference methods.

- Bracewell (1999) is still a standard and very readable textbook on the Fourier transform. There is also an interesting section on the life of Joseph Fourier.

- Trefethen (2015) is a great book on spectral methods built around Matlab programs.

- Stein and Shakarchi (2003) is the first of a more recent three-volume account of Fourier analysis and its applications.

- Fichtner (2010) provides a detailed discussion of Fourier-derived difference operators on staggered and central grids.

....

EXERCISES

Comprehension questions

(5.1) Explain the concept of *exact* interpolation behaviour in the context of pseudospectral methods. What are cardinal functions?

(5.2) Explain the meaning of the term *pseudospectral*. What is so *spectral* about the pseudospectral method?

(5.3) Through which concepts can the *exact* interpolation or derivative be achieved? What is the price for this accuracy?

(5.4) Motivate the use of function approximations (e.g. Fourier series, Chebyshev polynomials) for Earth science problems.

(5.5) What are the main differences between Fourier and Chebyshev approaches? Give application examples.

(5.6) Discuss pros and cons of the Fourier method compared with the finite-difference method. Give examples where you would prefer one over the other. What is the role of computer architecture?

(5.7) Are there fundamental differences between the numerical dispersion behaviour of the finite-difference and pseudospectral methods? If so, why?

(5.8) What is the meaning of the convolution theorem and what is its significance concerning numerical differentiation?

(5.9) The pseudospectral method appears simple, elegant, and very accurate. Why is it not the preferred method of choice today? Could the pseudospectral concept in 2(3)D be combined with the finite-difference method (for space derivatives)?

Theoretical problems

(5.10) How is the orthogonality of functions defined? Show the orthogonality of $sin(nx)$ for $n > 0$ evaluating

$$\int_{-\pi}^{\pi} \sin(jx)\sin(kx)\,dx.$$

(5.11) The Fourier coefficients for an odd function can be obtained by

$$b_n = \frac{2}{L}\int_0^L f(x)\sin\left(\frac{n\pi x}{L}\right)dx.$$

What is the meaning of the Fourier coefficients? Calculate the coefficients $n = 1, 2, \ldots$ for $f(x) = x$ and $L = 1$ and plot the approximate function using

$$g_N(x) = \sum_{n=1}^{N} b_n \sin\frac{n\pi x}{L}. \tag{5.87}$$

(5.12) Derive the Fourier series for $f(x) = x^2$ in the interval $x \in [0, 2\pi]$ to recover Eq. 5.15.

(5.13) In general, the spectrum $F(k)$ of the derivative of a function $f(x)$ is given by

$$F(k) = \frac{1}{\sqrt{2\pi}} \int_{-\infty}^{\infty} f(x)e^{-ikx} dx.$$

Use integration by parts to show that (only by) assuming $f(x)$ vanishes if $x \to \pm\infty$ we obtain the extremely useful result that $F^{(n)}(k) = (ik)^n F(k)$ is the spectrum of the nth derivative of $f(x)$.

(5.14) The fact that we discretize space with dx implies that our wavenumber space is limited by the Nyquist wavenumber $k_{max} = \pi/dx$. Derive the analytical form of the difference operator $d(x)$ by an inverse transform

$$d(x) = \int_{-k_{max}}^{k_{max}} (ik)e^{ikx} dk.$$

Hint: Use integration by parts. You need to recover Eq. 5.58.

(5.15) Use the concept of the previous exercise to derive the exact interpolation operator

$$d(x) = \int_{-k_{max}}^{k_{max}} e^{ikx} dk.$$

(5.16) Derive the dispersion relation for the Fourier pseudospectral approximation of the 1D acoustic wave equation applying the von Neumann analysis. Hint: Start with the wave equation and insert discrete plane wave trial functions. You want to recover Eq. 5.38.

(5.17) Use the definition

$$\begin{aligned} \cos[(n+1)\phi] + \cos[(n-1)\phi] \\ = 2\cos(\phi)\cos(n\phi) \end{aligned} \tag{5.88}$$

to recover the first five Chebyshev polynomials $T_n(\cos(\phi)) = \cos(n\phi) = T_n(x)$, $n = 1 \ldots 5$ with $x = \cos(\phi)$.

Programming exercises

For the following exercises you can make use of the codes in the supplementary electronic material.

(5.18) Define an arbitrary function (e.g. a Gaussian) and initialize its derivative on the same regular spatial grid. Calculate the numerical derivative using the Fourier method and the difference to the analytical derivative. Vary the wavenumber content of the analytical function. Does it make a difference? Is the derivative always exact to machine precision?

(5.19) Calculate a program that initializes the Chebyshev differentiation matrix and perform the same task as in the previous exercise. Note that you need to use the Chebyshev collocation points for the spatial grids. Increase the number of grid points and discuss the difference of grid-point distance at the centre and the boundary of the physical domain.

(5.20) Code a Fourier pseudospectral approximation to the 1D acoustic wave equation from scratch and compare with the analytical solution. Use parameters given in the examples.

(5.21) Code a Chebyshev pseudospectral approximation to the 1D acoustic wave equation from scratch and compare with the analytical solution. Calculate the derivatives using matrix–vector multiplication as discussed in the text.

(5.22) Determine numerically the stability limit of the Fourier (and/or Chebyshev) method applied to the 1D (2D) acoustic wave equation by varying the Courant criterion.

(5.23) Implement a positive velocity discontinuity of 50% at the centre of the 1D domain. Observe the reflection as a function of dominant wavelength (i.e. change the dominant frequency of the source wavelet).

(5.24) Keep the physical and numerical parameters of the Fourier simulation constant and vary the number of grid points *nx*. Add a statement in the code that measures the run time (it makes sense to only log the time extrapolation loop). Plot the run time as a function of *nx* keeping the size of the physical domain constant. Do the same with the Chebyshev code. Compare the required time step as a function of *nx*.

(5.25) Code the analytical solution to the acoustic wave problem in 1D. Compare the numerical result of the Fourier method with the analytical solution in an appropriate frequency band. Do the same using the basic 1D finite-difference code. Fix the accuracy of the final solution keeping the dominant frequency and propagation distance the same (e.g. 5%). Find numerical parameters for the Fourier method and finite-difference method that lead to the defined accuracy. Compare and discuss the computational set-ups in terms of memory requirements, number of time steps, Courant criterion, and computation time (compare with Fig. 5.14).

The Finite-Element Method

6

The numerical methods encountered so far (the finite-difference and pseudospectral methods) were based on purely mathematical concepts. The problem posed was how one can deal with space or time derivatives in partial differential equations when both space and time are discretized on (regular) grids. The *finite-element* method is a numerical approach that first originated in solid mechanics and structural engineering and was later put on solid mathematical foundations, in part with concepts that were developed in the nineteenth century. The naming of some of the elementary ingredients of the numerical method (e.g. *stiffness* and *mass* matrices) indicate the origin in mechanics.

What are the basic principles? We always have to bring the problems described with continuous partial differential equations into some discrete form, in order to allow realistic problems to be solved. The engineering approach was to subdivide a mechanical structure into beams (i.e. elements) that are behaving in an identical physical way, to link them at appropriate points (e.g. element corners), and finally to join all of them into a complete system (assembly). Obviously, in engineering, mechanical structures are usually geometrically complicated, which is why this requirement was the point of departure for the finite-element method, rather than the exception. Other techniques (e.g. the finite-difference and pseudospectral methods) were originally introduced for regular grids and later tweaked towards complex geometries, with more or less success.

In terms of physics and governing equations, the path from structural engineering to seismology is a very short one. So it was natural that seismologists sought to apply the finite-element method to seismic wave propagation. Today, finite-element-type methods are particularly used for problems where geometrically complex structures like surface topography or internal structures like sedimentary layers have to be included (see Fig. 6.1). The finite-element method has been tremendously successful in the engineering world. Many commercial and non-commercial finite-element frameworks are available, which can be used to solve a huge class of problems.

The book series on the finite-element method by O. C. Zienkiewicz and co-authors (Zienkiewicz and Taylor, 1989; Zienkiewicz et al., 2013), which has been running for half a century, is probably the most successful literature on any numerical method. The books have an awesome clarity and I highly recommend them for readers especially interested in this method. They characterize the finite-element method as *a general discretization procedure of continuum mechanics problems posed by mathematically defined statements*.[1] Also, they note that the

[1] And further define two major steps: (1) The continuum is divided into a finite number of parts (elements), the behaviour of which is specified by a finite number of parameters; and (2) the solution of the complete system as an assembly of its elements follows precisely the same rules as those applicable to *standard discrete problems*.

Computational Seismology. First Edition. Heiner Igel.
© Heiner Igel 2017. Published in 2017 by Oxford University Press.

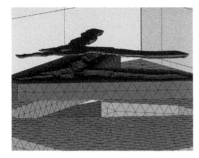

Fig. 6.1 *Tetrahedral finite-element mesh for the seismic velocity structure of the Grenoble basin. The various domains (sedimentary basin, bedrock) are meshed separately and matched. With this approach the geometry of internal interfaces, topography, and/or fault surfaces can be honoured. Figure courtesy of M. Käser.*

finite-element method is a unique approach which contains other methods (like finite-differences) as special cases.

It is in this spirit that the finite-element method will be presented in this chapter, focusing on the mathematical steps that lead to the solution system. However, the book on *Basis and Fundamentals of the Finite-Element Method* (Zienkiewicz et al., 2013) has 700 pages! Therefore it is clear that we can only scratch the surface here. Suggestions for further reading are given at the end of the chapter.

6.1 History

We will focus on applications of the finite-element method to problems in seismology. As the free-surface boundary condition is implicitly fulfilled, the method lends itself to the study of surface wave propagation (Lysmer and Drake, 1972; Schlue, 1979). Seismic scattering problems were simulated with the method presented in the dissertation of Day (1977). Its use was also investigated for problems in exploration seismics (Marfurt, 1984). It was found at the time that the method was more compute-intensive than other low-order methods such as finite differences. Also, in addition to the physical propagation modes, parasitic modes for high-order implementations were found (Kelly and Marfurt, 1990). Later, Seron et al. (1990) further developed the low-order finite-element methods, making them more efficient for seismic exploration problems.

At the same time the mathematical and engineering foundations of finite elements were further developed in the classic books by Strang (1988), Zienkiewicz and Taylor (1989), and Zienkiewicz et al. (2013). Parallel implementations on the legendary CM-2 massively parallel supercomputer were presented by Li et al. (1994). Finite-element principles are also the basis for the so-called direct solution method (DSM), which was introduced by R. Geller and co-workers (Cummins et al., 1994a; Cummins et al., 1994b; Cummins et al., 1997).

Many applications of the finite-element method to problems in seismic shaking hazards and engineering seismology were conducted by the group of J. Bielak and co-workers. Examples include modelling of ground motion simulation (Bielak et al., 1998), with specific applications to the response in alluvial valleys, e.g. Bielak and Xu (1999). The methods were later extended to the problem of full waveform inversion (Askan and Bielak, 2008; Epanomeritakis et al., 2008).

Hybrid methods that make use of advantages of both finite-difference and finite-element methods (e.g. for the implementation of free-surface boundary conditions) were presented by Moczo et al. (2010b).

To my knowledge, the classic low-order finite-element method was not at first as widely used by the seismological community (or in seismic exploration) as finite-difference methods. This is most likely related to the more simple mathematical concepts underlying finite-difference methods and the ease with which algorithms could be adapted to specific problems. Another point is the fact that finite-difference solutions are more easily implemented on parallel computers.

Low-order finite-element methods require the solution of an often very large linear system of equations with global communication requirements.

It took developments that started in the eighties, leading to high-order variations of the finite-element method using Lagrange polynomials or Chebyshev polynomials as basis functions (termed *spectral elements*), that initiated the success of this approach as we know it today (see next chapter).

In my view the classic low-order finite-element method deserves its place as a separate chapter in this volume despite the fact that the spectral element approach in the following chapter is a straightforward extension. This allows the introduction of the fundamental concepts of Galerkin-type methods at a more fundamental level, starting with the static problem well known from engineering applications.

6.2   Finite elements in a nutshell

We start by stating the 1D elastic wave equation with space-dependent density ρ, shear modulus μ, and forcing term f:

$$\rho \partial_t^2 u = \partial_x \, \mu \, \partial_x u + f. \tag{6.1}$$

We seek solutions to the displacement field $u(x, t)$. A fundamental contrast with the finite-difference method is that we do not solve for the displacement field u directly but instead replace it by a finite sum over (initially linear) basis functions φ_i:

$$u(x) \approx \overline{u}(x) = \sum_{i=1}^{N} u_i(t)\varphi_i(x). \tag{6.2}$$

Our unknowns are the coefficients of the basis functions φ_i, which we term u_i as they actually correspond to the discrete displacement values at node points x_i.

Furthermore, we formulate a so-called *weak* form of the wave equation, multiplying the original strong form by a test function φ_j with the same basis (this is the Galerkin principle), followed by an integration over the entire physical domain. This leads to a linear system of equations for domain D of the form

$$\int_D \rho \, \partial_t^2 \, \overline{u} \, \varphi_j \, dx + \int_D \mu \, \partial_x \, \overline{u} \, \partial_x \, \varphi_j \, dx = \int_D f \, \varphi_j \, dx, \tag{6.3}$$

where we seek to find the approximate displacement field $\overline{u}$ for given model parameters and forcing. With appropriate definitions of the vectors and matrices of this linear system, we can replace the time derivative of the solution field with finite differences and formulate an extrapolation problem as with all other numerical methods discussed in this volume.

Given appropriate initial conditions (everything is at rest, $\mathbf{u}(t = 0) = 0$) the solution at the next time step $\mathbf{u}(t + dt)$ can be found by the following matrix–vector equation:

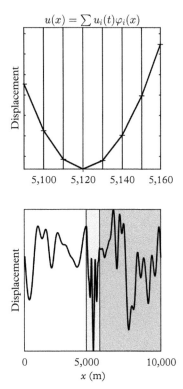

$$u(x) = \sum u_i(t)\varphi_i(x)$$

Displacement

5,100   5,120   5,140   5,160

Displacement

0   5,000   10,000

x (m)

Fig. 6.2 *Finite element method in a nutshell. **Bottom:** Snapshot of a finite-element simulation of 1D elastic wave propagation in a medium with three different velocities (grey shading).   **Top:** Detail at the centre of the domain. The vertical lines indicate the element boundaries and the + signs at the locations at which the displacement values are evaluated. Inside the elements the displacement field is described by a linear function. The solution field $u(x)$ is approximated by a sum over basis functions φ_i appropriately weighted.*

[2] Because of its potentially huge size and sparsity, it is hardly ever initialized as a matrix in realistic applications.

$$\mathbf{u}(t + dt) = dt^2 (\mathbf{M}^T)^{-1} \left[\mathbf{f} - \mathbf{K}^T \mathbf{u} \right] + 2\mathbf{u}(t) - \mathbf{u}(t - dt), \tag{6.4}$$

where $\mathbf{M}$ and $\mathbf{K}$ are *mass* and *stiffness* matrices, respectively.

One of the most important aspects of the finite-element method is the fact that these are global matrices in the sense that if a physical domain is discretized with N points, then the matrix shape is $N \times N$. Note also that one of the matrices has to be inverted.[2] In general, the matrix inversion has to be performed using sophisticated solution strategies, depending on the specific problem, grid type, and spatial dimensions. An example of a 1D finite-element simulation is shown in Fig. 6.2.

In our example the mass matrix $\mathbf{M}$ consists of elements of the form $\int_D \rho \varphi_i \varphi_j dx$ and the stiffness matrix $\mathbf{K}$ is built up with elements of the form $\int_D \mu \nabla \varphi_i \nabla \varphi_j dx$. These integrals can be computed in an elegant way for each element by mapping the physical space to a local reference space. If the parameters μ and ρ are constant inside the elements, these integrals can be computed analytically. The most attractive feature of the finite-element method is the fact that it can be formulated for arbitrary element shapes. This allows the simulation of models with highly complex geometric features and is the reason why this method is so successful for engineering problems.

In seismology, wide use of the finite-element method was hampered by the necessity to solve a huge system of equations. This situation can be substantially improved by a specific choice of basis functions and a numerical integration scheme, leading to the spectral-element method discussed in the next chapter.

6.3 Static elasticity

To make the introduction to the finite-element method as easy as possible, we start with the case of static elasticity. Departing from the 1D elastic wave equation

$$\rho(x)\partial_t^2 u(x, t) = \partial_x \mu(x)\partial_x u(x, t) + f(x, t), \tag{6.5}$$

we assume that the displacement does not depend on time ($\partial_t^2 u(x, t) = 0$). Here, μ is the shear modulus, u is the displacement field, and f is external forcing. Furthermore, the elastic properties of our 1D medium (e.g. a string or a bar) are independent of space (homogeneous, $\mu(x) = $ const.), so that we arrive at the following differential equation:

$$-\mu \partial_x^2 u = f. \tag{6.6}$$

This has the mathematical form of a Poisson equation (if we assume $\mu = 1$ or bring it to the right-hand side). The problem we are solving here corresponds to the question of how much a string is displaced if you pull with a certain force (see Fig. 6.3).

In the following we proceed with the mathematical steps that lead to the classic discrete finite-element solution scheme. The first step is to transform the *strong* form of the differential equation into the *weak* form by multiplying the equation with an arbitrary space-dependent (real, well-behaved) test function that we denote as $v \to v(x)$. Then we integrate the equation on both sides over the entire physical domain D with $x \in D$ and obtain (omitting space dependencies)

$$-\int_D \mu \partial_x^2 u\, v\, dx = \int_D f\, v\, dx. \tag{6.7}$$

Note that we have not changed the solution to this equation. Our unknown field is $u(x)$ given constant μ and our choice of forcing $f(x)$. We carry out an integration by parts of the left side of the previous equation:

$$-\int_D \mu \partial_x^2 u\, v\, dx = \int_D \mu\, \partial_x u\, \partial_x v\, dx - [\mu \partial_x u\, v]_{x_{min}}^{x_{max}}, \tag{6.8}$$

where the last term is an anti-derivative.

The next step and the argument involved is of fundamental importance and leads to one of the most attractive features of finite- (or spectral-) element methods. Remember that a free-surface condition implies that the stress vanishes at some boundaries of our physical domain. In 1D the stress is given as $\sigma = \mu \partial_x u$. As the anti-derivative is evaluated at the domain boundaries, this implies—given free-surface conditions—that this term vanishes. From a computational point of view this means we get the free-surface boundary conditions for free (it is implicitly solved correctly). For several other methods (like the finite-difference method and the pseudospectral method) the correct implementation of the free-surface condition with the same accuracy as the wavefield inside the domain is a big headache. Thus, assuming we have a free surface at the edges of our domain D, we obtain the weak form as

$$\mu \int_D \partial_x u\, \partial_x v\, dx = \int_D f\, v\, dx, \tag{6.9}$$

which is still a continuous description.

To enter the discrete world we replace our exact solution $u(x)$ by $\overline{u}$, a sum over some basis functions φ_i that we do not yet specify:[3]

$$u \approx \overline{u}(x) = \sum_{i=1}^{N} u_i \varphi_i. \tag{6.10}$$

If our original weak equation holds for u we also expect it to hold for $\overline{u}$, and replacing u we obtain

$$\mu \int_D \partial_x \overline{u}\, \partial_x v\, dx = \int_D f\, v\, dx. \tag{6.11}$$

The next step is another fundamental concept of numerical analysis, known as the *Galerkin* method or principle. As a choice for our test function $v(x)$ we use the same set of basis functions. Thus $v \to \varphi_j(x)$.

Fig. 6.3 *Static elasticity. A string with homogeneous properties (density and shear modulus) is pulled with a certain force. The Poisson equation determines the displacement of the string given appropriate boundary conditions. Don't overdo this experiment, particularly if you have old strings.*

[3] By developing the weak form of the wave equation we are reducing the dimensionality of our original problem from infinite- to finite-dimensional, solving the differential equation in a *subspace* using linear algebra. The most important property of the Galerkin method is the fact that the error is orthogonal to this subspace.

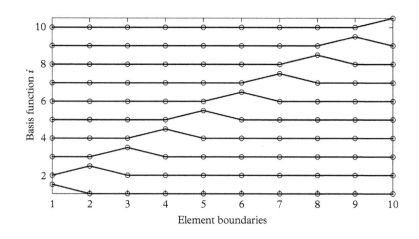

Fig. 6.4 *Linear basis functions for the finite-element method. A 1D domain is discretized with n − 1 elements having n = 10 element boundaries (open circles). The basis functions* φ_i = 1 *at* $x = x_i$. *With this basis an arbitrary function can be exactly interpolated at the element boundary points* x_i. *For the finite-element system of equations, integrals over the basis functions and their derivatives have to be evaluated separately for each element.*

What is the simplest choice for our basis functions φ_i? Denoting $x_i, i = 1, 2, ..., N$ as the boundaries of our elements we define our basis functions such that $\varphi_i = 1$ at $x = x_i$ and zero elsewhere. Inside the elements our solution field is described by a linear function. In mathematical terms this is obtained by

$$\varphi_i(x) = \begin{cases} \frac{x-x_{i-1}}{x_i-x_{i-1}} & \text{for } x_{i-1} \le x \le x_i \\ \frac{x_{i+1}-x}{x_{i+1}-x_i} & \text{for } x_i \le x \le x_{i+1} \\ 0 & \text{elsewhere,} \end{cases} \tag{6.12}$$

but more easily grasped by looking at Fig. 6.4. Recall the definition of our approximate solution field $u \approx \bar{u}(x) = \sum_{i=1}^{N} u_i\varphi_i$. It immediately becomes clear why we denoted the coefficients of the basis functions as u_i. Because of the unit value of basis function φ_i at $x = x_i$ the coefficients *are* the solutions of the displacement field, defined at the element boundaries. Inside the finite elements the solution field is interpolated by a linear function. It is important to note (and this is a major difference to finite-volume or discontinuous Galerkin methods) that adjacent elements share the same value at the boundaries. This finally leads to the requirement to solve a large global linear system of equations.

We are ready to assemble our discrete version of the weak form by replacing the continuous displacement field by its approximation and applying the Galerkin principle. Putting Eq. 6.10 into Eq. 6.9 we obtain for each $j = 1, \dots, N$

$$\mu \int_D \partial_x \left(\sum_{i=1}^{N} u_i\varphi_i \right) \partial_x \varphi_j \, dx = \int_D f \varphi_j \, dx$$

$$\sum_{i=1}^{N} u_i \, \mu \int_D \partial_x \varphi_i \, \partial_x \varphi_j \, dx = \int_D f \varphi_j \, dx, \tag{6.13}$$

which is a system of N equations since we project the solution on the basis functions φ_j with $j = 1, ..., N$. In the second equation we switched the sequence of integration and sum. The discrete system thus obtained can be written using matrix–vector notation. We define the solution vector **u** (i.e. the coefficients of our basis functions) as

$$\mathbf{u} = \begin{pmatrix} u_1 \\ u_2 \\ \vdots \\ u_N \end{pmatrix} \tag{6.14}$$

corresponding to the values at the left and right element boundaries. The source vector **f** can be written as

$$\mathbf{f} = \begin{pmatrix} \int_D f\, \varphi_1\, dx \\ \int_D f\, \varphi_2\, dx \\ \vdots \\ \int_D f\, \varphi_N\, dx \end{pmatrix} \tag{6.15}$$

and the matrix containing the integral over the basis function derivatives **K** reads

$$\mathbf{K} \rightarrow K_{ij} = \mu \int_D \partial_x \varphi_i\, \partial_x \varphi_j\, dx. \tag{6.16}$$

This system of equations can be written in component form as

$$u_i\, K_{ij} = f_j, \tag{6.17}$$

where we use the Einstein summation convention and in matrix–vector notation

$$\mathbf{K}^T \mathbf{u} = \mathbf{f}. \tag{6.18}$$

Note that here we encounter for the first time an important matrix in finite-element analysis: **K** is called the *stiffness* matrix, because in its original form it contains elastic parameters under the integration sign (here we assume they are constant). This system of equations has as many unknowns as equations. Provided that the matrix is positive definite, we can determine its inverse. In that case the solution to our problem is finally

$$\mathbf{u} = (\mathbf{K}^T)^{-1}\mathbf{f}. \tag{6.19}$$

Again, we encounter a fundamental characteristic of finite-element-type problems. To find solutions, a linear system of equations has to be solved. The size of the system matrix is in general the number of degrees of freedom squared, or N^2 if the domain is discretized with N points (the corresponding number of elements is then $N - 1$). We will shortly illustrate this with an example.

What are the consequences of having to invert a *global* system matrix? It implies that we look into solving a gigantic system of linear equations for realistic cases in more than one dimension! We cannot in general expect that the matrix is diagonal (which implies trivial inversion, see the chapter on spectral-element methods) or simply banded. Therefore, the entire toolbox of linear algebra concerning the solution of such systems has to be invoked. Solving linear systems of equations on parallel computers has become an independent field of research. Every new generation of parallel hardware is benchmarked against linear system solvers.[4]

Before we provide details on how to evaluate this system of equations we briefly look at the problem of special boundary conditions.

6.3.1   Boundary conditions

In case of a *free surface* (i.e. stress-free) boundary there is nothing to do, as it is implicitly fulfilled. Other boundary conditions can be implemented in a straightforward way. In case you would like to invoke specific values at the boundaries the approximate solution becomes

$$\bar{u} = u_1 \varphi_1 + \sum_{i=2}^{N-1} u_i \varphi_i + u_N \varphi_N, \tag{6.20}$$

where u_1 and u_N are the boundary values. Injecting this into the weak form we obtain, after rearranging terms,

$$\sum_{i=2}^{N-1} u_i \, \mu \int_D \partial_x \varphi_i \, \partial_x \varphi_j \, dx = \int_D f \, \varphi_j \, dx$$
$$+ u(x_{min}) \int_D \partial_x \varphi_1 \, \partial_x \varphi_j \, dx \tag{6.21}$$
$$+ u(x_{max}) \int_D \partial_x \varphi_N \, \partial_x \varphi_j \, dx,$$

observing that we have basically modified the right-hand side of the equation, and thus the source term. It is instructive to show this in a graphical way (see Fig. 6.5). For a finite-element system of size N we fix the boundary values and therefore reduce the number of unknowns to $N-2$. The system *feels* the boundary conditions through a modified source term that affects the solution everywhere.

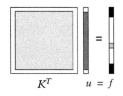

Fig. 6.5 *Static elasticity with boundary conditions. Graphical representation of the matrix–vector system with boundary conditions. The modified global system matrix has $N-2 \times N-2$ elements. The point in the middle of the source vector corresponds to a forcing inside the medium. The black top and bottom elements denote the boundary condition modifying the source term f.*

6.3.2   Reference element, mapping, stiffness matrix

When we introduced the basis functions earlier they were defined in the entire physical domain D. Another characteristic feature of the finite-element method that substantially simplifies the calculations performed on a computer is the fact that for all elements basically the same calculations have to be performed, with

[4] In my view this is one reason why in seismology the finite-element method did not take off as much as the finite-difference method when computational resources began to make 2D and 3D calculations practical.

only slight modifications. Therefore, a standard procedure is to map the physical domain to a reference element. In our case we centre the local coordinate system denoted as ξ at point x_i and obtain

$$\begin{aligned} \xi &= x - x_i \\ h_i &= x_i - x_{i-1}, \end{aligned} \tag{6.22}$$

where h_i denotes the size of element i defined in the interval $x \in [x_i, x_{i+1}]$. After this coordinate transform the definition of the local basis functions becomes

$$\varphi_i(\xi) = \begin{cases} \frac{\xi}{h_i} + 1 & \text{for } -h_i \leq \xi \leq 0 \\ 1 - \frac{\xi}{h_i} & \text{for } 0 \leq \xi \leq h_i \\ 0 & \text{elsewhere,} \end{cases} \tag{6.23}$$

and their derivatives

$$\partial_\xi \, \varphi_i(\xi) = \begin{cases} \frac{1}{h_i} & \text{for } -h_i \leq \xi \leq 0 \\ -\frac{1}{h_i} & \text{for } 0 \leq \xi \leq h_i \\ 0 & \text{elsewhere.} \end{cases} \tag{6.24}$$

We can now proceed with calculating the elements of the stiffness matrix $\mathbf{K}$ defined as

$$K_{ij} = \mu \int_D \partial_x \varphi_i \, \partial_x \, \varphi_j \, dx, \tag{6.25}$$

with the corresponding expression in local coordinates ξ

$$K_{ij} = \mu \int_{D_\xi} \partial_\xi \varphi_i \, \partial_\xi \varphi_j \, d\xi. \tag{6.26}$$

Before solving these integrals, let us pause for a moment and reflect on what the structure of this matrix might look like. First, we consider the basis functions and their derivatives in the local coordinate system. This is illustrated in Fig. 6.6. We note that they overlap only for adjacent indices, leading to a banded matrix structure. Because the basis functions and their derivatives are not differentiable at the element boundaries x_i the integrals have to be evaluated for each elemental domain separately. Let us calculate some of the elements of matrix K_{ij}, starting with the diagonal elements. For example, for K_{11} we obtain

$$\begin{aligned} K_{11} &= \mu \int_D \partial_x \varphi_1 \, \partial_x \varphi_1 \, dx \\ &= \mu \int_0^h \frac{-1}{h} \frac{-1}{h} d\xi = \frac{\mu}{h^2} \int_0^h d\xi = \frac{\mu}{h}, \end{aligned} \tag{6.27}$$

where we only have to perform the integration over element $i = 1$, since $\partial_x \, \varphi_1 = 0$ elsewhere. For the next diagonal element K_{22} we see that the derivatives overlap in

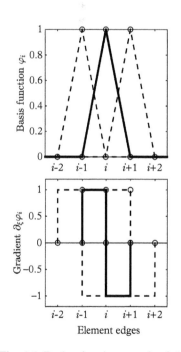

Fig. 6.6 *Basis functions and their derivatives.* **Top:** *The basis function φ_i (thick solid line) is shown along with the neighbouring functions $\varphi_{i\pm1}$ (thin dashed lines).* **Bottom:** *The same for their derivatives with respect to the space coordinate ξ.*

both elements 1 and 2, implying that in the local coordinate system the integration has to be performed for the interval $\xi \in [-h, h]$ (assuming for simplicity here that the elements are all of the same size).

$$
\begin{aligned}
K_{22} &= \mu \int_D \partial_x \varphi_2 \, \partial_x \varphi_2 \, dx \\
&= \mu \int_{-h}^{0} \partial_\xi \varphi_2 \, \partial_\xi \varphi_2 \, d\xi + \mu \int_{0}^{h} \partial_\xi \varphi_2 \, \partial_\xi \varphi_2 \, d\xi \\
&= \frac{\mu}{h^2} \int_{-h}^{0} d\xi + \frac{\mu}{h^2} \int_{0}^{h} d\xi = \frac{2\mu}{h}.
\end{aligned} \tag{6.28}
$$

Equivalently, the off-diagonal terms overlap only in one element, for example

$$
\begin{aligned}
K_{12} &= \mu \int_D \partial_x \varphi_1 \, \partial_x \varphi_2 \, dx \\
&= \mu \int_{0}^{h} \partial_\xi \varphi_1 \, \partial_\xi \varphi_2 \, d\xi = \mu \int_{0}^{h} \frac{-1}{h} \frac{1}{h} d\xi \\
&= \frac{-\mu}{h^2} \int_{0}^{h} d\xi = \frac{-\mu}{h} \\
K_{21} &= K_{12}.
\end{aligned} \tag{6.29}
$$

Finally, the stiffness matrix for an elastic physical system with constant shear modulus μ and uniform element size h (see Section 6.4.2 for the general case) reads

$$
K_{ij} = \frac{\mu}{h} \begin{pmatrix} 1 & -1 & & & \\ -1 & 2 & -1 & & \\ & & \ddots & & \\ & & -1 & 2 & -1 \\ & & & -1 & 1 \end{pmatrix} \tag{6.30}
$$

and these numbers probably ring some bells. It might not come as a surprise that the space-dependent terms in our linear system are proportional to the three-point operator matrix for a second finite-difference derivative. This will be discussed in more detail shortly. Let us turn this into a computer program.

Table 6.1 *1D static elastic problem with the finite-element method. Homogeneous case.*

Parameter	Value
x_{max}	1
nx	20
μ	1
h	0.0526
$u(0)$	0.15
$u(1)$	0.05

6.3.3   Simulation example

We demonstrate the finite-element solution to the static elastic problem with a simple toy problem. The parameters are given in Table 6.1. The physical domain is defined in the interval $x \in [0, 1]$ and we apply a unit forcing at $x = 0.75$ at one of the element boundary points.

The following Python code fragment presents a possible implementation of the finite-element solution and the calculation of the stiffness matrix in the case of constant element size and shear modulus.

```
# [...]
# Basic parameters
nx = 20            # Number of boundary points
u = zeros(nx)      # Solution vector
f = zeros(nx)      # Source vector
mu = 1             # Constant shear modulus
# Element boundary points
x = linspace(0, 1, nx) # x in [0,1]
h = x[2] - x[1] # Constant element size
# Assemble stiffness matrix K_ij
K = zeros((nx, nx))
for i in range(1, nx-1):
    for j in range(1, nx-1):
        if i == j:
            K[i, j] = 2 * mu/h
        elif i == j + 1:
            K[i, j] = -mu/h
        elif i + 1 == j:
            K[i, j] = -mu/h
        else:
            K[i, j] = 0
# Souce term is a spike at i = 15
f[15] = 1
# Boundary condition at x = 0
u[0] = 0.15 ; f[1] = u[0]/h
# Boundary condition at x = 1
u[nx-1] = 0.05 ; f[nx-2] = u[nx-1]/h
# finite element solution
u[1:nx-1] = linalg.inv(K[1:nx-1, 1:nx-1]) @ f[1:nx-1].T
# [...]
```

At this point we make a short detour and ask how this static problem might be solved with finite differences for comparison. Starting with the Poisson equation $-\mu \partial_x^2 u = f$, omitting space dependencies, we replace the left-hand side with a finite-difference approximation and obtain

$$-\mu \frac{u(x-h) - 2u(x) + u(x+h)}{h^2} = f, \tag{6.31}$$

and, after rearranging,

$$u(x) = \frac{u(x-h) + u(x+h)}{2} + \frac{h^2}{2\mu}f. \tag{6.32}$$

How should we interpret this equation? An update at point x is obtained by averaging the two surrounding points plus some scaled source term. We should not expect this to lead to the correct solution in one step. But this equation can be used as an iterative procedure with an initial guess for the unknown field u. With discretization of $u_i = u(x_i)$ and iteration step k this can be written as

$$u_i^{k+1} = \frac{u_{i+1}^k + u_{i-1}^k}{2} + \frac{h^2}{2\mu} f_i, \qquad (6.33)$$

with initial guess $u_i^{k=1} = 0$. This is implemented in the following script, and we compare it to the finite-element solution. This approach is called a *relaxation method*.

```
# [...]
# Forcing
f[15] = 1/h   # force vector
for it in range(nt):
    # Calculate the average of u (omit boundaries)
    for i in range(1, nx-1):
        du[i] = u[i+1] + u[i-1]
    u = 0.5 * (f * h**2/mu + du)
    u[0] = 0.15      # Boundary condition at x=0
    u[nx-1] = 0.05   # Boundary condition at x=1
# [...]
```

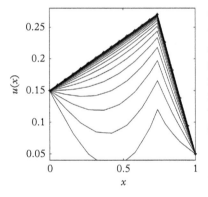

Fig. 6.7 *Static elasticity with boundary conditions. Simulation example comparing the finite-element solution (thick solid line) with a finite-difference-based relaxation method (thin lines) that iteratively converges to the correct solution (see text for details).*

The results are shown in Fig. 6.7. The finite-element solution reaches the solution to the problem in one step involving the inversion of the stiffness matrix K. In this example 500 iterations were employed for the relaxation method and the solution is shown after every 25 iterations. Obviously, the finite-element method is much faster. An interesting point is that the force term in the finite-element method is divided by the element width h by means of an integral over the derivative of the basis function. In the finite-difference approximation the division by h comes from the requirement to have the final solution independent of the grid distance (see Section 4.4 on source injection in Chapter 4 on the finite-difference method).

It is instructive to extend this simplest case to arbitrary element sizes and space-varying elastic properties (see exercises). In the next section we add time dependence to our system.

6.4 1D elastic wave equation

Let us apply what we have learned about the finite-element approach and the Galerkin principle to the 1D elastic wave equation

$$\rho \partial_t^2 u = \partial_x \mu \, \partial_x u + f, \qquad (6.34)$$

where again we omit space and time dependencies. From now on we assume that the properties of the medium, density ρ, and shear modulus μ, are both space-dependent. We obtain a weak form of the wave equation by integrating over the entire physical domain D and at the same time multiplying the original equation with an arbitrary basis function φ_j:

$$\int_D \rho \, \partial_t^2 \, u \, \varphi_j \, dx = \int_D \partial_x \, \mu \, \partial_x \, u \, \varphi_j \, dx + \int_D f \, \varphi_j \, dx. \qquad (6.35)$$

Integration by parts of the term containing the space derivatives leads to

$$\int_D \partial_x \, \mu \, \partial_x \, u \, \varphi_j \, dx = \left[\mu \partial_x \, u \, \varphi_j \right] - \int_D \mu \, \partial_x \, u \, \partial_x \, \varphi_j \, dx, \qquad (6.36)$$

and again we drop the term with the antiderivative that is evaluated at the edges of the physical domain. This is equivalent to a stress-free boundary condition as discussed above. Reinjecting this result into the weak form, we obtain

$$\int_D \rho \, \partial_t^2 \, u \, \varphi_j \, dx + \int_D \mu \, \partial_x \, u \, \partial_x \, \varphi_j \, dx = \int_D f \, \varphi_j \, dx, \qquad (6.37)$$

where u is the continuous unknown displacement field. We replace the exact displacement field by an approximation $\bar{u}$ of the form

$$u(x, t) \rightarrow \bar{u}(x, t) = \sum_{i=1}^{N} u_i(t) \, \varphi_i(x), \qquad (6.38)$$

where the coefficients u_i are expected to correspond to a discrete representation of the solution field at the element boundaries, following earlier the discussion on linear basis functions. The approximate displacement field is of course constrained by the same wave equation, thus

$$\int_D \rho \, \partial_t^2 \, \bar{u} \, \varphi_j \, dx + \int_D \mu \, \partial_x \, \bar{u} \, \partial_x \, \varphi_j \, dx = \int_D f \, \varphi_j \, dx. \qquad (6.39)$$

Injecting Eq. 6.38 into Eq. 6.39, we can turn the continuous weak form into a system of linear equations:

$$\int_D \rho \, \partial_t^2 \left(\sum_{i=1}^{N} u_i(t) \, \varphi_i \right) \varphi_j \, dx$$
$$+ \int_D \mu \, \partial_x \left(\sum_{i=1}^{N} u_i(t) \, \varphi_i \right) \partial_x \, \varphi_j \, dx \qquad (6.40)$$
$$= \int_D f \, \varphi_j \, dx,$$

where we highlight the fact that only the coefficients $u_i(t)$ are time-dependent. Changing the order of integration and summation, we obtain

$$\sum_{i=1}^{N} \partial_t^2 u_i(t) \int_D \rho \, \varphi_i \, \varphi_j \, dx + \sum_{i=1}^{N} u_i(t) \int_D \mu \, \partial_x \, \varphi_i \, \partial_x \, \varphi_j \, dx = \int_D f \, \varphi_j \, dx, \qquad (6.41)$$

using the fact that the unknown coefficients u_i only depend on time.

As in the static elastic case we proceed using matrix–vector notation, with the following definitions for the time-dependent solution vector of displacement values $\mathbf{u}(t)$, mass matrix $\mathbf{M}$ (sensible name, as it contains density), the already well-known stiffness matrix $\mathbf{K}$, and the source vector $\mathbf{f}$:

$$\mathbf{u}(t) \rightarrow u_i(t)$$

$$\mathbf{M} \rightarrow M_{ij} = \int_D \rho \, \varphi_i \, \varphi_j \, dx$$

$$\mathbf{K} \rightarrow K_{ij} = \int_D \mu \, \partial_x \varphi_i \, \partial_x \varphi_j \, dx \qquad (6.42)$$

$$\mathbf{f} \rightarrow f_j = \int_D f \, \varphi_j \, dx.$$

Thus we can write the system of equations as

$$\partial_t^2 \mathbf{u}\mathbf{M} + \mathbf{u}\mathbf{K} = \mathbf{f}, \qquad (6.43)$$

or with transposed system matrices as

$$\mathbf{M}^T \partial_t^2 \mathbf{u} + \mathbf{K}^T \mathbf{u} = \mathbf{f}. \qquad (6.44)$$

For the second time derivative we use a standard finite-difference approximation:

$$\partial_t^2 \mathbf{u} \approx \frac{\mathbf{u}(t+dt) - 2\mathbf{u}(t) + \mathbf{u}(t-dt)}{dt^2}, \qquad (6.45)$$

replacing the original partial derivative with respect to time to obtain

$$\mathbf{M}^T \left[\frac{\mathbf{u}(t+dt) - 2\mathbf{u}(t) + \mathbf{u}(t-dt)}{dt^2} \right] = \mathbf{f} - \mathbf{K}^T \mathbf{u}. \qquad (6.46)$$

This system can be extrapolated in an analoguous fashion to the other numerical methods we have already encountered. Starting from an initial state $\mathbf{u}(t = 0) = 0$ we can determine the displacement field at time $t + dt$ by

$$\mathbf{u}(t+dt) = dt^2 (\mathbf{M}^T)^{-1} \left[\mathbf{f} - \mathbf{K}^T \mathbf{u} \right] + 2\mathbf{u}(t) - \mathbf{u}(t-dt). \qquad (6.47)$$

This equation will be implemented in the computer program. Again, like in the static case, the solution depends on the inversion of a global system matrix, in this case the mass matrix. In the general case, for 2D or 3D problems this huge matrix might be sparse but there is no way around a global solution scheme unless we find a scheme for a diagonal mass matrix. This is in fact possible, with the right choice of basis functions (Lagrange polynomials) and a corresponding numerical integration scheme (Gauss integration). This is a specific form of the spectral-element method that is discussed in Chapter 7. Another observation is that in this formulation the mass and stiffness matrices do not depend on time. For the mass matrix this implies that it can be inverted once and for all prior to the time extrapolation as an initialization step. What remains to be done before presenting examples is to discuss how the system matrices can be calculated.

6.4.1   The system matrices

To calculate the entries of the system matrices we transform the space coordinate into a local system, as we have done in the static case

$$\xi = x - x_i$$
$$h_i = x_{i+1} - x_i. \tag{6.48}$$

However, now we allow the element size h_i to vary. With the definition above element i is defined in the interval $x \in [x_i, x_{i+1}]$. In the local coordinate system the basis functions are defined by

$$\varphi_i(\xi) = \begin{cases} \frac{\xi}{h_{i-1}} + 1 & \text{for } -h_{i-1} \leq \xi \leq 0 \\ 1 - \frac{\xi}{h_i} & \text{for } 0 \leq \xi \leq h_i \\ 0 & \text{elsewhere,} \end{cases} \tag{6.49}$$

with the corresponding derivatives

$$\partial_\xi \, \varphi_i(\xi) = \begin{cases} \frac{1}{h_{i-1}} & \text{for } -h_{i-1} \leq \xi \leq 0 \\ -\frac{1}{h_i} & \text{for } 0 \leq \xi \leq h_i \\ 0 & \text{elsewhere.} \end{cases} \tag{6.50}$$

An example of basis functions and their derivatives for an irregular grid with varying element sizes is shown in Fig. 6.8. We are ready to assemble our system matrices.

The mass matrix

Looking at the global definition of the mass matrix **M** with components

$$M_{ij} = \int_D \rho \, \varphi_i \, \varphi_j \, dx \tag{6.51}$$

and considering the specific nature of our basis functions (see Fig. 6.8), we realize that the only non-zero entries are around the diagonal and are of components $M_{i,i-1}$ $M_{i,i}$ and $M_{i,i+1}$ for $i = 2, ..., N - 1$. Elements M_{11} and M_{NN} have to be treated separately. For the diagonal elements we obtain

$$M_{ii} = \int_D \rho \, \varphi_i \, \varphi_i \, dx = \int_{D_\xi} \rho \, \varphi_i \, \varphi_i \, d\xi \tag{6.52}$$

in the local coordinate system. Integration has to be carried out over the elements to the left and right of the boundary points x_i. We thus obtain

$$M_{ii} = \rho_{i-1} \int_{-h_{i-1}}^0 \left(\frac{\xi}{h_{i-1}} + 1 \right)^2 d\xi + \rho_i \int_0^{h_i} \left(1 - \frac{\xi}{h_i} \right)^2 d\xi$$
$$= \frac{1}{3} \left(\rho_{i-1} \, h_{i-1} + \rho_i \, h_i \right), \tag{6.53}$$

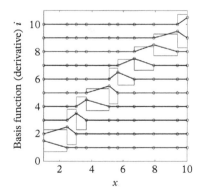

Fig. 6.8 *Basis functions and derivatives for irregular mesh. Example of a finite-element domain with irregular element sizes h_i. The basis functions (thick solid lines) are illustrated with the normalized derivatives (thin solid lines).*

where we assume that the density is constant inside each element (otherwise we would have to use numerical integration). For the off-diagonal elements the basis functions overlap only in one element, for example

$$M_{i,i-1} = \rho_{i-1} \int_{-h_{i-1}}^{0} \left(\frac{\xi}{h_{i-1}} + 1 \right) \frac{-\xi}{h_{i-1}} d\xi = \frac{1}{6} \rho_{i-1} h_{i-1} \qquad (6.54)$$

or

$$M_{i,i+1} = \rho_i \int_{0}^{h_i} \left(1 - \frac{\xi}{h_i} \right) \frac{\xi}{h_i} d\xi = \frac{1}{6} \rho_i h_i. \qquad (6.55)$$

It is instructive to sketch the integration intervals and the relevant basis functions using Eq. 6.49 to visualize the above formulation (see Fig. 6.8).

The banded nature of the mass matrix, assuming constant element size h and density ρ, reads

$$\mathbf{M} = \frac{\rho h}{6} \begin{pmatrix} \ddots & & & & 0 \\ 1 & 4 & 1 & & \\ & 1 & 4 & 1 & \\ & & 1 & 4 & 1 \\ 0 & & & & \ddots \end{pmatrix}. \qquad (6.56)$$

In the general case, with varying element size the mass matrix is not symmetric.

The stiffness matrix

The same concepts apply to the stiffness matrix. We move to the local coordinate system by

$$K_{ij} = \int_{D} \mu \, \partial_x \varphi_i \, \partial_x \varphi_j \, dx = \int_{D_\xi} \mu \, \partial_\xi \varphi_i \, \partial_\xi \varphi_j \, d\xi \qquad (6.57)$$

and obtain for a diagonal element, assuming constant shear modulus μ inside each element,

$$\begin{aligned} K_{ii} &= \mu_{i-1} \int_{-h_{i-1}}^{0} \left(\frac{1}{h_{i-1}} \right)^2 d\xi + \mu_i \int_{0}^{h_i} \left(-\frac{1}{h_i} \right)^2 d\xi \\ &= \frac{\mu_{i-1}}{h_{i-1}} + \frac{\mu_i}{h_i}, \end{aligned} \qquad (6.58)$$

and for the off-diagonal elements

$$\begin{aligned} K_{i,i+1} &= \mu_i \int_{0}^{h_i} \left(-\frac{1}{h_i} \right) \left(\frac{1}{h_i} \right) d\xi = -\frac{\mu_i}{h_i} \\ K_{i,i-1} &= \mu_{i-1} \int_{-h_{i-1}}^{0} \left(\frac{1}{h_{i-1}} \right) \left(-\frac{1}{h_{i-1}} \right) d\xi = -\frac{\mu_{i-1}}{h_{i-1}}, \end{aligned} \qquad (6.59)$$

while all other elements of the stiffness matrix are zero. For example, assuming constant shear modulus and element size the stiffness matrix reads

$$\mathbf{K} = \frac{\mu}{h} \begin{pmatrix} \ddots & & & & 0 \\ -1 & 2 & -1 & & \\ & -1 & 2 & -1 & \\ & & -1 & 2 & -1 \\ 0 & & & & \ddots \end{pmatrix}. \qquad (6.60)$$

At this point we have all the ingredients to assemble a program simulating 1D elastic wave propagtion in a heterogeneous medium with irregular elements.

6.4.2 Simulation example

Let us initialize a 1D physical domain and simulate elastic wave propagation with the algorithm developed above. We start with a homogeneous domain on a regular grid and compare with the finite-difference method. The parameters for the simulation are given in Table 6.2. Before showing the results we would like to illustrate the striking similarity of the final algorithm to the finite-difference method. Despite the fact that we introduced the finite-difference approach through a *local* view, it can also be formulated in matrix–vector form (we have used this already in the case of the Chebyshev pseudospectral method).

It is easy to show (see exercises) that if we initialize a differentiation matrix D for the second derivative based on finite differences we obtain (according to the wave equation we multiply by μ)

$$\mathbf{D} = \frac{\mu}{dt^2} \begin{pmatrix} -2 & 1 & & & \\ 1 & -2 & 1 & & \\ & & \ddots & & \\ & & 1 & -2 & 1 \\ & & & 1 & -2 \end{pmatrix} \qquad (6.61)$$

and, defining a diagonal mass matrix *Minv* containing the inverse densities, we can extrapolate the finite-difference scheme with the algorithm shown in the following Python code fragment:

```
# Time extrapolation
for it in range(nt):
    # Finite Difference Method
    unew = (dt**2)*Minv @ (D @ u + f/dx*src[it])\
        + 2*u - uold
    uold, u = u, unew
[...]
```

It is instructive to compare the matrix structures with the finite-element method in a graphical way. This is shown in Fig. 6.9. As already indicated, the stiffness matrix is basically equivalent to the matrix form of the second derivative operator. Only the inverse mass matrices have different structure. While in the

Table 6.2 *Simulation parameters for 1D elastic simulation. Homogeneous case.*

Parameter	Value
x_{max}	10,000 m
nx	1,000
v_s	3,000 m/s
ρ	2,500 kg/m^3
h	10 m
eps	0.5
f_0	20 Hz

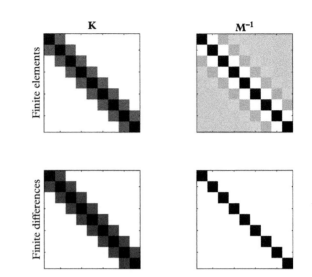

Fig. 6.9 *System matrices. The structure of the system matrices for the finite-element method are compared with the finite-difference method formulated with matrix–vector operations.* **Top row:** *Stiffness and inverse mass matrix for the finite-element method.* **Bottom row:** *Stiffness (differential) matrix and diagonal mass matrix for the finite-difference method.*

finite-difference method it is diagonal, the finite-element-based mass matrix is banded around the diagonal.

The following code example illustrates the calculation of the mass matrix for the general case of varying element size:

```
# Mass matrix M_ij
M = zeros((nx, nx))
for i in range(1, nx-1):
    for j in range(1, nx-1):
        if i == j:
            M[i, j] = (ro[i-1] * h[i-1]
                       + ro[i] * h[i]) / 3
        elif j == i + 1:
            M[i, j] = ro[i] * h[i]/6
        elif j == i - 1:
            M[i, j] = ro[i-1] * h[i-1]/6
        else:
            M[i, j] = 0
# Corner elements
M[0,0] = ro[0] * h[0] / 3
M[nx-1, nx-1] = ro[nx-1] * h[nx-2] / 3
```

The implementation of the finite-element time extrapolation of the coefficients is almost identical to the finite-difference algorithm just presented. The mass matrix is inverted before the time extrapolation starts. The source is injected at one boundary element point. The source time function is a first derivative of a Gaussian.

```
# Time extrapolation
for it in range(nt):
    # Finite Element Method
    unew = (dt**2)*Minv @ (f*src[it] - K @ u)
            + 2*u - uold
    uold, u = u, unew
# [...]
```

In all listings above h corresponds to the element size, *src* contains the source time function, u, *new*, *uold* are the displacement fields at $t \pm dt$, and f is the force vector.

Results of the simulation are shown in Fig 6.10 for various propagation distances and compared with the finite-difference method. It is interesting to note that in this case the finite-difference and the finite-element methods have basically the same dispersive behaviour, but with opposite sign (i.e. in the finite-element method the high frequencies arrive earlier, like in the pseudospectral method).

Finally, let us explore one of the most important advantages of the finite-element method; the simplicity with which the element size can vary. This is also called h-adaptivity (referring to the adaptation of element size h in our case due to geometrical features or the velocity model).

h-adaptivity

Think about an Earth model in which the seismic velocities have strong variations. This is more the rule than the exception. An example is the Earth's mantle. When including the P-velocity in the oceans (1.5 km/s) the seismic velocities span almost an entire magnitude given the maximum P-velocities above the core–mantle boundary (13 km/s). Any numerical scheme with globally constant element size has to be accurate for the shortest wavelength. Therefore, regions with higher velocities might be substantially oversampled, reducing computational efficiency.

The numerical methods we have encountered so far (finite differences and pseudospectral methods) are not really able to solve this problem in the most general case. The finite-element method with options for deformed hexahedral or tetrahedral elements offers this flexibility. We demonstrate this in the 1D case with a strongly heterogeneous velocity model in which the number of grid points per wavelength is kept constant in the entire physical domain. The parameters for this model with three different subdomains are given in Table 6.3. The model mimics (in a slightly exaggerated way) the situation in a fault zone with a central low-velocity zone (damage zone) with different material properties on the two sides of the fault (this situation is examined with the 2D finite-difference code; see exercises).

The seismic velocities differ by a factor of 4 (in seismological terms this is indeed a dramatic velocity contrast). We keep the number of grid points constant across the model by adapting the element size accordingly in each domain. A source is injected in the centre of the low-velocity zone. The results of the simulation are shown for the entire domain as a function of time in Fig. 6.11. The boundaries act as free surfaces, and thus reflect the entire wavefield with reversed

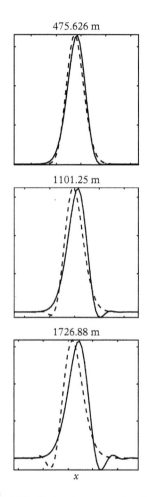

Fig. 6.10 *Finite-element simulation. Snapshots of the displacement wavefield calculated with the finite-element method (solid line) are compared with the finite-difference method (dotted line) at various distances from the source, using the same parameters. The length of the window is 500 m.*

Table 6.3 *Simulation parameters for 1D elastic simulation. Heterogeneous case.*

	Left	Middle	Right
x	4,600 m	1,000 m	4,600 m
v_s	6,000 m/s	1,500 m/s	3,000 m/s
dx	40 m	10 m	20 m
ρ	2,500 kg/m^3	2,500 kg/m^3	2,500 kg/m^3

Parameter	Value
nt	18,000
dt	3.3 ms
f_0	5 Hz
eps	0.5

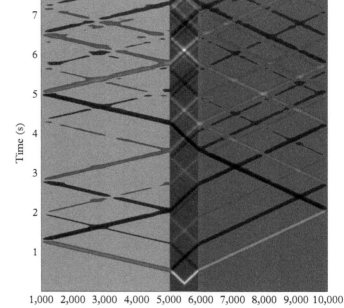

Fig. 6.11 *Finite-element simulation with varying element size. Snapshots of displacement values are shown as a function of time. Where displacement amplitudes are below a threshold, the velocity model is shown in greyscale. The parameters for this simulation are given in Table 6.3. The element size is defined such that the number of grid points per wavelength is constant in the entire physical domain. Note the polarity change of the reflections at the boundaries and the slope of the signals in the x – t plane indicating their velocities.*

polarity. The results illustrate the reverberating effect of the low-velocity zone. The dominant frequency in this case is sampled with 30 elements per wavelength. Note that in this case the *h*-adaptive mesh also leads to an equal sampling of the wavefield in time within each subdomain (see exercises).

Let us have a close-up look at one of the boundaries where the element size changes. This is shown in Fig. 6.12. There are several points to notice. First, because of the large velocity change the wavelength in the left domain is substantially longer. Second, note the continuous, but non-differentiable behaviour of the wavefield at the boundary. The advantage here is that we do not have to take space derivatives across the domain boundary as would be the case with the finite-difference method.

The theory and the applications presented so far were based on the simplest linear-basis functions; in terms of polynomial order this corresponds to order 1. In the following section, we briefly illustrate how this *classical* ansatz can be extended to higher orders in space.

6.5 Shape functions in 1D

How can we formally derive basis functions for finite-element discretizations? Is there a way to improve accuracy in space by moving to high orders?

Let us recall how we replaced the originally continuous unknown field $u(x)$ by a sum over some basis functions φ_i:

$$u(x) = \sum_{i=1}^{N} c_i \varphi_i(x), \tag{6.62}$$

where $x \in D$ is defined in the entire physical domain. Here, we denote the coefficients of the basis functions by c_i. This facilitates the notation when going to higher orders. As mentioned before, a standard procedure in finite-element analysis is to map all elements to a local coordinate system, which makes life easier as all integrals can be calculated in the same way (except for some constants). We define

$$\xi = \frac{x - x_i}{x_{i+1} - x_i}, \tag{6.63}$$

where our reference element is defined with $\xi \in [0, 1]$. In the following we present a formal procedure to derive the so-called shape functions that are used to describe the solution field. Even though we have encountered the linear form of these functions already, this procedure will in principle allow extensions to any order.

Linear shape functions

We put ourselves at element level and assume that our unknown function $u(\xi)$ is linear:

$$u(\xi) = c_1 + c_2 \xi, \tag{6.64}$$

where c_i are real coefficients. Each element has two node points, namely the element boundaries at $\xi_{1,2} = 0, 1$. This leads to the following conditions and solutions for coefficients c_i

$$\begin{aligned} u_1 &= c_1 \rightarrow c_1 = u_1 \\ u_2 &= c_1 + c_2 \rightarrow c_2 = -u_1 + u_2 \end{aligned}. \tag{6.65}$$

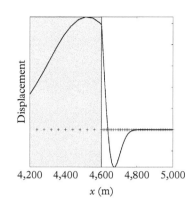

Fig. 6.12 *Finite element simulation, domain boundary. Detail of the finite-element simulation (snapshot of the displacement field) with varying element size at one of the domain boundaries. The crosses indicate the element boundaries and the changing element size. Note the continuous but non-differentiable behaviour of the displacement field at the interface.*

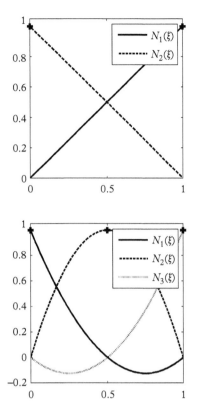

Fig. 6.13 *Finite-element shape functions.* **Top:** *Linear shape functions as used in the development of the finite-element solution to static and dynamic elastic problems. Node points are indicated by crosses.* **Bottom:** *Quadratic shape functions requiring one more node point at the centre of the element.*

This can be written in matrix notation, which will help us when dealing with high-order systems. We obtain

$$\begin{bmatrix} u_1 \\ u_2 \end{bmatrix} = \begin{bmatrix} 1 & 0 \\ 1 & 1 \end{bmatrix} \begin{bmatrix} c_1 \\ c_2 \end{bmatrix}, \tag{6.66}$$

and, using matrix inversion,

$$\begin{bmatrix} c_1 \\ c_2 \end{bmatrix} = \begin{bmatrix} 1 & 0 \\ -1 & 1 \end{bmatrix} \begin{bmatrix} u_1 \\ u_2 \end{bmatrix}. \tag{6.67}$$

With appropriate matrix and vector definitions this can be written as

$$\mathbf{u} = \mathbf{Ac} \rightarrow \mathbf{c} = \mathbf{A}^{-1}\mathbf{u}, \tag{6.68}$$

implying that to obtain coefficients $\mathbf{c}$ we need to calculate the inverse of $\mathbf{A}$. Putting the coefficients thus obtained back into the original equation we obtain

$$\begin{aligned} u(\xi) &= u_1 + (-u_1 + u_2)\xi \\ &= u_1(1-\xi) + u_2\xi \\ &= u_1 N_1(\xi) + u_2 N_2(\xi), \end{aligned} \tag{6.69}$$

where we introduced a novel concept, the shape functions $N_i(\xi)$ with the following form

$$N_1(\xi) = 1 - \xi, \quad N_2(\xi) = \xi. \tag{6.70}$$

This is a fundamental concept holding both for finite- and spectral-element methods in this *nodal* form: One elemental shape function is multiplied by a value of the solution field (e.g. the displacement field) at a specific node point. The sum over the weighted shape function of general order N

$$u(\xi) = \sum_{i=1}^{N} u_i N_i(\xi) \tag{6.71}$$

gives the approximate *continuous* representation of the solution field $u(\xi)$ inside the element. The shape functions are illustrated in Fig. 6.13.

Quadratic shape functions

Extending these concepts to higher orders is straightforward. Describing our solution field by quadratic functions requires

$$u(\xi) = c_1 + c_2\xi + c_3\xi^2, \tag{6.72}$$

where we added one more node point at the centre of the element $\xi_{1,2,3} = 0, 1/2, 1$. With these node locations we obtain

$$\begin{aligned} u_1 &= c_1 \\ u_2 &= c_1 + 0.5c_2 + 0.25c_3 \\ u_3 &= c_1 + c_2 + c_3, \end{aligned} \tag{6.73}$$

and after inverting the resulting system matrix $\mathbf{A}$

$$\mathbf{A}^{-1} = \begin{bmatrix} 1 & 0 & 0 \\ -3 & 4 & -1 \\ 2 & -4 & 2 \end{bmatrix} \tag{6.74}$$

we can represent the final quadratic solution field inside the element with

$$\begin{aligned} u(\xi) &= c_1 + c_2\xi + c_3\xi^2 \\ &= u_1(1 - 3\xi + 2\xi^2) + \\ &\quad u_2(4\xi - 4\xi^2) + \\ &\quad u_3(-\xi + 2\xi^2), \end{aligned} \tag{6.75}$$

resulting in the following shape functions

$$\begin{aligned} N_1(\xi) &= 1 - 3\xi + 2\xi^2 \\ N_2(\xi) &= 4\xi - 4\xi^2 \\ N_3(\xi) &= -\xi + 2\xi^2, \end{aligned} \tag{6.76}$$

illustrated in Fig. 6.13 (bottom). The extension to cubic shape functions is straightforward but in that case derivative information at the boundaries is necessary to constrain the coefficients.

6.6 Shape functions in 2D

Shape functions start getting interesting with more dimensions. The most frequently used element shapes in 2D are triangles (e.g. after Delauney triangulation of arbitrary point clouds) and rectangles. We limit ourselves to the linear case.

Triangular shape functions

We start with a triangle of arbitrary shape defined somewhere in x–y space. This is illustrated in Fig. 6.14. To perform the integration operations when calculating the system matrices we move to the local coordinate system $\xi, \eta \in [0, 1]$ (sometimes the reference space is chosen to be $[-1, 1]$), through

$$\begin{aligned} x &= x_1 + (x_2 - x_1)\xi + (x_3 - x_1)\eta \\ y &= y_1 + (y_2 - y_1)\xi + (y_3 - y_1)\eta. \end{aligned} \tag{6.77}$$

We seek to describe a linear function inside our triangle; therefore

$$u(\xi, \eta) = c_1 + c_2\xi + c_3\eta. \tag{6.78}$$

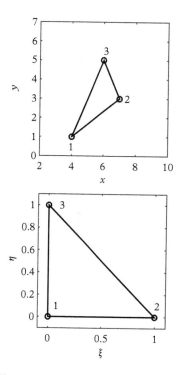

Fig. 6.14 *Triangular elements. Mapping of physical coordinates $(x, y$, top) to a local reference frame (bottom) with coordinates ξ, η.*

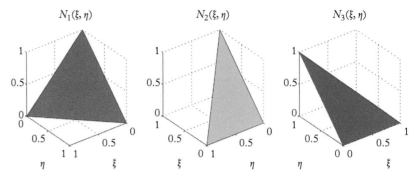

Fig. 6.15 *Triangular elements, shape functions. The three corner nodes lead to an equivalent number of shape functions $N_i(\xi, \eta)$ with unit value at one of the corners.*

We only know our function at the corners of the reference triangle; therefore the constraints for coefficients c_i are

$$
\begin{aligned}
u_1 &= u(0,0) = c_1 \\
u_2 &= u(1,0) = c_1 + c_2 \\
u_3 &= u(0,1) = c_1 + c_3.
\end{aligned}
\tag{6.79}
$$

This leads, using the same matrix inversion approach described above, to the following shape functions for triangular elements:

$$
\begin{aligned}
N_1(\xi, \eta) &= 1 - \xi - \eta \\
N_2(\xi, \eta) &= \xi \\
N_3(\xi, \eta) &= \eta.
\end{aligned}
\tag{6.80}
$$

These shape functions are illustrated in Fig. 6.15. Again, extension to higher orders is possible.

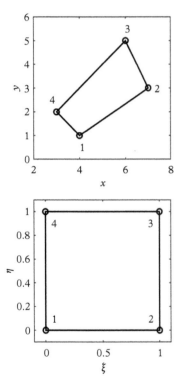

Rectangular shape functions

Similarly, shape functions can be derived for general quadrilateral elements. An example is shown in Fig. 6.16. We map space to a local coordinate system through

$$
\begin{aligned}
x &= x_1 + (x_2 - x_1)\xi + (x_4 - x_1)\eta + (x_3 - x_2)\xi\eta \\
y &= y_1 + (y_2 - y_1)\xi + (y_4 - y_1)\eta + (y_3 - y_2)\xi\eta.
\end{aligned}
\tag{6.81}
$$

Requiring linear behaviour of the function inside the element

$$
u(\xi, \eta) = c_1 + c_2\xi + c_3\eta + c_4\xi\eta
\tag{6.82}
$$

we obtain the following shape functions:

Fig. 6.16 *Quadrilateral elements. Mapping of physical coordinates (x, y, top) to a local reference frame (bottom) with coordinates ξ, η.*

$$
\begin{aligned}
N_1(\xi, \eta) &= (1 - \xi)(1 - \eta) \\
N_2(\xi, \eta) &= \xi(1 - \eta) \\
N_3(\xi, \eta) &= \xi\eta \\
N_4(\xi, \eta) &= (1 - \xi)\eta.
\end{aligned}
\tag{6.83}
$$

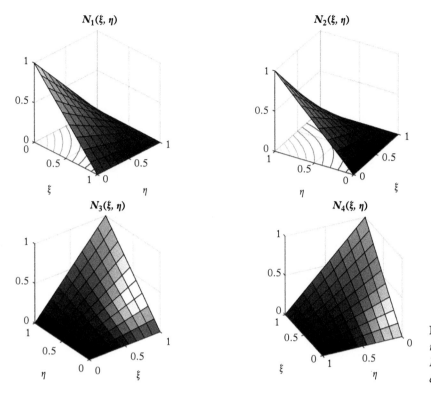

$N_1(\xi, \eta)$ $N_2(\xi, \eta)$

$N_3(\xi, \eta)$ $N_4(\xi, \eta)$

Fig. 6.17 *Rectangular reference elements. The four linear shape functions $N_i(\xi, \eta)$ with unit value at one of the corners.*

These four shape functions are illustrated in Fig. 6.17. There are many more aspects to shape functions, in particular in connection with numerical integration. The reader is referred to the extensive coverage in Zienkiewicz et al. (2013). Numerical integration will be discussed in Chapter 7 in connection with the spectral-element method.

6.7 The road to 3D

The extension of the simple finite-element codes presented here to 2D and 3D is substantially more involved than for 3D finite-difference or pseudospectral aproaches. The reason is that in 2D the global system matrices already become huge, and handling those requires getting into linear algebra algorithms. This goes far beyond the scope of this introductory text (even Zienkiewicz et al. (2013) only touches the problem). However, usually this is handled by libraries (like LAPACK) that are specifically designed and optimized to solve sparse systems. Those interested in descriptions of 3D implementations and strategies for parallelization are referred to Bao et al. (1996) and Bielak et al. (1998).

Recent extensions include the introduction of spatially adaptive meshes using the octree approach (Bielak et al., 2005).

To my knowledge it is fair to say that the classic low-order finite-element method plays a minor role today for large-scale computational problems in seismology, exploration seismics, and rupture dynamics, in comparison with the finite-difference or the spectral-element methods. The reason is, on one hand, that extensions of the finite-element method towards higher orders using a combination of Lagrange polynomials and Gauss integration leads to a global system of equations that does not require linear algebra tools for handling huge matrix inversion (this is called the spectral-element method, see Chapter 7). On the other hand, the resulting explicit extrapolation algorithm can be parallelized using domain decomposition in a very efficient way, and does not need to rely on linear algebra libraries.

Other flavours, like the finite-element discontinuous Galerkin method, have recently been introduced to seismic wave propagation, provoking a lot of interest, in particular for dynamic rupture problems and wave propagation through media with highly complex geometrical features. Because of the importance of these developments in current research, each of these methods receives its own chapter here.

Chapter summary

- The finite-element method was originally developed for static structural engineering problems.
- The *element* concept relates to describing the solution field in an analogous way inside each element, thereby facilitating the required calculations of the system matrices.
- The finite-element approach can in principle be applied to elements of arbitrary shape. Most used shapes are triangles (tetrahedral geometries) or quadrilaterals (hexahedral geometries).
- The finite-element method is a series expansion method. The continuous solution field is replaced by a finite sum over (not necessarily orthogonal) basis functions.
- For static elastic problems or the elastic wave-propagation problem, finite-element analysis leads to a (large) system of linear equations. In general, the matrices are of size $N \times N$ where N is the number of degrees of freedom.
- Because of the specific interpolation properties of the basis functions, their coefficients take the meaning of the values of the solution field at specific node points.

- In an initialization step, the global stiffness and mass matrices have to be calculated. They depend on integrals over products of basis functions and their derivatives.

- If equation parameters (e.g. elastic parameters, density) vary inside elements, numerical integration has to be performed.

- The stress-free surface condition is implicitly solved. This is a major advantage compared to other methods (e.g. the finite-difference method), particularly in the presence of surface topography.

- The classic finite-element method plays a minor role in seismology compared with its higher-order extension. The spectral-element method leads to a fully explicit scheme, which is easier to implement on parallel hardware.

FURTHER READING

- Zienkiewicz et al. (2013) provides an excellent introduction to the basics of finite-element analysis, with an engineering focus.

- Durran (1999) contains a compact introduction to the finite-element method. He discusses several other numerical methods for wave-propagation problems in a comparative way.

- Strang (1988) and later editions is another classic book on finite-element analysis with very clear descriptions of simple 1D problems.

- A nice introduction to high-performance computing with examples in linear algebra (solution of large linear systems) is given in the recent book by Eijkhout (2015) (freely available as a PDF).

EXERCISES

Comprehension questions

(6.1) In which community was the finite-element method primarily developed? Give some typical problems.

(6.2) What are weak and strong forms of partial differential equations? Give examples.

(6.3) Discuss the pros and cons of the finite-element method vs. low-order finite-difference methods.

(6.4) Present and discuss problem classes that can be handled well with the finite-element method. Compare with problems better handled with other methods.

(6.5)   Compare the spatial discretization strategies of finite-element and finite-difference methods.

(6.6)   Describe the derivation strategy of finite-element shape functions.

(6.7)   Discuss qualitatively (use sketches) the use of basis functions. Compare with the interpolation properties of the pseudospectral method.

(6.8)   Is the finite-element method a global or a local scheme? Explain.

(6.9)   Why does the finite-element method require the solution of a (possibly huge) system of linear equations? What is the consequence for parallel computing?

(6.10) Why is the classic linear finite-element method not used so much for seismological research today?

(6.11) Explain the benefits of the finite-element method with respect to Earth models with complex geometries.

Theoretical problems

(6.12) The advection equation is

$$\partial_t q(x, t) + c(x)\partial_x q(x, t) = 0,$$

where $q(x, t)$ is the scalar quantity to be advected and $c(x)$ is the advection velocity. Write down the weak form of this equation and perform integration by parts. What happens to the anti-derivative? Does it cancel out at the boundaries as in the 1D elastic wave equation? Note: This is the point of departure for the discontinuous Galerkin method.

(6.13) Are the linear basis functions

$$\varphi_i(x) = \begin{cases} \frac{x-x_{i-1}}{x_i-x_{i-1}} & \text{for } x_{i-1} < x \le x_i \\ \frac{x_{i+1}-x}{x_{i+1}-x_i} & \text{for } x_i < x < x_{i+1} \\ 0 & \text{elsewhere} \end{cases}$$

orthogonal?

(6.14) Derive the formulae for the calculation of the mass matrix elements (Eq. 6.54 and Eq. 6.55) by sketching the integration interval, and the corresponding basis functions.

(6.15) Calculate all entries of the stiffness matrix $K_{ij} = \int_D \mu \partial_x \varphi_i \partial_x \varphi_j$ for a static elastic problem with $\mu = 70$ GPa and $h = 1$ m for a problem with $n = 5$ degrees of freedom.

(6.16) A finite-element system has the following parameters: Element sizes $h = [1, 3, 0.5, 2, 4]$, density $\rho = [2, 3, 2, 3, 2]$ kg/m^3. Calculate the entries of the mass matrix given by $M_{ij} = \int_D \rho \, \varphi_i \, \varphi_j \, dx$ using linear basis functions.

(6.17) *h*-adaptivity. For the simulation with varying velocities and element size with parameters given in Table 6.3, calculate the time step required for $\epsilon = 0.5$ in each of the subdomains. Remember that the stability criterion is $\epsilon = c_{max}dt/dx$ where c_{max} is the maximum velocity in the entire physical domain. Discuss the result.

(6.18) Follow the approach of the derivation of shape functions and derive the cubic case in 1D: $u(x) = c_1 + c_2\xi + c_3\xi^2 + c_4\xi^3$. What are key differences compared to quadratic and linear cases?

(6.19) Derive the quadratic shape functions $N(\xi, \eta)$ for 2D triangles with the following node points:

$$P_1(0,0), P_2(1,0), P_3(0,1),$$
$$P_4(1/2,0), P_5(1/2,1/2), P_6(0,1/2).$$

Note: Use Python (or another program) to solve the linear system of equations.

(6.20) Derive the quadratic shape functions $N(\xi, \eta)$ for 2D rectangles with the following node points:

$$P_1(0,0), P_2(1/2,0), P_3(1,0), P_4(1,1/2),$$
$$P_5(1,1), P_6(1/2,1), P_7(0,1), P_8(0,1/2).$$

(6.21) Derive the derivative matrix $\mathbf{D}$ for the finite-difference-based second derivative (Eq. 6.61). Show that when applied to a vector $\mathbf{u}$ that contains an appropriate function (e.g. a Gaussian, sin function) you obtain an approximation of its derivative.

Programming exercises

(6.22) Write a computer program that solves the 1D static elasticity problem (Eq. 6.19) using finite elements. Also code the finite-difference-based relaxation problem (Eq. 6.33) and compare the results. Reproduce Fig. 6.7. Extend the formulation to arbitrary element sizes.

(6.23) Code the 1D elastic wave equation using finite elements (Eq. 6.47). Determine numerically the stability limit and compare with the finite-difference solution. Implement the analytical solution for the homogeneous case (note: it is the same as the 1D acoustic wave equation). Compare the numerical dispersion behaviour of the finite-element method with the corresponding low- (or high-) order finite-difference method.

(6.24) Initialize a strongly heterogeneous velocity model with spatially varying element size. Try to match the results with a regular-grid finite-difference implementation of the same model. Discuss the two approaches in terms of time step, run time, and memory usage.

(6.25) Plot the high-order 2D shape functions derived in the theoretical problems above.

(6.26) Derive a finite-difference-based centred differentiation matrix for the first derivative. Implement the 1D elastic wave equation in matrix form. Compare with the 1D finite-element implementation.

7 The Spectral-Element Method

The spectral-element method[1] is currently one of the most widely used numerical approaches for seismic wave-propagation problems. Let us briefly compare it with the other methods discussed so far and outline why this might be the case.

The finite-difference method suffers in particular from the difficulties in accurately implementing free-surface boundary conditions in the case of realistic topography. The elegant pseudospectral method exploited for the first time the spectral convergence of function interpolation (and derivative) for specific choices of basis functions. This will be a central ingredient of the spectral-element method, but at a local elemental level. The global communication requirements of the pseudospectral method prevent good scaling on parallel hardware, and it is thus unattractive for 3D problems. Furthermore, adaptation to models with complex geometry is difficult.

The classic low-order finite-element approach only fixes part of the problem. A major advantage it has is the fact that the free-surface boundary condition comes for free as it is implicitly solved. In addition, unstructured tetrahedral or hexahedral grids are possible, allowing geometrically complex model features. However, a large linear system of equations has to be solved, and this is cumbersome to implement efficiently on parallel hardware.

So what makes the spectral-element method so powerful? By using a specific set of basis functions inside the elements—Lagrange polynomials—combined with an interpolation scheme based upon the Gauss–Lobatto–Legendre (GLL) collocation points, the mass matrix that needs to be inverted in the finite-element formulation becomes diagonal (note that this only works with rectangular grids in 2D or hexahedral grids in 3D). This implies that the scheme can be explicitly extrapolated just like finite-difference or pseudospectral implementations, without the need to solve a large linear system of equations. In combination with the interpolation properties of Lagrange polynomials, this makes the algorithm extremely efficient and lends itself to implementation on parallel hardware. As is the case with other finite-element-type schemes, the geometrical flexibility comes with the requirement to generate a computational mesh that is numerically stable (see Fig. 7.1).

Following a brief section on its history, the focus will be on illustrating the various ingredients that make up the power of the spectral-element method.

[1] The term *spectral* should not give the impression that the method is based on solutions of the wave equation in the spectral domain. This is not the case. As with all other previous methods, the solutions are sought in the space–time domain.

7.1 History

The spectral-element method was born out of ideas developed within the framework of pseudospectral methods, which used the concepts of exact interpolation on collocation points with spectral convergence properties.[2] The similarity with the concept of basis functions in the classic finite-element method is obvious and the idea was to take advantage of the exponential convergence properties of the spectral basis functions. The first appearances of the combined approach were in Patera (1984) and Maday and Patera (1989) in fluid dynamics. These authors were also the first to use the term 'spectral elements'.

Spectral-element formulations for elastic wave problems were first published by Priolo et al. (1994), Seriani and Priolo (1994), and Faccioli et al. (1996). In these early applications Chebyshev polynomials were used to approximate the unknown fields. This improved the dispersion properties compared to classic finite-element methods. However, the necessity to invert a large linear system of equations still inhibited wide use of the approach. Building on the work by Maday and Patera (1989), the breakthrough in seismology came with the work of Komatitsch and Vilotte (1998), which introduced the combination of Lagrange polynomials as interpolants and an integration scheme based on Gauss quadrature defined on the GLL points for the elastic wave equation. This led to a diagonal mass matrix that can be trivially inverted. As a consequence, a fully explicit scheme is possible that is easy to parallelize and has the desired properties described in the previous section.

A further milestone was the implementation of the spectral-element method for global wave propagation by Chaljub (2000) and Chaljub et al. (2003) using the cubed-sphere concept by Ronchi et al. (1996). This allowed for the first time the simulation of the complete wavefield in a 3D *heterogeneous* spherical Earth and triggered the development of the *specfem3d* community code (Komatitsch and Tromp, 2002*a*; Komatitsch and Tromp, 2002*b*; see Fig. 7.2), which is, at the time of writing, undoubtedly one of the best-engineered openly accessible wave-propagation simulation tools. Fichtner and Igel (2008) implemented a spectral-element method for wave propagation in spherical coordinates (3D spherical sections) that led to the first ever application of the adjoint inversion method to regional earthquake data. A recent review paper by Peter et al. (2011) demonstrates the great flexibility of the spectral-element method for both modelling and inversion.

The geometrical flexibility of this approach of course comes with the necessity to prepare a potentially complex computational mesh.[3]

As indicated, the advantages of the spectral-element approach concerning the diagonal mass matrix are restricted to hexahedral elements in 3D. This implies that the preparation of a mesh from given geophysical data such as the free-surface topography, and the curved, discontinuous internal boundaries, can be a task that takes weeks to months of hard work, and expertise in the use of meshing software (e.g. CUBIT, ANSYS) is required. As indicated in the introduction,

Fig. 7.1 *A spectral-element mesh to model soil–structure interactions with a hexahedral grid implementation. Note the variations in element size and the deformation of the hexahedral elements with curved boundaries. Figure courtesy of M. Stuppazzini.*

[2] This is related to the question of how fast an approximation converges to the exact function. In (pseudo-) spectral methods the rate of convergence has an exponential form, provided the function is sufficiently smooth.

[3] Several other spectral-element developments are discussed in Chapter 10 on applications. Freely available spectral-element codes are listed in the Appendix.

Fig. 7.2 *Specfem3d: Logo of the well-known community code with a snapshot of global wave propagation for a simulation of the devastating M9.1 earthquake near Sumatra in December 2004. The figure was used as the title page of Science on 20 May 2005. Reprinted with permission.*

the efficient meshing of specific Earth models for simulation purposes both for hexahedral and tetrahedral meshes is still an open issue.

In the following section we look at some of the basic concepts of spectral elements. Subsequently, we develop the complete mathematical formulation for the solution of the 1D elastic wave equation.

7.2 Spectral elements in a nutshell

Let us set the stage for assembling the various ingredients of the spectral-element method applied to the wave-propagation problem. Again, we start with the classic 1D elastic wave equation:

$$\rho \partial_t^2 u = \partial_x \left(\mu \partial_x u \right) + f, \tag{7.1}$$

in which the displacement u, external force f, mass density ρ, and shear modulus μ depend on x, and u and f also on t. In the following these dependencies are implicitly assumed. An important boundary property that has to obeyed is the stress-free condition that occurs at the Earth's surface. This condition is expressed as the vanishing of the traction perpendicular to the free surface, thus its normal vector n_j

$$\sigma_{ij} \, n_j = 0, \tag{7.2}$$

where σ_{ij} is the symmetric stress tensor. Using the stress–strain relation and adapting to our elastic 1D problem, we obtain

$$\mu \, \partial_x \, u(x, t) \Big|_{x=0,L} = 0, \tag{7.3}$$

where our spatial boundaries are at $x = 0, L$ and the stress-free condition applies at both ends (other conditions like absorbing boundaries are also possible).

Before we dive into the details of numerical analysis, let us illustrate some basic concepts graphically. In Fig. 7.3 a snapshot of a 1D displacement wavefield simulated with the spectral-element method is shown for a medium with a random distribution of elastic parameters. The top figure is a zoomed-in view of one of the elements, the central concept of finite-element-type techniques. Inside each element we approximate the unknown function u by a sum over a set of basis functions (thin solid lines). In this case the unknown function is approximated by a sum over Lagrange polynomials of a specific order. The order determines the number of points inside the elements (black squares in Fig. 7.3, top) at which the solution is *exactly* interpolated. This could in principle be achieved for a set of regularly spaced points. However, for various reasons that will become apparent, it is preferable to use a specific set of collocation points known as the GLL points.

As is obvious from Fig. 7.3, these points are unevenly spaced. When higher-order polynomials are used, the grid points densify even more towards the element boundaries. This implies that we have to decrease time steps, while keeping everything else the same. As presented in Chapter 6 on the finite-element method, we also need to integrate basis functions, their derivatives, and elastic parameters over the elements. Thus we need to come up with a numerical integration scheme, as an analytical integration is not in general possible. The method of choice is a special case of the Gauss quadrature approach, the Gauss–Lobatto–Legendre quadrature, that consists of a sum over the function to be integrated, evaluated at the GLL points appropriately weighted. These points are equivalent to those illustrated in Fig. 7.3, top. It is precisely this fact that leads to a global system of matrix equations that can be solved with high efficiency as no global system matrix inversion is required.

Thus, in the following sections, we (1) formulate the weak form of the wave equation, and (2) provide the transformation of the equation down to the elemental level, introducing the concept of the Jacobian. The discretization of our system comes with (3) the approximation of our unknown function u using Lagrange polynomials as interpolants. The formulation also requires (4) the evaluation of the first derivatives of the Lagrange polynomials, the calculation of which requires the Legendre polynomials. (5) The numerical integration scheme based on GLL quadrature allows us to calculate all system matrices at elemental level, which are then (6) assembled in a final step to obtain the global system of equations that is extrapolated in time using a simple finite-difference scheme.

7.3 Weak form of the elastic equation

Despite the equivalence to the finite-element method, in order to keep the chapter independent we recall briefly the weak form as a starting point for the specific spectral-element discretization.

We multiply both sides of Eq. (7.1) by a time-independent test function $v(x)$. This may be any function of the set of functions that are, together with their first derivative, square integrable[4] over the integration domain D (i.e. a continuous and 'well-behaved' function). D is here the complete computational domain defined with $x \in D = [0, L]$.

$$\int_D v \, \rho \, \partial_t^2 u \, dx - \int_D v \, \partial_x \left(\mu \, \partial_x u \right) \, dx = \int_D v f \, dx. \tag{7.4}$$

We integrate the second term of the left-hand side of the above equation by parts, to obtain

$$\int_D v \, \rho \, \partial_t^2 u \, dx + \int_D \mu \, \partial_x v \, \partial_x u \, dx = \int_D v f \, dx, \tag{7.5}$$

where we made use of the boundary condition

$$\partial_x u(x, t) \big|_{x=0} = \partial_x u(x, t) \big|_{x=L} = 0. \tag{7.6}$$

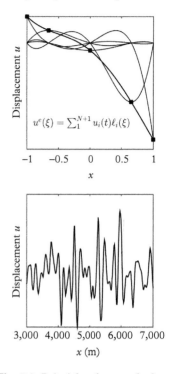

$$u^e(\xi) = \sum_1^{N+1} u_i(t) \ell_i(\xi)$$

Fig. 7.3 *Principle of spectral element discretization.* **Bottom:** *Snapshot of the displacement field u during a simulation in a strongly heterogeneous 1D medium.* **Top:** *Close-up of the displacement field inside one element discretized with order N = 4 collocation points, at which the solution is exactly interpolated using Lagrange polynomials (ℓ_i, grey lines). These points are also used for the numerical integration scheme. The equation describes the interpolation scheme using Lagrange polynomials.*

---

[4] If you take the integral of the square of the absolute values of a *square-integrable* function, it is finite.

Comparing equations (7.6) and (7.3) we note the equivalence of these conditions. This implies that the physical free-surface boundary condition, as in the classic finite-element approach, is implicitly fulfilled; an extremely attractive feature for 3D problems involving the Earth's surface! We are now left with the problem of finding solutions of the displacement field u for arbitrary space-dependent test functions v.

At this point we are still in the continuous world in which solutions are sought by analytical means. As we seek to simulate wave propagation in Earth models with heterogeneous distributions of elastic parameters we need to find appropriate discrete representations of the seismic wavefield u with which we can find solutions by numerical means. This can be achieved by the Galerkin method. Here, we approximate the exact solution for $u(x, t)$ by a finite superposition of n basis functions $\varphi_i(x)$ with $i = 1, \ldots, N_p$ weighted by time-dependent coefficients $u_i(t)$.[5] We expect the test functions and the solution to be continuous. Note that we do not specify the basis functions here and we are still considering the complete spatial domain D (this will be changed later when we go down to the element level). The approximate displacement field is denoted by $\overline{u}(x, t)$:

$$u(x, t) \approx \overline{u}(x, t) = \sum_{i=1}^{N_p} u_i(t) \, \varphi_i(x). \tag{7.7}$$

We expect that the accuracy of this approximation will depend on the specific choice of basis function and the number of functions superimposed (i.e. N_p). In the following we restrict ourselves to the problem of finding solutions to Eq. (7.5) for our approximate displacement field $\overline{u}(x, t)$. In addition, we make another important step by using as test functions the same functions that are used to approximate our unknown fields (Galerkin principle), obtaining

$$\int_D \varphi_i \, \rho \, \partial_t^2 \overline{u} \, dx \, + \int_D \mu \, \partial_x \varphi_i \, \partial_x \overline{u} \, dx = \int_D \varphi_i \, f \, dx, \tag{7.8}$$

with the requirement that the medium is at rest at $t = 0$. Combining Eqs. 7.7 and 7.8 leads to an equation for the unknown coefficients $u_i(t)$

$$\sum_{i=1}^{N_p} \left[\partial_t^2 u_i(t) \int_D \rho(x) \, \varphi_j(x) \, \varphi_i(x) \, dx \right]$$

$$+ \sum_{i=1}^{N_p} \left[u_i(t) \int_D \mu(x) \, \partial_x \varphi_j(x) \, \partial_x \varphi_i(x) \, dx \right] \tag{7.9}$$

$$= \int_D \varphi_i \, f(x, t) \, dx$$

for all basis functions φ_j with $j = 1, \ldots, n$. This is the well-known equation for finite-element problems, which can be written in matrix notation:

$$\mathbf{M} \partial_t^2 \mathbf{u}(t) \, + \mathbf{K} \mathbf{u}(t) = \mathbf{f}(t), \tag{7.10}$$

[5] The use of the letter u_i for the expansion coefficients might be surprising here. N_p is the number of required basis functions for a specific polynomial order N. However, it will turn out that because of the choice of the basis functions these coefficients will actually correspond to the discrete values of the displacement field at grid points at element boundaries and/or inside the elements.

with implicit matrix–vector operations. The mass matrix—here defined over the entire domain—is

$$M_{ji} = \int_D \rho(x)\, \varphi_j(x)\, \varphi_i(x)\, dx, \tag{7.11}$$

the stiffness matrix

$$K_{ji} = \int_D \mu(x)\, \partial_x\varphi_j(x)\, \partial_x\varphi_i(x)\, dx, \tag{7.12}$$

and the vector containing the volumetric forces $f(x, t)$

$$f_j(t) = \int_D \varphi_i\, f(x, t)\, dx. \tag{7.13}$$

This simple matrix equation has to be solved for the space-independent but time-dependent coefficients **u**. The vector of coefficients will take the meaning of the actual displacement values at a global set of points imposed by the specific basis functions to be introduced shortly (and must not be confused with the classic three-component displacement vector in the 3D elastic wave equation). This system of equations is illustrated graphically in Fig. 7.4, with a hint towards the final solution structure. The mass matrix is diagonal, thus its inversion is trivial. The stiffness matrix has a banded structure in this case with the bandwidth depending on the number of basis functions that are required inside each element.

A simple centred finite-difference approximation of the second derivative in Eq. 7.10 and the following mapping

$$\mathbf{u}^{new} \rightarrow \mathbf{u}(t + dt)$$
$$\mathbf{u} \rightarrow \mathbf{u}(t) \tag{7.14}$$
$$\mathbf{u}^{old} \rightarrow \mathbf{u}(t - dt)$$

leads us to the solution for the coefficient vector $\mathbf{u}(t + dt)$ for the next time step as already well known from the other solution schemes in previous chapters:

$$\mathbf{u}^{new} = dt^2 \left[\mathbf{M}^{-1} (\mathbf{f} - \mathbf{K}\ \mathbf{u}) \right] + 2\mathbf{u} - \mathbf{u}^{old} \tag{7.15}$$

It is important to note that so far nothing differs from the classic finite-element approach. The problems that remain to be solved are finding appropriate basis

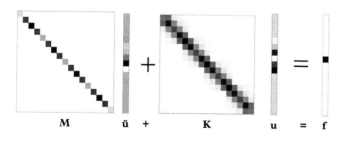

$$\mathbf{M} \qquad \ddot{\mathbf{u}} \quad + \qquad \mathbf{K} \qquad \mathbf{u} \quad = \quad \mathbf{f}$$

Fig. 7.4 *Symbolic and mathematical representation of the global system that has to be solved. The unknown acceleration $\partial_t^2 \mathbf{u}$ is found by a simple temporal finite-difference approximation. The solution requires the inversion of mass matrix* **M** *which is trivial because it is diagonal. This is the key feature of the spectral-element method.*

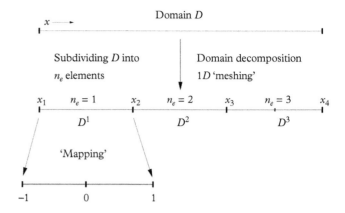

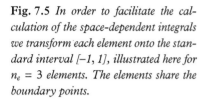

Fig. 7.5 *In order to facilitate the calculation of the space-dependent integrals we transform each element onto the standard interval [−1, 1], illustrated here for n_e = 3 elements. The elements share the boundary points.*

functions and integration schemes to efficiently calculate mass matrix, stiffness matrix, and forces.

7.4 Getting down to the element level

The solution of the elastic wave equation given in Eq. 7.9 is of a global nature, that is, u represents the complete physical domain. One level of discretization came through the approximation of the unknown field u by a finite sum over some basis functions φ_i. However, a further level of discretization is required to facilitate the final solution. We divide the domain D into subdomains D_e (which do not need to be of the same size). This is illustrated in Fig. 7.5 for an example of a 1D spatial domain D divided into n_e = 3 subdomains (the elements). This allows the introduction of discontinuities in material parameters, leading to discontinuity of the displacement gradient ∇u. This step, without further specifying the basis functions φ_i, leads to

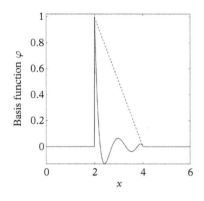

Fig. 7.6 *Illustration of local basis functions. By defining basis functions only inside elements, the integrals can be evaluated in a local coordinate system. The graph assumes three elements n_e = 3 with equal size h = 2. A finite-element-type linear-basis function (dashed line) is shown alongside a spectral-element-type Lagrange polynomial-basis function of degree N = 5 (solid line). Compare with the linear local-basis functions introduced in Chapter 6 on the finite-element method (Fig. 6.4).*

$$
\begin{aligned}
&\sum_{i=1}^{N_p}\left[\partial_t^2 u_i(t)\sum_{e=1}^{n_e}\int_{D_e}\rho(x)\varphi_j(x)\varphi_i(x)\,dx\right] \\
&+\sum_{i=1}^{N_p}\left[u_i(t)\sum_{e=1}^{n_e}\int_{D_e}\mu(x)\partial_x\varphi_j(x)\partial_x\varphi_i(x)\,dx\right] \\
&=\sum_{e=1}^{n_e}\int_{D_e}\varphi_j(x)f(x,t)\,dx,
\end{aligned}
\tag{7.16}
$$

representing a linear system of N_p equations for each j.

It is important to note that the coefficients u_i depend on a sum over all elements. To avoid this global dependence we introduce (as was the case in the classic finite-element method) basis functions that are only defined on the subdomains D_e. This is illustrated in Fig. 7.6. Here a linear-basis function is compared with a spectral-element-type non-linear-basis function. Mathematically this locality has important beneficial consequences. Instead of defining basis functions in D (as is the case in pseudospectral methods) we now restrict them to reside inside the elements D_e. We thus end up with the approximation

$$\bar{u}(x, t)\Big|_{x \in D_e} = \sum_{i=1}^{N_p} u_i^e(t) \varphi_i^e(x), \tag{7.17}$$

where N_p denotes the number of basis functions for polynomial order N to be summed up.[6]

As a consequence, the integrals are now local to one specific element D_e and to obtain the solution inside we need only to sum over all basis functions with the appropriate coefficients:

$$\sum_{i=1}^{N_p} \partial_t^2 u_i^e(t) \int_{D_e} \rho(x) \varphi_j^e(x) \varphi_i^e(x) \, dx$$

$$+ \sum_{i=1}^{N_p} u_i^e(t) \int_{D_e} \mu(x) \partial_x \varphi_j^e(x) \partial_x \varphi_i^e(x) \, dx \tag{7.18}$$

$$= \int_{D_e} \varphi_j^e(x) f(x, t) \, dx.$$

As was the case for the global system in Eq. 7.9, we can use matrix notation to obtain

$$\mathbf{M}^e \partial_t^2 \mathbf{u}(t) + \mathbf{K}^e \mathbf{u}^e(t) = \mathbf{f}^e(t), \qquad e = 1, \dots, n_e. \tag{7.19}$$

Here $\mathbf{u}^e$, $\mathbf{K}^e$, $\mathbf{M}^e$, and $\mathbf{f}^e$ are (1) the coefficients of the unknown displacement inside the element, (2) stiffness and (3) mass matrices with information on the density and elastic parameters, and (4) the forces, respectively, and n_e is the number of elements. Matrix–vector multiplications are implicit.[7]

To facilitate the mathematical operations under the integrals it is useful to map the spatial coordinates of an element to a reference interval. In principle this can be any interval but for our specific choice of basis function we use the interval $[-1, 1]$ as indicated in Fig. 7.5. If we want to integrate an arbitrary function in our reference interval $[-1, 1]$ we have to apply a coordinate transformation from our global system $x \in D$ to our local coordinates which we denote $\xi \in F_e$. This transformation can be written as:

$$F_e \colon [-1, 1] \rightarrow D_e, \qquad x = F_e(\xi),$$
$$\xi = \xi(x) = F_e^{-1}(x), \qquad e = 1, \dots, n_e, \tag{7.20}$$

where n_e is the number of elements, and $\xi \in [-1, 1]$. Thus the physical coordinate x can be related to the local coordinate ξ via

---

[6] It turns out that for the specific basis functions we use order N will require $N_p = N + 1$ functions to be summed.

[7] The sizes of the elemental vectors and matrices in this system are:

$$\mathbf{u}^e \rightarrow N_p$$
$$\mathbf{K}^e \rightarrow N_p \times N_p$$
$$\mathbf{M}^e \rightarrow N_p \times N_p$$
$$\mathbf{f}^e \rightarrow N_p,$$

where N_p is the number of basis functions inside the elements.

$$x(\xi) = F_e(\xi) = h_e \frac{(\xi + 1)}{2} + x_e, \qquad (7.21)$$

where x_e is the coordinate of the left side of the element (see Fig. 7.5) and h_e is the element size. The length h_e may vary for each element, allowing the adaptation of computational meshes. The inverse mapping is given as

$$\xi(x) = 2\frac{(x - x_e)}{h_e} - 1. \qquad (7.22)$$

From Eq. 7.18 we expect to have to solve integrals of products of basis functions, their derivatives, and elastic parameters. A coordinate change $x \to \xi$ leads to

$$\int_{D_e} f(x)\,dx = \int_{-1}^{1} f(\xi)\frac{dx}{d\xi}\,d\xi, \qquad (7.23)$$

where the integrand has to be multiplied by the Jabobian[8] $\mathcal{J}$ defined as

$$\mathcal{J} = \frac{dx}{d\xi} = \frac{h_e}{2}. \qquad (7.24)$$

The inverse Jacobian is also required, when derivatives of the basis functions need to be integrated, thus

$$\mathcal{J}^{-1} = \frac{d\xi}{dx} = \frac{2}{h_e}. \qquad (7.25)$$

Finally, we can assemble our system of equations inside each element as

$$\sum_{i=1}^{N_p} \partial_t^2 u_i^e(t) \int_{-1}^{1} \rho\,[x(\xi)]\,\varphi_j^e\,[x(\xi)]\,\varphi_i^e\,[x(\xi)]\,\frac{dx}{d\xi}\,d\xi$$

$$+ \sum_{i=1}^{N_p} u_i^e(t) \int_{-1}^{1} \mu\,[x(\xi)]\,\partial_\xi \varphi_j^e\,[x(\xi)]\,\partial_\xi \varphi_i^e\,[x(\xi)] \left(\frac{d\xi}{dx}\right)^2 \frac{dx}{d\xi}\,d\xi \qquad (7.26)$$

$$= \int_{-1}^{1} \varphi_j^e\,[x(\xi)]\,f\,[(x(\xi)), t]\,\frac{dx}{d\xi}\,d\xi.$$

Note that this is a system of N_p equations for each index j corresponding to one particular basis function for which the wave equation needs to hold. Equation 7.26 is a *semi-discrete* weak form of the elastic wave equation for one element only, as we have not yet discretized the time axis. What remains to be done is to find a choice of basis functions φ_i and a numerical integration scheme such that the calculation of the integrals, and the assembly and solution of the final global system, become as efficient and accurate as possible.

[8] Germany-born Carl Jacobi (1804–1851) was considered a mathematical *wunderkind* and one of the greatest mathematicians of the nineteenth century; he is mostly known for his contributions to the theory of elliptic functions.

7.4.1 Interpolation with Lagrange polynomials

It is about time we disclosed what basis functions we will be choosing to approximate (i.e. 'interpolate') our unknown displacement field and why. Remember we seek to approximate $u(x, t)$ by a sum over space-dependent-basis functions φ_i weighted by time-dependent coefficients $u_i(t)$:

$$u(x, t) \approx \overline{u}(x, t) = \sum_{i=1}^{N_p} u_i(t) \, \varphi_i(x). \qquad (7.27)$$

As interpolating functions we finally choose the Lagrange polynomials[9] and use ξ as the space variable representing our elemental domain:

$$\varphi_i \rightarrow \ell_i^{(N)}(\xi) := \prod_{j \neq i}^{N+1} \frac{\xi - \xi_j}{\xi_i - \xi_j}, \qquad i, j = 1, 2, \ldots, N + 1, \qquad (7.28)$$

where ξ_i are (in general arbitrary, separated) fixed points in the interval $[-1, 1]$. Let us look at this definition in more detail. Writing the sum explicitly we obtain

$$\ell_i^{(N)}(\xi) = \frac{\xi - \xi_1}{\xi_i - \xi_1} \frac{\xi - \xi_2}{\xi_i - \xi_2} \cdots \frac{\xi - \xi_N}{\xi_i - \xi_N} \frac{\xi - \xi_{N+1}}{\xi_i - \xi_{N+1}}, \qquad (7.29)$$

which is well defined as long as $k \neq i$ since only separate points are allowed. We further observe that for specific points ξ_j

$$\ell_{i \neq j}^{(N)}(\xi_j) = \frac{\xi_j - \xi_1}{\xi_i - \xi_1} \cdots \frac{\xi_j - \xi_j}{\xi_i - \xi_j} \cdots \frac{\xi_j - \xi_{N+1}}{\xi_i - \xi_{N+1}} = 0 \qquad (7.30)$$

and

$$\ell_i^{(N)}(\xi_i) = \prod_{j \neq i}^{N+1} \frac{\xi_i - \xi_j}{\xi_i - \xi_j} = 1. \qquad (7.31)$$

We have just demonstrated the orthogonality of the Lagrange polynomials, which can be expressed as

$$\ell_i^{(N)}(\xi_j) = \delta_{ij}, \qquad (7.32)$$

where δ_{ij} is the Kronecker symbol, which is 1 if $i = j$ and 0 otherwise.

If you cannot picture the Lagrange polynomials by looking at the above equations you are probably not alone. But before plotting them, we need to resolve another issue. Which points in the ξ_i interval $[-1, 1]$ should we choose for our simulation scheme? We use the so-called *Gauss–Lobatto–Legendre*[10] (GLL) points (see Fig. 7.7), a choice for which there are several reasons.

[9] The first appearance of the spectral-element method was with Chebyshev polynomials. However, this choice of basis function does not lead to the diagonal mass matrix that makes inversion so efficient!

[10] Did Gauss, Lobatto, and Legendre ever have a drink together? Probably not, but Mozart and Haydn did get together regularly for jam sessions at around that time in Vienna. Gauss (1777–1855) lived in Göttingen, Legendre (1752–1833) in Paris, and Lobatto (1797–1866) in Amsterdam.

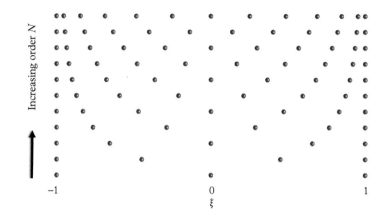

Fig. 7.7 *Gauss–Lobatto–Legendre po-ints. Illustration of their spatial distri-bution in the interval [–1, 1] for polyno-mial order N = 2 to N = 12 (from bottom to top), corresponding to $N_p = N + 1$ collocation points. The distribution of points is symmetric around the origin. Note the decreasing distance between collocation points towards the element boundaries (compare with Chebyshev polynomials).*

First of all, with this set of points the basis functions are defined such that

$$\ell_i^{(N)}(\xi_i) = 1 \quad \text{and} \quad \partial_\xi \ell_i^{(N)}(\xi_i) = 0. \tag{7.33}$$

This implies

$$|\ell_i^{(N)}(\xi)| \le 1, \qquad \xi \in [-1, 1], \tag{7.34}$$

minimizing the interpolation error in between the collocation points due to nu-merical inaccuracies. The densification of points towards the boundaries avoids overshooting of the interpolated function near the boundaries (similar to the Gibbs phenomenon discussed in Chapter 5 on the pseudospectral method). An-other important aspect is that an integration scheme exists (with the same name) that uses precisely this point set, leading to a diagonal mass matrix.

Fig. 7.8 illustrates $N + 1$ Lagrange polynomials of degree $N = 2$ and 6. The order N is equivalent to the number of intervals inside each element. This implies that with $N = 2$ we would obtain a spectral-element discretization with even spacing of collocation points. Note also that with $N = 1$ we exactly recover the classic linear finite-element scheme! As the order increases, the difference between the distance of collocation points increases in a linear way (see Fig. 7.7). The GLL points are the roots of the first derivative of the Legendre polynomials L_N of degree N (definition below).

Let us look at an example of function interpolation using Lagrange polynomi-als. In Fig 7.9 we approximate a known function (here a sum over sine functions) by Lagrange polynomials of various orders. According to

$$u^e(\xi) = \sum_{i=1}^{N+1} u^e(\xi_i)\ell_i(\xi) \tag{7.35}$$

the function approximation is given as a sum over $N + 1$ polynomials weighted with the values of the function at the collocation points ξ_i. The superscript (N) of the Lagrange polynomials is omitted from now on as the order is indicated by the summation limits. This equation provides the reason why it makes sense to call the coefficients u_i as they correspond exactly to the continuous function u at the collocation points. Note the decreasing misfit between the approximation and the original function for increasing order in Fig 7.9 (see also exercises). The accuracy obviously increases with order. However, it is important to note that this does not mean the highest possible orders should be used for the final algorithm. A strategy will be discussed when we assemble the complete spectral-element algorithm.

With the definition of Eq. 7.28 we are now able to express the general finite-element system Eq. 7.26 with our choice of basis function to obtain

$$\sum_{i=1}^{N+1} \partial_t^2 u_i^e(t) \int_{-1}^{1} \rho(\xi) \ell_j(\xi) \ell_i(\xi) \frac{dx}{d\xi} d\xi$$

$$+ \sum_{i=1}^{N+1} u_i^e(t) \int_{-1}^{1} \mu(\xi) \partial_\xi \ell_j(\xi) \partial_\xi \ell_i(\xi) \left(\frac{d\xi}{dx}\right)^2 \frac{dx}{d\xi} d\xi \qquad (7.36)$$

$$= \int_{-1}^{1} \ell_j(\xi) f(\xi, t) \frac{dx}{d\xi} d\xi.$$

Again, it is worth pausing for a moment and clarifying the structure of this system of equations. We have equations for each index $j = 1, \ldots, N + 1$ while the summation is over $i = 1, \ldots, N + 1$. Everything is known (the analytical Lagrange polynomials, the models of density and shear modulus); only the displacement and acceleration values u_i and $\partial_t^2 u_i$ are unknown.

To simplify notation we use the following mapping for density ρ, elastic constant μ, and forces f:

$$\rho(\xi) := \rho[x(\xi)], \qquad \mu(\xi) := \mu[x(\xi)], \qquad f(\xi) := f[x(\xi)]. \qquad (7.37)$$

In practice that means we initiate the density and shear moduli at each collocation point. Note that assuming constant density and shear moduli inside elements would simplify the problem here, since we could take them out of the integrals and solve the integrals analytically. However, the fact that we can allow the geophysical parameters to vary smoothly inside elements is an attractive feature for seismic wave-propagation problems. As a consequence, the integrals in Eq. 7.36 unfortunately cannot be evaluated analytically, because we want to keep this flexibility for strongly heterogeneous models. Therefore, we are forced to resort to numerical integration, the topic of the next section.

7.4.2   Numerical integration

As illustrated above we have to find an efficient way of solving the integrals over the element domain D_e given in Eq. 7.36. Like finding the derivatives of a function numerically, the approximation of integrals is a vast field of its own. A fundamental principle that is often applied is the concept of replacing the function to be

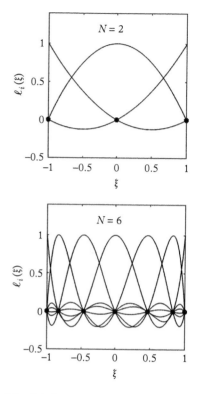

Fig. 7.8 *Lagrange Polynomials.* **Top:** *Family of $N + 1$ Lagrange polynomials for $N = 2$ defined in the interval $\xi \in [-1, 1]$. Note their maximum value over the whole interval does not exceed unity.* **Bottom:** *Same for $N = 6$. The domain is divided into N intervals of uneven length. When using Lagrange polynomials for function interpolation the values are exactly recovered at the Gauss–Lobatto–Legendre collocation points (squares).*

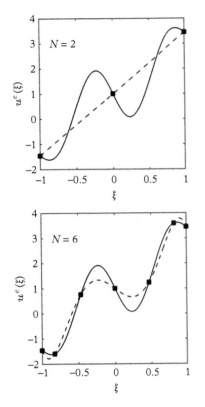

Fig. 7.9 *Interpolation with Lagrange Polynomials. The function to be approximated is given by the solid lines. The approximation is given by the dashed line exactly interpolating the function at the GLL points (squares). Top: Order N = 2 with three grid points. Bottom: Order N = 6 with seven grid pints.*

integrated $f(x)$ by a polynomial approximation that can be integrated analytically. A well-known example is the classic Gauss quadrature. It can be shown that a degree $2N + 1$ polynomial can be integrated exactly with only $N + 1$ collocation points. The problem is that the corresponding collocation points lie inside the element (and not at the boundaries). The requirement that the boundaries of the integrals are included leads to the so-called Gauss–Lobatto–Legendre quadrature (well, we heard these three names before) using the GLL points for integration.

As interpolating functions we use again the Lagrange polynomials and obtain the following integration scheme for an arbitrary function $f(x)$ defined in the interval $x \in [-1, 1]$:

$$\int_{-1}^{1} f(x)\,dx \approx \int_{-1}^{1} P_N(x)\,dx = \sum_{i=1}^{N+1} w_i f(x_i), \tag{7.38}$$

with

$$P_N(x) = \sum_{i=1}^{N+1} f(x_i)\ell_i^{(N)}(x), \tag{7.39}$$

and the integration weights are calculated with

$$w_i = \int_{-1}^{1} \ell_i^{(N)}(x)\,dx. \tag{7.40}$$

Examples of these integration weights are given in Table 7.1. Note the decreasing values towards the element boundaries compensating for the narrowing intervals. For the numerical integration the same principle applies as was the case for the interpolation scheme. The higher the order, the more accurate the integration is (see exercises). Another point is the specific form of the integrand $f(x)$. In principle, the smoother this function is, the more accurate the approximation; that is, the faster the numerical result converges to the exact solution as the polynomial order increases.

Let us take an example illustrating how numerical integration works. We initialize an arbitrary function, here a sum of sinusoidal functions

$$f(\xi) = \sum_{i=1}^{5} \sin\left(\frac{\pi}{a_i}\xi + a_i\right) \tag{7.41}$$

with $a = [0.5, 1, -3, -2, -5, 4]$, which can be easily integrated analytically. Using the GLL weights from Table 7.1 we integrate this function numerically for varying order N and compare with the analytical solution. Examples are shown in Fig. 7.10. This is a fairly tough problem for the interpolator, but a decent result is obtained with integration order $N \geq 6$. It is remarkable how different the approximation is compared to the original function, yet the numerical integral calculation is very accurate.

At this point we can draw an important conclusion: Within the realm of the polynomial representation of our displacement field, the density, elastic constants, and forces, the only error we are accumulating in the spatial domain is by the numerical integration scheme just discussed. The only other error in the complete spectral-element algorithm comes from the finite-difference approximation of the time derivatives. The use of the GLL integration scheme given in Eq. 7.38 to evaluate the integrals in Eq. 7.36 has important consequences. We now introduce another level of discretization by replacing the continuous integration over the elements by a sum over (again) $N + 1$ weighted functional values located at our well-known GLL points, the same locations at which we interpolate our unknown function u. With this integration scheme leading to an additional sum over k, we obtain at element level

$$\sum_{i,k=1}^{N+1} \partial_t^2 u_i^e(t) w_k \rho(\xi) \ell_j(\xi) \ell_i(\xi) \frac{dx}{d\xi}\bigg|_{\xi=\xi_k}$$

$$+ \sum_{i,k=1}^{N+1} w_k u_i^e(t) \mu(\xi) \partial_\xi \ell_j(\xi) \partial_\xi \ell_i(\xi) \left(\frac{d\xi}{dx}\right)^2 \frac{dx}{d\xi}\bigg|_{\xi=\xi_k} \quad (7.42)$$

$$\approx \sum_{k=1}^{N+1} w_k \ell_j(\xi) f(\xi, t) \frac{dx}{d\xi}\bigg|_{\xi=\xi_k}.$$

We can make use of the cardinal interpolation property of the Lagrange polynomial $\ell_i^{(N)}(x_j) = \delta_{ij}$ to arrive at the solution equation for our spectral-element system *at the element level* using matrix notation:

$$\sum_{i=1}^{N+1} M_{ji}^e \partial_t^2 u_i^e(t) + \sum_{i=1}^{N+1} K_{ji}^e u_i^e(t) = f_j^e(t), \quad e = 1, \ldots, n_e \quad (7.43)$$

$$M_{ji}^e = w_j \rho(\xi) \frac{dx}{d\xi} \delta_{ij}\bigg|_{\xi=\xi_j}$$

$$K_{ji}^e = \sum_{k=1}^{N+1} w_k \mu(\xi) \partial_\xi \ell_j(\xi) \partial_\xi \ell_i(\xi) \left(\frac{d\xi}{dx}\right)^2 \frac{dx}{d\xi}\bigg|_{\xi=\xi_k} \quad (7.44)$$

$$f_j^e = w_j f(\xi, t) \frac{dx}{d\xi}\bigg|_{\xi=\xi_j}.$$

To some extent the above three equations for mass matrix, stiffness matrix, and sources constitute the core of the preparatory computations that need to be done before starting the time extrapolation for the wavefield.[11] But wait a minute! There is one item that needs to be considered further. Note that to calculate the stiffness matrix we need to know the derivatives of the Lagrange polynomials evaluated at the GLL collocation points. As our intention is to provide you with all equations necessary to solve (at least) the 1D problem using spectral elements, we will deviate briefly to present how these derivatives are calculated.

Table 7.1 *Collocation points and integration weights of the GLL quadrature for order $N = 2, \ldots, 4$.*

N	ξ_i	ω_i
2:	0	4/3
	± 1	1/3
3:	$\pm\sqrt{1/5}$	5/6
	± 1	1/6
4:	0	32/45
	$\pm\sqrt{3/7}$	49/90
	± 1	1/10

[11] Note the δ_{ij} in the equation for the mass matrix M_{ji}^e, implying a diagonal structure. As a consequence, finding its inverse, required for the solution of the global system of equations, is trivial.

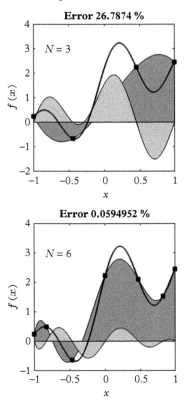

Fig. 7.10 *Gauss integration. The exact function (thick solid line) is approximated by a Lagrange polynomials (thin solid line) that can be integrated analytically. Thus, the integral of the true function (thick solid) is replaced by an integral over the polynomial function (dark grey). The difference between the true and approximate functions is given in light grey. Top: N = 3. Bottom: N = 6.*

7.4.3   Derivatives of Lagrange polynomials

From Eq. 7.26 it is clear that we also need the derivatives of our basis functions, that is, each Lagrange polynomial, as they are part of the integrands. Thus we seek (if possible) analytical solutions to $\partial_\xi \ell_i(\xi)$. Here we present a common scheme for its calculation using a recursive formula.

The GLL integration scheme implies that we do not need to know the derivatives everywhere in $[-1, 1]$ but only at the GLL collocation points. It turns out that these derivatives can be efficiently calculated using Legendre polynomials. They are also defined in the interval $\xi \in [-1, 1]$ and are given as:

$$L_N(\xi) = \frac{1}{2^N N!} \frac{d^N}{d\xi^N} \left(\xi^2 - 1 \right)^N, \tag{7.45}$$

where N denotes the polynomial degree. The Legendre polynomials can be calculated using the following recursive formula:

$$
\begin{aligned}
L_0(\xi) &= 1 \\
L_1(\xi) &= \xi \\
L_{n \geq 2}(\xi) &= \frac{1}{n}[(2n-1)\, \xi\, L_{n-1}(\xi) - (n-1)L_{n-2}(\xi)].
\end{aligned}
\tag{7.46}
$$

An illustration of the Legendre polynomials is given in Fig. 7.11. Following Funaro (1993), the derivatives of the Lagrange polynomials can be calculated using:

$$\partial_\xi \ell_k(\xi_i) = \sum_{j=0}^{N} d_{ij} \ell_k(\xi_j)\,, \quad k = 0, \ldots, N, \tag{7.47}$$

with

$$
d_{ij} = \begin{cases}
-\frac{1}{4} N(N+1) & \text{if } i = j = 0 \\[2mm]
\frac{L_N(\xi_i)}{L_N(\xi_j)} \frac{1}{\xi_i - \xi_j} & \text{if } 0 \leq i \leq N,\ 0 \leq j \leq N,\ i \neq j \\[2mm]
0 & \text{if } 1 \leq i = j \leq N - 1 \\[2mm]
\frac{1}{4} N(N+1) & \text{if } i = j = N.
\end{cases}
\tag{7.48}
$$

For a spectral-element simulation of a specific order N, a matrix with the derivatives $\partial_\xi \ell_k(\xi_i)$ for each polynomial k at all $N+1$ collocation points ξ_i is precalculated and used to evaluate the integrals (see Matlab/Python routines given in the supplementary material). Note that these precalculated derivatives of Lagrange polynomials can be used to approximate the derivative of an arbitrary function $u(\xi)$ defined in $\xi \in [-1, 1]$ at the GLL collocation points ξ_i:

$$\partial_\xi u^e(\xi) = \sum_{i=1}^{N+1} u^e(\xi_i) \partial_\xi \ell_i(\xi). \qquad (7.49)$$

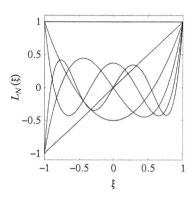

At this point we have almost finished! At least, we have prepared everything that we need to express our approximate solutions inside one element. What remains to be done is to link the elemental results and formulate the complete solution over the entire physical domain. This process is called *assembly*.

7.5   Global assembly and solution

Eqs. 7.44 provided us with solutions of the wave equation inside an element without interaction with the outside. To understand how to assemble the global solution it is instructive to recall the original global solution we obtained (Eq. 7.9). We have done nothing other than to divide up the entire domain D into ne elements and describe the solutions at an elemental level. As indicated earlier the classic finite- (spectral-) element method assumes continuity of the solution fields at the element boundaries. Therefore, we simply need to add up the elemental solutions at the corresponding boundary collocation points. In the spectral-element, method each element boundary thus only has *one* value.[12] Before we present the calculation of the global system of equations, let us discuss the dimensions of the resulting vectors and matrices, that is, the overall *number of degrees of freedom* in our spectral-element system.

Comparing with Fig. 7.5 it is straightforward to see that, for a system with ne elements and given polynomial order N, the global number of collocation points n_g of our system is $n_g = ne \times N + 1$. This is illustrated in Fig. 7.12 for a physical domain of size 10 km and elements of equal size.

We have derived a diagonal elemental mass matrix, which implies that we can store its entries as a vector. Nevertheless we present its global shape for illustrative purposes in Fig. 7.13, highlighting two of the elemental matrices inside the domain that make up the global system. Note that the elemental matrices in general differ from each other as they depend on elastic parameters and density. In addition, their Jacobians depend on element size.

We denote the global matrices (vectors) with subscript g and illustrate their mathematical form for a system with $n_e = 3$ elements and order $N = 2$ Lagrange polynomials. As the mass matrix is diagonal we show it in vector form and obtain

Fig. 7.11 *Illustration of the Legendre polynomials up to order $N = 6$. The Legendre polynomials are used to calculate the first derivatives of the Lagrange polynomials. They can also be used to calculate the integration weights of the GLL quadrature.*

[12] This is the main difference with the discontinuous Galerkin method, where every element has its own boundary values and information between elements is transmitted by flux terms.

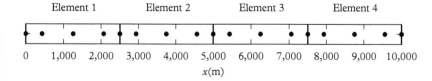

Element 1   Element 2   Element 3   Element 4

0   1,000   2,000   3,000   4,000   5,000   6,000   7,000   8,000   9,000   10,000

x(m)

Fig. 7.12 *Global GLL collocation points for ne = 4 elements, order N = 4 polynomials, and a physical domain with $x \in [0; 10,000]$ m.*

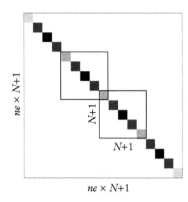

ne × N+1

N+1

N+1

ne × N+1

$$
\mathbf{M}_g = \begin{pmatrix} M^{(1)}_{1,1} \\ M^{(1)}_{2,2} \\ M^{(1)}_{3,3} \\ \\ \\ \\ \\ \end{pmatrix} + \begin{pmatrix} \\ \\ M^{(2)}_{1,1} \\ M^{(2)}_{2,2} \\ M^{(2)}_{3,3} \\ \\ \\ \end{pmatrix} + \begin{pmatrix} \\ \\ \\ \\ M^{(3)}_{1,1} \\ M^{(3)}_{2,2} \\ M^{(3)}_{3,3} \end{pmatrix} = \begin{pmatrix} M^{(1)}_{1,1} \\ M^{(1)}_{2,2} \\ M^{(1)}_{3,3} + M^{(2)}_{1,1} \\ M^{(2)}_{2,2} \\ M^{(2)}_{3,3} + M^{(3)}_{1,1} \\ M^{(3)}_{2,2} \\ M^{(3)}_{3,3} \end{pmatrix} \quad (7.50)
$$

Fig. 7.13 *Global mass matrix. The diagonal structure of the mass matrix is illustrated for a 1D example with ne = 4 elements and order N = 4 Lagrange polynomials. Elemental matrices are overlapping and summed at the global mass matrix entries representing the element boundaries (compare with Eq. 7.50).*

where the corner entries of the elemental matrices are summed (here every third element). The upper indices in brackets denote the elements. Note that we have $N + 1 = 3$ collocation points inside each element (see Fig. 7.13).

For the global stiffness matrix we obtain in an analogous way $\mathbf{K_g} =$

$$
\begin{pmatrix}
K^{(1)}_{1,1} & K^{(1)}_{1,2} & K^{(1)}_{1,3} & & & & \\
K^{(1)}_{2,1} & K^{(1)}_{2,2} & K^{(1)}_{2,3} & & & 0 & \\
K^{(1)}_{3,1} & K^{(1)}_{3,2} & K^{(1)}_{3,3}+K^{(2)}_{1,1} & K^{(2)}_{1,2} & K^{(2)}_{1,3} & & \\
 & & K^{(2)}_{2,1} & K^{(2)}_{2,2} & K^{(2)}_{2,3} & & \\
 & & K^{(2)}_{3,1} & K^{(2)}_{3,2} & K^{(2)}_{3,3}+K^{(3)}_{1,1} & K^{(3)}_{1,2} & K^{(3)}_{1,3} \\
 & 0 & & & K^{(3)}_{2,1} & K^{(3)}_{2,2} & K^{(3)}_{2,3} \\
 & & & & K^{(2)}_{3,1} & K^{(2)}_{3,2} & K^{(2)}_{3,3}
\end{pmatrix} \quad (7.51)
$$

with each element represented by an $(N + 1) \times (N + 1)$ matrix. Inside the domain the first and last diagonal element are summed up with the value from the adjacent element. For a 1D grid, and also with varying element size, the stiffness matrix has a banded structure (Fig. 7.14).

Note that for general irregular grids (hexahedral or tetrahedral) a connectivity has to be defined and the stiffness matrix structure can be arbitrarily filled. An example of a stiffness matrix for an irregular grid is shown in Fig. 7.15.

Equivalently, the vector containing information on the source is given as

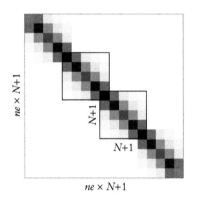

ne × N+1

N+1

N+1

ne × N+1

$$
\mathbf{f}_g = \begin{pmatrix} f^{(1)}_1 \\ f^{(1)}_2 \\ f^{(1)}_3 + f^{(2)}_1 \\ f^{(2)}_2 \\ f^{(2)}_3 + f^{(3)}_1 \\ f^{(3)}_2 \\ f^{(3)}_3 \end{pmatrix}. \quad (7.52)
$$

Fig. 7.14 *Global stiffness matrix. The diagonal structure of the stiffness matrix is illustrated for a 1D example with ne = 4 elements and order N = 4 Lagrange polynomials. Elemental matrices are overlapping and summed at the global stiffness matrix entries representing the element boundaries (compare with Eq. 7.51).*

We end up with a system of equations for $n_g = n_e \times N + 1$ coefficients for the displacement $\mathbf{u}_g$, where N is the interpolation order and n_e is the number of elements. As illustrated above, the matrices $\mathbf{M}_g$ and $\mathbf{K}_g$ have dimensions $n_g \times n_g$. Because of its diagonal structure, $\mathbf{M}_g$ is obviously never initialized as a matrix. To save memory it is stored as a vector of the diagonal elements. The force vector $\mathbf{f}_g$ also has n_g elements. The time-dependent coefficients $\mathbf{u}_g$ are extrapolated with a simple centred finite-difference scheme to the next time step $t + dt$:

$$\mathbf{u}_g(t + dt) = dt^2 \left[\mathbf{M}_g^{-1} \left(\mathbf{f}_g(t) - \mathbf{K}_g \mathbf{u}_g(t) \right) \right]$$
$$+ 2\mathbf{u}_g(t) - \mathbf{u}_g(t - dt). \qquad (7.53)$$

In this final algorithm only the coefficients $\mathbf{u}_g$ are updated as a response to the time-dependent forces that are injected through $\mathbf{f}_g$ at predefined locations after the system was at rest at $t = 0$ (our initial condition). Even though in principle mass and stiffness matrices could be modified during the time extrapolation (e.g. when the elastic parameters or density are time-dependent or the computational mesh is adapted), this case will not be considered here. In many seismological applications the Earth model and the mesh remain constant.

Before turning this algorithm into a computer program let us throw some light on the forces that are injected.

7.6 Source input

In many research applications in seismology it is sufficient to treat the seismic source as acting at a single point. In addition, any finite source representation can be obtained by summing over many point sources according to the superposition principle. What happens if we activate a force at a single collocation point inside (or at the edge of) an element? This is illustrated in Fig. 7.16. As is the case in the finite-difference method, injecting a source at a single collocation point is not a problem. In fact, due to the Galerkin approach the integral correctly represents a delta-function.

If the source is not located directly at a collocation point, a possible solution is to use (smooth) spatially limited functions to spread the point source to adjacent grid points. Note that, depending on its spatial extension, this may lead to a low-pass filtering of the injected source time function. The injection of sources is described in more detail in Fichtner (2010) for both spectral-element and finite-difference methods. In the vicinity of the point sources the solution may be erroneous as the near-field terms are not properly represented. This is discussed in Nissen-Meyer et al. (2007). However, the authors note that the wavefields are accurate when records are taken more than two elements away from the physical source location.

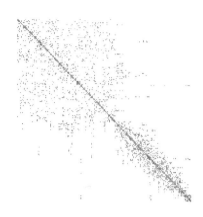

Fig. 7.15 *Graphical illustration of a sparse global matrix for the case of an irregular mesh. The non-zero entries are shown in black.*

7.7 The spectral-element method in action

7.7.1 Homogeneous example

We can now proceed to implement Eq. 7.53 in a computer program and illustrate some of the spectral-element-specific features with code fragments written in Python. Compared to the finite-difference method the preparatory steps for the spectral-element extrapolation are substantially more involved, even though the final extrapolation is very similar. Thus we illustrate the workflow schematically

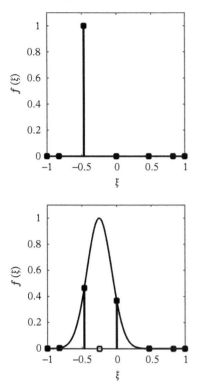

Fig. 7.16 *Illustration of the seismic source input into a spectral-element algorithm.* **Top:** *A point source representation at a collocation point (black bar).* **Bottom:** *Sources not collocated with collocation points (open square) can be input by spreading them with appropriately scaled spatially limited functions (here, a Gaussian).*

---

in Fig. 7.17 (which, with a few alterations, is also representative of the nodal discontinuous Galerkin method).

For any finite-element-type method, the calculation of the system matrices (stiffness and mass) constitutes the most important preparatory step before the time-loop is started to extrapolate the initial conditions. The following code snippet shows that the mathematical developments described above lead to an extremely dense algorithm for the calculation of the *elemental* stiffness and mass matrices:

```
# Elemental Mass matrix
# stored as a vector since it's diagonal
Me = zeros(N+1)
for i in range(0,N+1):
    Me[i] = rho * w[i] * J
# [...]
# Elemental Stiffness Matrix
Ke = zeros([N+1,N+1])
for i in range(0,N+1):
    for j in range(0, N+1):
        for k in range(0, N+1):
            Ke[i,j] = Ke[i,j] + mu*w[k]*Ji*l1d[i,k]
                *l1d[j,k]
# [...]
```

Here the elemental mass matrix *Me* is initialized in vector form because of its diagonal structure. We integrate density multiplied by the product of our basis functions (i.e. unity). Because of their orthogonality the resulting δ-function leads to this extremely simple formulation.[13] In this case the parameters μ and ρ are constant inside the element.

The integration weights *w* are initialized from precalculated tables (see Table 7.1). In 1D the Jacobian $\mathcal{J}$ (and its inverse $\mathcal{J}i$) is a scalar and in this example kept constant (i.e. all elements have the same size). This can easily be changed to allow space-dependent element size, one of the most attractive features of element-based techniques.

The elemental stiffness matrix *Ke* is calculated by integrating over the elastic coefficients *mu* (shear modulus) and the product of the derivatives of the basis functions (initialized here through a function *l1d* implementing Eqs. 7.47 and 7.48), multiplied by the inverse Jacobian (one inverse Jabocian and one Jacobian cancel each other out in the 1D case).

As indicated in Eq. 7.44, the spatial source function also has to be projected onto the basis functions and integrated. In general, the force vector will vary in each element (zero except for the source point or region). Therefore, it is initialized as a matrix of size $ne \times N + 1$ with *ne* the number of elements and *N* the polynomial order. The matrix *s* represents the spatial source function, which can contain several source locations (e.g. a finite source) to be injected at collocation points. The code fragment below describes the force initialization in the general

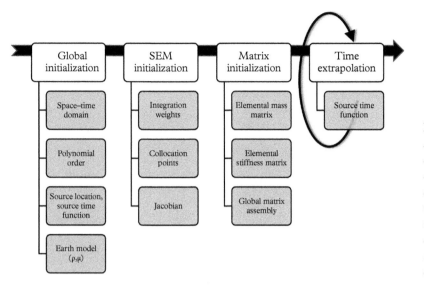

Fig. 7.17 *Schematic workflow of the spectral-element (SEM) solution of the elastic wave equation. A substantial part consists of preparing the interpolation and integration procedures required to initialize the global mass and stiffness matrices. The final time extrapolation is extremely compact and does not require the inversion of a global matrix (whereas classic finite-element methods do).*

case with an example of a point source input in the central element $ne/2$ at the first collocation point (i.e. at the boundary between element $ne/2 - 1$ and $ne/2$).

```
# Initialization
fe = zeros((ne,N+1))
s = zeros((ne,N+1))
# Point source
s[int(ne/2),1] = 1
# Force vector
for k in range(ne):
    for i in range(N+1):
        fe[k,i] = s[k,i] * w[i] * J
```

Before the time extrapolation, the elemental force vectors are assembled to a global form (see Section 7.5) and reshaped to a vector with $n_g = n_e \times (N + 1)$ elements.

As an example of a global matrix assembly we present a code fragment initializing the global stiffness matrix K of size $ng \times ng$ from elemental stiffness matrices Ke of size $N + 1 \times N + 1$. Here, $i0$ and $j0$ are the elements in the global matrix from which the elemental stiffness matrices are added. Note that here we assume equal stiffness matrices for each element. It is straightforward to make these elements dependent (see exercises and supplementary material).

```
# Global Stiffness Matrix
K = zeros([ng,ng])
# Values except at element boundaries
for k in range(1,ne+1):
```

```
i0 = (k-1) * N+1
j0 = i0
for i in range(-1,N):
    for j in range(-1,N):
        K[i0+i,j0+j] = Ke[i+1,j+1]
# Values at element boundaries
for k in range(2,ne+1):
    i0 = (k-1) * N
    j0 = i0
    K[i0,j0] = Ke[0,0] + Ke[N,N]
```

Following the assembly of the global matrices as described in the previous section, the vector of coefficients *u* (which does in fact correspond to the displacement values due to the clever choice of our basis functions) can be extrapolated to the next time step *unew* by a simple finite-difference approximation. Here, for illustrative purposes, *Minv* (inverse mass matrix) is initialized as a diagonal matrix, *f* contains the forcing, and *uold* the coefficients at the previous time step.

Remapping of the coefficient vector allows the subsequent extrapolation to the next time step in each iteration step using implicit matrix–vector operations. Here, *nt* is the global number of time steps that depends on the desired seismogram length and the minimum distance between collocation points and the choice of elastic parameters. The various other initializations that have to be done prior to the simulation are similar to the other numerical techniques presented in the previous chapters, and so are not further illustrated here (see supplementary electronic material).

```
# [...]
# Time extrapolation
for it in range(nt):
# Extrapolation
    unew = dt**2 * Minv @ (f - K @ u) + 2 * u - uold
    uold, u = u, unew
# [...]
```

Note that, as illustrated already in Chapter 6 on the finite-element method, this matrix–vector extrapolation scheme is formally identical to the finite-difference method, if the global stiffness matrix is replaced by a scaled finite-difference operator.

Finally, let us analyse examples with numerical solutions of the wave equation solved with the above algorithm (Eq 7.53). The parameters of a simulation in a homogeneous material are given in Table 7.2. The source time function has a dominant period of $T_{dom} = 0.15$ s and is initialized by

$$s(t) = -2a(t-t_0)e^{\frac{(t-t_0)^2}{a^2}} \tag{7.54}$$

Table 7.2 *Spectral-element simulation, homogeneous case.*

Parameter	Value
x_{max}	10 km
ne	250
v_s	2,500 m/s
ρ	2,000 kg/m^3
N	2–8
ϵ	0.8

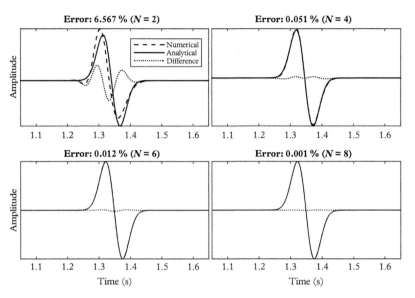

Fig. 7.18 *Spectral-element method, homogeneous medium. Comparison of numerical (dashed) and analytical (black) solution and their differences (dotted) of the 1D spectral-element implementation. The order of the interpolation (and integration) scheme is given at the top of the figures together with the overall energy misfit in %. Note the improvement of the numerical solution with increasing polynomial order N.*

with $a = 4/T_{dom}$. This is the first derivative of a Gaussian. The receiver is at a distance of 3 km from the source. In Fig. 7.18 we compare the numerical solution (dashed line) with the analytical solution (black line) and calculate the relative energy misfit in % as a function of the polynomial order used in the simulation, while everything else is kept constant.

The spectral-element algorithm as implemented above allows us to change the spatial accuracy with only one parameter N without any further modification. However, note that, assuming constant stability criterion $\epsilon = 0.8$, the time step dt is decreasing with increasing order as the minimum distance between collocation points is the decisive factor. This effect was discussed at length in connection with the Chebyshev pseudospectral method. The decreasing time step with polynomial order actually prevents us going to much higher orders. In realistic simulations (e.g. using specfem3d) the highest order is usually $N = 4$.

The results shown in Fig. 7.18 indicate how the simulation with polynomial order $N = 2$ (corresponding to a regular grid) would not lead to a sufficiently accurate solution for the chosen set-up. It is instructive to see how the solution accuracy dramatically improves when the order of the interpolation (and integration) scheme increases—of course at the cost of longer simulation time. This is due to the increase in the number of floating point operations per time step and the increasing number of time steps due to the decrease of smallest grid-point distance.

In many cases one aims to have an overall error of the numerical solution of $\leq 1\%$. It is interesting to analyse the behaviour of the numerical solutions for this simple case as a function of the spectral-element-specific parameters that enter the algorithm and compare with the other methods discussed in this volume. This can be done using the extensive electronic supplementary material. In the homogeneous case we are fortunate to have an analytical solution to compare with. This is not the case for general heterogeneous models. The next section presents a heterogeneous model and discusses a strategy for benchmarking complex models.

7.7.2   Heterogeneous example

In the spectral-element method the parameters λ and μ can vary at each collocation point. However, beware what this really means. Similar to the problem discussed in the previous section, any function that is described on the collocation points is multiplied by the Lagrange polynomials. This implies that a sharp discontinuity is replaced by a smooth representation and the Gibbs phenomenon occurs (explore this with the codes given in the supplementary material).

From an algorithmic point of view the extension to the heterogeneous case is straightforward. Mass and stiffness matrices have to be initialized separately for each element, with λ and μ varying at each collocation point. That means, when increasing the order of a scheme, the space-dependent parameters are interpolated in a slightly different way even though the results should converge.

We demonstrate this convergence behaviour with a random 1D velocity model. The parameters for the simulations are shown in Table 7.3. The constant velocity model is perturbed with a random perturbation[14]. The resulting velocity model is shown in Fig. 7.19 (top).

A source is injected at the centre of the model with a dominant period $T_{dom} = 1.2$ s (see Eq.7.54). The wavefield propagated in both directions through

Table 7.3 *Spectral-element simulation, heterogeneous case.*

Parameter	Value
x_{max}	10 km
ne	250
v_s^{mean}	2,500 m/s
ρ^{mean}	2,000 kg/m^3
N	2–12
ϵ	0.5
T_{dom}	0.12 s

[14] In practice we initialize a vector with random numbers, low-pass filter with the desired corner wavelength (here 500 m), and scale with the perturbation amplitude (here 25%).

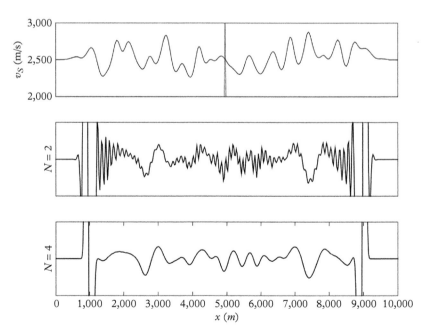

Fig. 7.19 *Spectral element method, heterogeneous case.* **Top:** *Shear velocity model with smooth random perturbation (solid line). The spatial source function is indicated by a dashed line.* **Middle:** *Displacement wavefield at t = 1.7 s for a simulation with order N = 2.* **Bottom:** *Simulation with order N = 4.*

the random model and is captured after a simulation time of t = 1.7 s in Fig. 7.19 for orders $N = 2, 4$. The parameters have been chosen such that we can illustrate the improvement of the solution with increasing order. In the figure the direct wave has been clipped so that the effect of the accuracy improvement becomes more visible. The simulation with order $N = 2$ shows strongly dispersive behaviour. Increasing to order $N = 4$ reduces substantially the high-frequency numerical noise.

But still ... how do we know our result is correct? This is a situation that always occurs when simulating heterogeneous models. One possible approach is to gradually increase the order (accuracy) of the scheme to see whether the results still change. A trick often used is to carry out one simulation at very high resolution (which might be expensive in the 3D case) and use this as a reference solution. That is fine if you are absolutely confident that there is no bug in your code. An alternative is to compare your results with other codes, possible using other numerical methods. Projects with various benchmark models are discussed in the Appendix.

The convergence behaviour in our heterogeneous example is documented in Fig. 7.20 for orders $N = 2, \ldots, 12$. The root-mean-square (rms) difference to the simulation with order $N = 12$ is shown as a function of order. At the same time we record the elapsed time of the extrapolation part. Visually the results do not change much for orders $N > 4$ but the error steadily decreases. This comes at the expense of a steady increase in computation time (here with respect to the simulation for $N = 2$). This is because, while keeping everything else constant, (1) the number of grid points and thus the floating-point-operations per time step increase, and (2) due to the decreasing minimal grid distance, the time step is decreasing accordingly. This effect is of course even more pronounced in 2D or 3D.

7.8 The road to 3D

The power of the spectral-element method lies in its efficient extension to 2D and 3D while keeping the explicit time marching scheme possible through the diagonal mass matrix when using hexahedral grids. Curved elements do not cause any problems, allowing efficient and accurate implementation of curved free surfaces (rather than approximating them as straight-line segments, see Fig. 7.21). In this section we merely give hints as to which articles can be useful in gaining an understanding of the spectral-element concepts applied to 3D. Many more articles *using* this method for research are presented in Chapter 10 on applications.

The basic spectral-element algorithms using Lagrange polynomials for 2D or 3D elastic wave propagation were presented in Komatitsch (1997), Komatitsch and Vilotte (1998), Komatitsch et al. (1999), and Komatitsch and Tromp (1999). These studies also document improvements using high-order time-extrapolation schemes (e.g. the Newmark scheme). The solution to fluid-solid media—highly relevant for marine exploration problems or global wave propagation—was presented by Komatitsch et al. (2000b). Further extensions include anisotropic

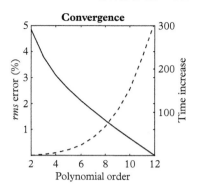

Fig. 7.20 *Convergence with increasing order. Convergence behaviour of the simulation shown in Fig. 7.19. The left axis shows the decrease of the rms error (%) as the polynomial order of the scheme increases. In the right axis the increase of the elapsed simulation time (for the time extrapolation part only) is shown.*

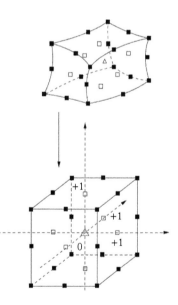

Fig. 7.21 *In 3D, elements might be skewed and have curved boundaries. In analogy to the 1D case, curved hexahedra are mapped to a reference interval through the Jacobian transformation. Figure courtesy of B. Schuberth.*

media (Komatitsch et al., 2000*a*), and poroelastic media (Boxberg et al., 2015). Further verification against analytical solutions was presented by Martin (2011).

The great success of the spectral-element method is also linked to the fact that in combination with the cubed-sphere approach it is currently the only *production* code for 3D global wave propagation. This was made possible through the work of Chaljub et al. (2003), Komatitsch and Tromp (2002*a*), and Komatitsch and Tromp (2002*b*), with many subsequent applications.

Regular-grid spectral-element implementations also have their merits, in particular when Earth models with moderate and smooth velocity perturbations are the target. Detailed algorithms for the case of wave propagation on a regional scale were presented in Fichtner and Igel (2008), Fichtner (2009), and Fichtner (2010). Further attractive applications are the axisymmetric approach for global wave propagation of Nissen-Meyer et al. (2007) and Nissen-Meyer et al. (2014), with recent online facilities developed by van Driel et al. (2015*b*).

The spectral-element method is now routinely used for forward problems of all kinds (including rupture problems), as well as full waveform inversion. The flexibility with hexahedral meshes led to the possibility of generating meshes with substantially varying element sizes (i.e. *h*-adaptivity). However, models with *very* discontinuous behaviour, or extremely complex geometries, still cause problems because of the difficulties in generating hexahedral meshes.

Attempts had been made to extend the spectral-element concepts to triangular meshes (Mercerat et al., 2006) and recently to tetrahedral meshes (May et al., 2016, see Fig. 7.22). The unstructured grids do not allow a diagonal mass matrix, so that, as in the classic finite-element methods, linear algebra libraries are required to solve the global linear system of equations. It will be interesting to see how these new methods compare with the discontinuous Galerkin-type methods.

The problems with finding appropriate meshes with hexahedral element shapes motivated the search for methods that (1) are based on tetrahedral (or arbitrarily shaped) elements that are easier to adapt to complex geometries, and (2) can better handle discontinuities in the solution field or the geophysical parameters. These problems are addressed with the remaining approaches; the finite-volume and the discontinuous Galerkin methods.

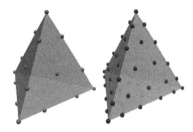

Fig. 7.22 *Spectral elements with tetrahedra. Two examples of a high-order nodal spectral-element discretization with tetrahedra. The mass matrix is no longer diagonal and has to be inverted numerically. From May et al. (2016).*

Chapter summary

- The spectral-element method combines the flexibility of finite-element methods with respect to computational meshes with the spectral convergence of Lagrange-basis functions used inside the elements.

- The enormous success of the spectral-element method is based upon the diagonal structure of the mass matrix that needs to be inverted to extrapolate the system in time combined with the spectral convergence of the basis functions. Due to the diagonality, no matrix inversion techniques need to be employed, allowing straightforward parallelization of the algorithm.

- The diagonal mass matrix is made possible by superimposing the collocation points of both interpolation and integration schemes (Gauss–Lobatto–Legendre integration).

- The errors of the spectral-element scheme accumulate from the (usually low-order finite-difference) time-extrapolation scheme and the numerical integration using Gauss–Lobatto–Legendre quadrature.

- In principle, the spectral-element method can also be formulated with other basis functions with similar (or even better) interpolation and integration properties (e.g. Chebyshev polynomials). However, then the mass matrix is not diagonal and a global system matrix needs to be inverted.

- Spectral-element solutions are usually formulated for hexahedral computational grids. For complex models (surface topography, internal curved boundaries) this might involve cumbersome mesh generation. Formulations for triangles or tetrahedra are in principle possible but the advantage of a diagonal mass matrix is lost.

- The spectral-element method is particularly useful for simulation problems where an uneven free surface plays an important role, and/or in which surface waves need to be accurately modelled. The reason is that the free-surface boundary is implicitly solved.

- Several well-engineered community codes are available for Cartesian and spherical geometries including basin scale, continental scale, and global Earth (or planetary scale) calculations.

FURTHER READING

- Fichtner (2010) provides further mathematical details on the spectral-element method (interpolation and integration schemes) and discusses forward and inverse problems in 3D.

- Pozrikidis (2005) is perhaps the most exhaustive book on the spectral-element method, with many examples provided using Matlab. The mathematical background is explained in great detail.

- Peter et al. (2011) give an excellent review of the capabilities of the specfem3d code family for both forward and inverse modelling.

. . . .

EXERCISES

Comprehension questions

(7.1) What is the main difference between classical finite and spectral-element methods. What is the meaning of spectral in this context?

(7.2) What is the free-surface boundary condition? Explain qualitatively why this boundary condition is implicitly fulfilled in finite- (and spectral-) element methods. Which problems in seismology might benefit from this behaviour?

(7.3) The spectral-element method allows in principle arbitrary high-order polynomials inside the elements. Can you give a reason why in practice only low-order polynomials (usually $N \leq 4$) are used, even for large simulations with long propagation distances?

(7.4) Why can sin and cos functions not be used within the *spectral*-element framework, given that they are so efficient for the pseudo*spectral* method?

(7.5) Explain the concepts of *weak* form and *strong* form of the wave equation.

(7.6) What is meant by *exact* interpolation at collocation points? Does it mean the solution is exact everywhere inside an element?

(7.7) Do you know how the *mass* and *stiffness* matrices got their names? Hint: This has to do with the field in which the finite-element method was developed.

(7.8) Compare finite-difference and spectral-element methods in terms of their potential domains of application in the field of seismic wave propagation. Give arguments.

Theoretical problems

(7.9) In the spectral-element method each element has $N + 1$ collocation points including the boundaries, where N is the polynomial order. Derive the equation for ng the global number of degrees of freedom (i.e. collocation points) in the 1D case for a problem with ne elements.

(7.10) We want to find a set-up for a simulation task. Assume that you want to propagate 10,000 km with velocity of $c = 5$ km/s. The stability criterion is given by $c\, dt/dx < 0.5$. Assume that 10 points per wavelength are enough to achieve sufficient accuracy. The dominant frequency of your wavefield is 0.2 Hz (i.e. period 5 s like crustal surface waves). The elements are discretized by Gauss–Lobatto–Legendre points. Examples are given in Table 7.1. Calculate the required number of spectral elements and the time steps for orders $N = 2, 3$, and 4. How many time steps would you roughly expect for each simulation?

(7.11) The Lagrange polynomials of order N are given by

$$\ell_i^{(N)}(x) := \prod_{k=1,\ k \neq i}^{N+1} \frac{x - x_k}{x_i - x_k}, \qquad i = 1, 2, \ldots, N + 1.$$

Write down all polynomials $\ell_i^{(2)}(x)$ for $N = 2$ and general points x_k with $k = 1, 2, 3$. Show that with N = 1 you recover the definition of linear-basis functions introduced in Chapter 6.

(7.12) The function $f(x) = 1/2x^2 - 1/3x^5$ is defined in the interval $x \in [0, 1]$. Evaluate its integral analytically. Calculate the integral using GLL quadrature for orders N = 1–4 (see Table 7.1). Compare analytical and numerical results.

(7.13) Derive the elemental mass matrix with Lagrange polynomials (Eq. 7.44) starting with the general form given in Eq. 7.11.

(7.14) Use the recursion formula Eq. 7.46 and derive the Legendre polynomials for order N = 0–4. Plot the results in the interval $[-1, 1]$.

Programming exercises

(7.15) Use the information on the GLL collocation points in Table 7.1 to write a function *lagrange* that returns the Lagrange polynomials $i \in [0, N]$ for arbitrary $\xi \in [-1, 1]$ where N is the order (see equation in exercise 7.11).

(7.16) Define an arbitrary function $f(x)$ and use the *lagrange* routine of the previous problem (or the supplementary material) to calculate the interpolating function for $f(x)$. Show that the interpolation is exact at the collocation points. Compare the original function $f(x)$ and the interpolating function on a finely spaced grid. Vary the order of the interpolating polynomials and calculate the error as a function of order.

(7.17) We want to investigate the performance of the numerical integration scheme (Gauss integration). Based on Table 7.1, write a program that performs GLL integration on the GLL points. Define a function $f(x)$ of your choice and calculate analytically the integral $\int f(x)\,dx$ for the interval $[-1, 1]$. Perform the integration numerically and compare the results. Modify the function and the order of the numerical integration. Discuss the results. Note: The error of the spatial scheme in the spectral-element method comes only from this integration step.

(7.18) Use the 1D spectral-element code (supplementary material) to determine experimentally the stability limit as a function of the order N of the Lagrange interpolation.

(7.19) Increase the order of the scheme and observe the necessary decrease of the time step, keeping the Courant criterion constant.

(7.20) Modify the spectral-element code to allow for space-dependent elastic parameters and density. Introduce a low-velocity zone (–30%) at the centre of the model spanning 5 elements. Input the source inside this zone and discuss the resulting wavefield.

(7.21) Introduce h-adaptivity (each element may have different size h) to the numerical scheme by making the Jacobian element dependent. Generate a space-dependent mesh size (e.g. decreasing the element size gradually towards the centre). Generate a velocity model that keeps the number of points per wavelength approximately constant.

(7.22) Use the power of the 1D spectral-element scheme to implement a strongly heterogeneous computational mesh: (1) a low-velocity zone in the middle of the region (source in and outside this region); (2) vary the element size using a Gauss function; (3) vary the element size randomly within some bounds. Document the effect on the solution in the homogeneous case. Investigate the effects on the waveforms for the heterogeneous case. Make sure you choose the right time step!

(7.23) Define an arbitrary function in the interval $[-1, 1]$. Use the available GLL and lagrange routines to compare the interpolation behaviour of Lagrange polynomials on regular grids vs. GLL points. Plot the energy misfit as a function of polynomial order (Runge phenomenon).

The Finite-Volume Method

All numerical methods we have encountered so far work (reasonably) well for the solution of wave equations. Seismic wave-propagation problems are usually characterized by the fact that the solutions are sufficiently smooth, so that, in practice, we simulate band-limited wavefields. This holds in most cases even if the parameters of the elastic wave equation (i.e. the seismic velocity model) are discontinuous. But what if the solution is characterized by discontinuities (e.g. shock waves, gravity waves, transport problems, etc. see Fig. 8.1)? This question led to the development of the finite-volume method.

In many problems of physics (e.g. fluid flow, material transport, advection problems) that are described by partial differential equations, the initial conditions (or source terms) contain discontinuities. In terms of spectral content this implies that infinite frequencies are part of the solution. We have seen in Chapter 5 on pseudospectral methods that discontinuities cause problems (the Gibbs phenomenon) as soon as we have to limit the wavenumber range of our solution (e.g. due to spatial discretization).

To some extent, the finite-volume method is a way to avoid taking spatial derivatives of the solution fields, by replacing them with so-called flux terms. The finite-volume method naturally follows from conserving mass (or a tracer concentration, elastic energy, etc.) in a volume cell of an advective system, balancing it with the flux into and out of it. This leads to the classic advection equation of material transport, and in fact this simple principle can also be used to derive the acoustic wave equation (Leveque, 2002).

The advection equation is an expression of a hyperbolic conservation law that is fundamental for many branches of continuum physics. It turns out that the seismic wave equation can be cast in this mathematical form. This implies that we can transfer results concerning the numerical solution of scalar advection directly to problems in seismology.

The most important advantage of the finite-volume method is the allowance of in principle arbitrarily shaped computational cells. Despite this flexibility, finite-volume methods have so far not been used for large-scale seismic simulation problems. However, the flux concepts introduced here are fundamental for the understanding of the discontinuous Galerkin method, discussed in Chapter 9. In fact, in their lowest-order implementations, the finite-volume and the discontinuous Galerkin methods are identical.

This chapter is structured as follows. After a brief section on the history of the finite-volume method we present, as always, the method in a nutshell. This

Computational Seismology. First Edition. Heiner Igel.
© Heiner Igel 2017. Published in 2017 by Oxford University Press.

Fig. 8.1 *Simulating the breaking of gravity waves is a challenging task. The finite-volume method was developed with a view to solving problems with discontinuous solutions and complex geometrical features.*

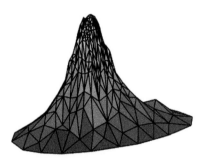

Fig. 8.2 *Topographic mesh. Triangulated topography of a seamount model lending itself to a volumetric discretization using tetrahedra. The finite-volume method offers an elegant solution to problems on tetrahedral meshes.*

is followed by a derivation of the numerical solution of the 1D scalar advection and the elastic wave equation using the finite-volume method from first principles. Finally, we analyse examples and compare the results with other methods encountered so far.

8.1 History

The story of finite volumes in seismology is quickly told. The method itself appeared in the early nineties with applications primarily in plasma physics (Hermeline, 1993), and computational fluid dynamics (Versteeg and Malalasekera, 1995). In an effort to investigate its use for seismic wave propagation, Dormy and Tarantola (1995) used the discrete version of the divergence theorem in an elegant way to derive a scheme that allowed the simulation of wave propagation through arbitrary cell shapes. Their approach can be considered an extension of the staggered-grid finite-difference method to arbitrary geometries. In its basic form it converges to the classic staggered-grid finite-difference method for regular grids (Virieux, 1986) and to the solution for minimal (hexagonal) grids (Magnier et al., 1994). The method was tested in two dimensions in its lowest-order form. A more mathematical treatment of the finite-volume method was presented in Eymard et al. (2000).

Käser et al. (2001) and Käser and Igel (2001) compared various numerical approaches, including the finite-volume concept of Dormy and Tarantola (1995), and quantified the accuracy of wave propagation modelled through unstructured grids in comparison with classic regular-grid methods. They concluded that, despite the flexibility of unstructured meshes, the price to pay is high in the sense that many more grid points per wavelength are necessary compared to classic techniques. This experience eventually led to the introduction of the high-order discontinuous Galerkin method to seismology.

High-order extensions were presented for advection-type problems by Ollivier-Gooch and Van Altena (2002) and Wang (2002) for general conservation laws. Harder and Hansen (2005) applied the finite-volume method to geophysical fluid flow, and Tadi (2004) developed an algorithm for 2D elastic wave propagation. In the comprehensive textbook by Leveque (2002) the finite-volume method is presented as a natural consequence of conservation laws. It is shown that the mathematical structure of elastic wave-propagation problems (in first-order form as a coupled system) is identical to the advection problem. This implies that the same numerical concepts developed for conservation laws can be directly applied to elastic wave propagation. This approach was adopted by Dumbser et al. (2007a) who presented the arbitrary high-order scheme (ADER) of the finite-volume method for seismic wave propagation. This removes the argument used by many against the finite-volume method that it cannot be easily extended to higher orders. Dumbser et al. (2007a) show that the finite-volume method might well be a competitive scheme, in particular for triangular or tetrahedral meshes (see Fig. 8.2). Further applications in seismology were presented by Benjemaa et al. (2007) and Benjemaa et al. (2009) for the dynamic rupture problem.

To my knowledge, the finite-volume method is not widely used today for large-scale problems in seismological research. As mentioned earlier, it can be understood as a special case of the discontinuous Galerkin method, and combined use (low- and high-order) might be a useful strategy for some applications.

8.2 Finite volumes in a nutshell

The finite-volume method was developed around the problem of transporting (advecting) material and conserving the integral quantity. As the first-order linear advection problem is formally equivalent to the elastic wave-propagation problem, all methods derived for the former equation equally apply. Therefore, we first present the solution to the scalar advection problem, later extending to more general cases.

The finite-volume method in its basic form takes an entirely local viewpoint, in the sense that the solution field $q(x, t)$ is tracked inside a representative cell of a finite volume. As the exact solution is not known, the field is approximated by an average quantity Q_i^n inside cell $\mathscr{C}$ as

$$Q_i^n = \frac{1}{dx} \int_\mathscr{C} q(x, t)\, dx. \qquad (8.1)$$

Here, as before, the lower index denotes cell $\mathscr{C}$ and the upper index denotes time level $t_n = ndt$. The cell $\mathscr{C}$ is centred at $x = x_i$, with left and right boundaries defined at $x_i - \frac{1}{2}dx$ and $x_i + \frac{1}{2}dx$, respectively (see Fig. 8.3).

Tracking the change of the values with time inside each cell implies that we equate the change of quantity Q_i from time step n to $n+1$ with the fluxes through the boundaries such that

$$\frac{Q_i^{n+1} - Q_i^n}{dt} = \frac{F_{i-1/2}^n - F_{i+1/2}^n}{dx}, \qquad (8.2)$$

where $F_{i\pm1/2}^n$ represent time integrals of the fluxes from time t_n to t_{n+1}. As information propagates with finite speed, it is reasonable to assume that, for example for the left boundary, the flux depends on the adjacent Q_i^n values only:

$$F_{i-1/2}^n = f(Q_{i-1}^n, Q_i^n). \qquad (8.3)$$

The requirement of conservation for a transport (advection) problem leads (entirely from basic principles) to the advection equation of the form

$$\partial_t q(x, t) + a\partial_x q(x, t) = 0, \qquad (8.4)$$

where a is a transport velocity. We will show in what follows that the elastic wave-propagation problem is formally equivalent to this equation and that all solution procedures derived for this simple advection problem can be used.

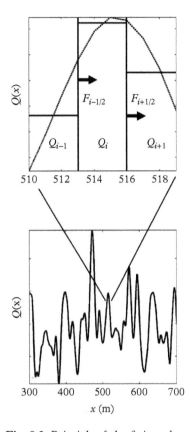

Fig. 8.3 *Principle of the finite-volume method illustrated for the scalar advection problem for a random initial field.* **Bottom:** *Snapshot in space of the solution field $Q(x)$.* **Top:** *Close-up of a detail of the solution field for cell volumes of size $dx = 3$. The values Q_i inside the volumes represent an average over the solution $q(x)$ (dotted line). Using knowledge about the advection equation (here: the positive scalar advection velocity (a), the values in each volume are updated by estimating the fluxes $F_{i+1/2}$ across the boundaries at $x = x_i \pm \frac{1}{2}dx$.*

For the constant-coefficient advection problem the flux terms are simply

$$F_{i-1/2}^n = aQ_{i-1}^n,$$
$$F_{i+1/2}^n = aQ_i^n \quad , \tag{8.5}$$

where a is the advection velocity. With these definitions we obtain a fully discrete extrapolation scheme as

$$Q_i^{n+1} = Q_i^n + a\frac{dt}{dx}(Q_{i-1}^n - Q_i^n). \tag{8.6}$$

Here we used the fact that the advection occurs in one direction only (upwind scheme). It is interesting to note—you might have seen it already—that the resulting numerical algorithm is equivalent to a simple finite-difference solution using a forward finite-difference scheme (i.e. upwind). However, this result was obtained with an entirely different approach. This will be further elaborated in the next section.

From a practical point of view the first-order scheme just introduced is useless, as it is highly dispersive. A considerably better solution is the Lax–Wendroff scheme, given as

$$Q_i^{n+1} = Q_i^n - \frac{adt}{2dx}(Q_{i+1}^n - Q_{i-1}^n) + \frac{1}{2}\left(\frac{adt}{dx}\right)^2 (Q_{i-1}^n - 2Q_i^n + Q_{i+1}^n), \tag{8.7}$$

which is second-order accurate and much less dispersive (Leveque, 2002).

From these very simple considerations a few important conclusions can be drawn. It appears that the finite-volume approach allows numerical schemes to be developed that are independent of cell shape, as long as appropriate fluxes are defined. Obviously, the assumption of constant values in each cell is suboptimal. However, it leads in principle to finite-difference-type algorithms. This assumption can be relaxed, and recently arbitrary high-order reconstructions inside grid cells have been proposed. The fundamental ingredient to finite-volume methods is the flux concept. The accurate calculation of the flux contribution to the cell update is related to the *Riemann problem* which considers the advection of a single discontinuity. Further interpretations and an alternative derivation of finite-volume concepts are given in Section 8.6.

8.3 The finite-volume method via conservation laws

Finite-volume methods were motivated by the challenge of finding solutions to problems with strongly heterogeneous parameters and possible discontinuities in the solution (e.g. shock waves). At discontinuities the partial-differential equations no longer hold. However, the integral equations, from which the finite-volume concepts are derived, are still valid, and this is a fundamental difference to other numerical approaches. From this point of view it is not obvious that the seismic wave-propagation problem is a good candidate for this method. However, there are situations in which classic numerical methods like finite differences break down. In addition, the theoretical concepts introduced here will form the basis for the discontinuous Galerkin method presented in Chapter 9.

To understand the fundamental concepts of finite-volume methods it is sufficient to consider the scalar advection equation that naturally follows from conserving mass in a flowing (advecting) system. In addition, the finite-volume concepts provide an alternative way to derive the equations for elastic wave motion from first principles. We will tightly follow the concepts presented in the excellent book by Leveque (2002).

We will then show that the problem of elastic wave propagation can be formulated as a coupled first-order system formally equivalent to the advection problem. In that sense the constant coefficient linear elastic wave equation can be viewed as a conservation law for stress and velocity (or elastic energy). In what follows we will restrict ourselves to methods that assume constant (average) values inside the cell volumes. Extensions to higher-order representations inside the cells are briefly discussed at the end of this chapter.

In the mid nineties an alternative way of presenting the finite-volume concept was introduced by Dormy and Tarantola (1995), taking Gauss's theorem as a starting point. As I find this approach very attractive it will also be briefly presented.

Let us start by posing a simple question regarding the transport (advection) of something (e.g. a tracer density in a flowing river, an isotope in an ocean, etc.). To describe what is happening we put ourselves into a *finite volume* cell that we denote as $\mathscr{C}$ (to keep it simple we stick here to 1D) and define the cell as limited by $x \in x_l, x_r$. We further assume a positive advection speed a. This set-up is illustrated in Fig. 8.4. The total mass of a quantity (e.g. tracer density, pressure) inside the cell is

$$\int_{x_l}^{x_r} q(x,t)\,dx \tag{8.8}$$

and a change in time can only be due to fluxes across the left and/or right cell boundaries. Thus

$$\partial_t \int_{x_l}^{x_r} q(x,t)\,dx = F_l(t) - F_r(t), \tag{8.9}$$

where $F_i(t)$ are rates (e.g. in g/s) at which the quantity flows through the left and right boundaries. If we assume advection with a constant transport velocity a this flux is given as a function of the values of $q(x,t)$ as

$$F \rightarrow f(q(x,t)) = aq(x,t); \tag{8.10}$$

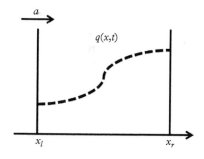

Fig. 8.4 *The finite-volume method is based on the concept of describing the evolution of a density (energy) field $q(x,t)$ inside a finite volume $\mathscr{C}$ over time.*

in other words

$$\partial_t \int_{x_l}^{x_r} q(x,t)\,dx = f(q(x_l,t)) - f(q(x_r,t)). \tag{8.11}$$

This is called the integral form of a hyperbolic conservation law. We can now make use of the definition of integration and antiderivates to obtain

$$\partial_t \int_{x_l}^{x_r} q(x,t)\,dx = -\int_{x_l}^{x_r} \partial_x f(q(x,t))\,dx$$
$$\int_{x_l}^{x_r} [\partial_t q(x,t) + \partial_x f(q(x,t))]\,dx = 0, \tag{8.12}$$

which leads to the well-known partial-differential equation of linear advection

$$\partial_t q(x,t) + \partial_x f(q(x,t)) = 0. \tag{8.13}$$

This simple and elegant derivation can also be developed for the problem of acoustic wave propagation, leading directly to the wave equation in first-order form. The interested reader is referred to Leveque (2002), Section 2.9.1. Leveque (2002) also makes the point that, despite the fact that the second-order wave equation is often described as the fundamental one in many books, it is in fact the first-order system that is the more fundamental equation. Also, efficient numerical methods are more easily derived for first-order systems than for second-order systems.

The upwind scheme

At this point let us start developing a numerical approximation (i.e. discretization) that we can solve on a computer. To do this, instead of working on the field $q(x,t)$ itself we approximate the integral of $q(x,t)$ over the cell $\mathscr{C}$ by

$$Q_i^n \approx \frac{1}{dx} \int_{\mathscr{C}} q(x,t^n)\,dx. \tag{8.14}$$

This is the average value of $q(x,t)$ inside the cell (see Fig. 8.5). In order to find an extrapolation scheme to approximate the future state of our finite-volume cells, we integrate Eq. 8.11 over time:

$$\int_{\mathscr{C}} q(x,t^{n+1})\,dx - \int_{\mathscr{C}} q(x,t^n)\,dx$$
$$= \int_{t_n}^{t_{n+1}} f(q(x_l,t))\,dt - \int_{t_n}^{t_{n+1}} f(q(x_r,t))\,dt, \tag{8.15}$$

where we rearranged terms and divided by dx in order to recover the average cell values, which we use in the numerical scheme below. Note that this equation is exact!

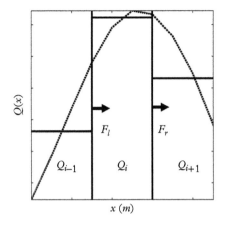

Fig. 8.5 *The finite-volume method: The cell averages (solid lines) are balanced by fluxes $F_{l,r}$ through the left and right boundaries. In this illustration there is advection to the right only (positive advection speed a).*

Finally, using the following terms for the fluxes at the boundaries,

$$F_{l,r}^n = \frac{1}{dt} \int_{t_n}^{t_{n+1}} f(q(x_{l,r}, t)) dt, \tag{8.16}$$

we obtain a first-order time-discrete scheme for the average values of our solution field $q(x, t)$ (see Eq. 8.14):

$$Q_i^{n+1} = Q_i^n - \frac{dt}{dx} (F_r^n - F_l^n), \tag{8.17}$$

where the upper index n denotes time level $t_n = ndt$ and the lower index i denotes cell $\mathscr{C}_i$ of size dx. It is worth noting that we reformulated our problem—defined as a space–time partial differential equation—without making use of space derivatives. The holy grail of finite-volume methods is an accurate representation of the flux terms in Eq. 8.16.

The simplest, and in practice most often-used, numerical flux is developed using the physics of the problem itself. We know that for hyperbolic problems the mass (tracer, energy, information) propagates along so-called characteristics. In seismological terms this is related to the question of how far a point of constant phase propagates in a time interval dt. This is illustrated in Fig. 8.6, here defining x_i to be the cell boundary coordinates and dx the constant cell size.

We thus seek to approximate the next cell update Q_i^{n+1}, knowing that

$$Q_i^{n+1} \approx q(x_i, t^{n+1}) = q(x_i - adt, t^n). \tag{8.18}$$

Information can only come from the cell to the left Q_{i-1}^n, and Q_i^{n+1} obviously will only change if the adjacent cells have different averages. Thus, we can predict the new cell average Q_i^{n+1} analytically by adding the appropriate mass flowing via the left boundary by interpolation. This comes down to simply calculating the

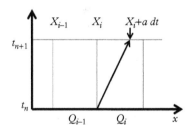

Fig. 8.6 *The upwind method. For the linear advection problem with positive speed a we can analytically predict where the tracer (information, seismic phase) will be located after time dt. The value of $q(x_i, t^{n+1})$ will be exactly the same as $q(x_i - adt, t^n)$. We can use this information to predict the new cell average Q_i^{n+1}.*

fraction of $Q_i^n - Q_{i-1}^n$ that makes it into the cell during time dt. We know velocity a and according to Fig. 8.6 this fraction is $(dx - a\,dt)/dx$.[1] We thus obtain

$$Q_i^{n+1} = Q_{i-1}^n + \frac{dx - adt}{dx}(Q_i^n - Q_{i-1}^n)$$

$$Q_i^{n+1} = Q_i^n\left(1 - \frac{adt}{dx}\right) + Q_{i-1}^n\frac{adt}{dx}. \tag{8.19}$$

After re-arranging we finally obtain a fully discrete scheme

$$Q_i^{n+1} = Q_i^n - \frac{adt}{dx}(Q_i^n - Q_{i-1}^n) \tag{8.20}$$

using only values from the direction where the 'wind' is coming from (here, from the left, see Fig. 8.7). This is the classic *upwind* scheme used in many advection-type problems (e.g. meteorology, ocean circulation). Referring to the semi-discrete integral Eq. 8.17 we can denote the right and left numerical fluxes as

$$F_{ir}^n = aQ_i^n$$

$$F_{il}^n = aQ_{i-1}^n. \tag{8.21}$$

Surprise, surprise! The experienced finite-difference modeller immediately recognizes this result. Eq. 8.20 is nothing but a finite-difference approximation of the scalar advection problem using a backward definition of the finite difference. The finite-volume approach does often lead to finite-difference-like algorithms, but to state that it is nothing but a simple finite-difference scheme is not correct. The 1D example illustrated here does not highlight the power of this concept for problems with arbitrary cell shapes and very strong discontinuities! More about this in Section 8.6.

The numerical scheme developed so far is further illustrated in Fig. 8.8 for both possible flow directions. Knowing the structure of the advection problem, the cell updates can be analytically calculated using only information from adjacent cells. In terms of large-scale computations this is important. Finite-volume schemes are always explicit and local schemes where the future of physical systems are estimated by only looking at the immediate neighbourhood. Such schemes lend themselves to efficient parallelization.

As demonstrated above, for the scalar advection problem the question of how cell averages have to be updated at both sides of a cell (element) boundary could be answered in a straightforward way using analytical solutions to the advection problem. For more complicated hyperbolic systems (like the elastic wave equation) the solution to this problem is more difficult. The general solution to this is called the *Riemann problem* and we will make use of the related concepts when discussing wave propagation in heterogeneous elastic media.

Finally, a word on stability. By looking at Eq. 8.19 we realize that this equation only makes sense if

$$\left|\frac{adt}{dx}\right| \le 1, \tag{8.22}$$

Fig. 8.7 *Looking upwind. Sometimes it is obvious where the wind is coming from (picture taken near South Point, Hawai'i—one of the windiest places on Earth).*

[1] Another way of writing this is

$$dx\left(1 - a\frac{dt}{dx}\right)$$

and this should ring a bell! If dt is too large we would advect beyond the next cell boundary invalidating our assumptions. Of course this corresponds to a stability criterion (see discussion below).

because otherwise the information (mass, tracer, energy, etc.) would propagate beyond the next cell boundary within the time interval dt, and the cell update would be wrong. Note also, as was the case with the lowest-order finite-difference method, a special case is $adt/dx = 1$, where the solution is exact to numerical precision, which, however, has no practical significance for realistic cases.

Without providing an analytical proof we note that the simple upwind scheme just developed is of first-order accuracy only and very dispersive. Thus it is not accurate enough to be of any use for actual simulation tasks. However, the methods based on constant cell averages can be extended to higher orders. This will be demonstrated in the next section.

The Lax–Wendroff scheme

In this section we will encounter a mathematical trick with which high-order extrapolation schemes can be developed. This approach was first presented by Lax and Wendroff (1960), making use of a concept called the Cauchy–Kowaleski procedure which replaces all the time derivatives by space derivatives using the original partial-differential equation.

Our goal is to find solutions to $\partial_t Q + a\partial_x Q = 0$. We start by using the Taylor expansion to extrapolate $Q(x, t)$ in time to obtain

$$Q(x, t^{n+1}) = Q(x, t^n) + dt\partial_t Q(x, t^n) + \frac{1}{2}dt^2\partial_t^2 Q(x, t^n) + \dots. \qquad (8.23)$$

From the governing equation we are also able to state by additional differentiations

$$\begin{aligned}\partial_t^2 Q &= -a\partial_x\partial_t Q \\ \partial_x\partial_t Q &= \partial_t\partial_x Q = \partial_x(-a\partial_x Q) \\ \partial_t^2 Q &= a^2\partial_x^2 Q,\end{aligned} \qquad (8.24)$$

noting that we just derived the second order form of the acoustic wave equation (space–time dependencies omitted for brevity). We can now proceed and replace the time derivatives in Eq. 8.23 with the equivalent expressions containing space derivatives only and obtain

$$Q(x, t^{n+1}) = Q(x, t^n) - dt\, a\partial_x Q(x, t^n) + \frac{1}{2}dt^2 a^2\partial_x^2 Q(x, t^n) + \dots. \qquad (8.25)$$

Using central differencing schemes for both space derivatives

$$\begin{aligned}\partial_x Q(x, t^n) &\approx \frac{Q_{i+1}^n - Q_{i-1}^n}{2dx} \\ \partial_x^2 Q(x, t^n) &\approx \frac{Q_{i+1}^n - 2Q_i^n + Q_{i-1}^n}{dx^2}\end{aligned} \qquad (8.26)$$

we finally obtain a fully discrete second-order scheme for the extrapolation of our cell average Q, with the upper index denoting time and the lower index denoting space discretization

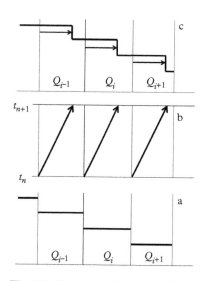

Fig. 8.8 *The finite-volume method, numerical fluxes.* **a:** *Average cell values Q_i^n at time step n.* **b:** *Information propagates along so-called characeristic curves a distance adt into adjacent cells.* **c:** *Differences in cell averages have propagated from left boundaries into adjacent cells and new cell averages can be updated analytically.*

$$Q_i^{n+1} = Q_i^n - \frac{adt}{2dx}(Q_{i+1}^n - Q_{i-1}^n) + \frac{1}{2}\left(\frac{adt}{dx}\right)^2 (Q_{i+1}^n - 2Q_i^n + Q_{i-1}^n), \qquad (8.27)$$

known as the Lax–Wendroff scheme.

Here we derived this scheme using standard finite-difference considerations without making use of flux concepts. However, this result can also be obtained by extending the finite-volume method towards higher order by approximating the solution inside the finite volume as a piecewise linear function of which the slope is determined by interpolation; see Leveque (2002) for details. The choice of slope considered (upwind, downwind, centred) then determines the specific second-order numerical scheme that evolves. The Lax–Wendroff scheme corresponds to the use of the downwind slope.

The Lax–Wendroff scheme can also be interpreted as a finite-volume method by considering the flux functions

$$\begin{aligned}
F_l^n &= \frac{1}{2}a(Q_{i-1}^n + Q_i^n) - \frac{1}{2}\frac{dt}{dx}a^2(Q_i^n - Q_{i-1}^n) \\
F_r^n &= \frac{1}{2}a(Q_i^n + Q_{i+1}^n) - \frac{1}{2}\frac{dt}{dx}a^2(Q_{i+1}^n - Q_i^n)
\end{aligned} \qquad (8.28)$$

that enter Eq. 8.17. For more details on high-order finite-volume scheme reader is referred to Leveque (2002). We are now ready to implement our first numerical finite-volume scheme for the scalar advection problem.

Table 8.1 *Simulation parameters for scalar 1D advection with the finite-volume method.*

Parameter	Value
x_{max}	75,000
nx	6,000
c	2,500 m/s
dt	0.0025 s
dx	12.5 m
$\epsilon = c\,dt/dx$	0.9
σ (Gauss)	200 m
x_0	1,000 m

8.4   Scalar advection in 1D

We proceed with implementing the two numerical schemes: (1) the upwind method and (2) the Lax–Wendroff scheme. Recalling their formulations,

$$Q_i^{n+1} = Q_i^n - \frac{adt}{dx}(Q_i^n - Q_{i-1}^n) \qquad (8.29)$$

and

$$Q_i^{n+1} = Q_i^n - \frac{adt}{2dx}(Q_{i+1}^n - Q_{i-1}^n) + \frac{1}{2}\left(\frac{adt}{dx}\right)^2 (Q_{i+1}^n - 2Q_i^n + Q_{i-1}^n), \qquad (8.30)$$

it is straightforward to initialize the discrete solution field Q and make an appropriate choice for the remaining parameters. An example is given in Table 8.1. To keep the problem simple we use a spatial initial condition, a Gauss function with half-width σ,

$$Q(x, t = 0) = e^{-1/\sigma^2 (x-x_0)^2}, \qquad (8.31)$$

which is advected with speed $c = 2,500$ m/s. The analytical solution to this problem is a simple translation of the initial condition to $x = x_0 + c\,t$, where $t = j\,dt$ is the simulation time at time step j.

```
# [...]
# Time extrapolation
for j in range(nt):
    # upwind
    if method == 'upwind':
        for i in range(1,nx-1):
            # Forward (upwind) (c>0)
            dQ[i] = (Q[i] - Q[i-1])/dx
        # Time extrapolation
        Q = Q - dt * c * dQ
    # Lax Wendroff
    if method == 'Lax-Wendroff':
        for i in range(1, nx-1):
            # Forward (upwind) (c>0)
            dQ1[i] = Q[i+1] - 2 * Q[i] + Q[i-1]
            dQ2[i] = Q[i+1] - Q[i-1]
        # Time extrapolation
        Q = Q - c/2*dt/dx*dQ2 + 0.5*(c*dt/dx)**2 *dQ1
    # Boundary condition
    # Periodic
    # Q[0] = Q[nx-2]
    # Absorbing
    Q[nx-1] = Q[nx-2]
# [...]
```

The code snippet above illustrates the Python implementation of the time extrapolation discussed earlier omitting the specification part. The fields $Q, dQ, dQ1, dQ2$ are vectors with nx elements. Before the time extrapolation, Q is initialized with the Gaussian function specified above.

Note that the spatial loop omits the boundary points; this allows us to implement specific boundary conditions (e.g. circular, absorbing, reflecting) by using so-called ghost cells. In this case the physical boundaries are the left and right limits of cells 2 and $nx - 1$, respectively. In the case of positive advection velocity c, we can implement periodic and absorbing boundary conditions with the statements

$$\text{Periodic: } Q_1^n = Q_{nx-1}^n$$
$$\text{Absorbing: } Q_{nx}^n = Q_{nx-1}^n \tag{8.32}$$

and according statements for negative advection speeds or propagation in both directions (see Fig. 8.9). We illustrate the algorithm for a homogeneous medium with constant scalar advection velocity c. The heterogeneous case is studied for the elastic wave-propagation problem.

The results of the simulation examples are shown in Fig. 8.10. Comparison with the analytical solution of this simple advection problem illustrates

Fig. 8.9 *Boundary conditions. Absorbing or circular boundary conditions can be implemented by using ghost cells outside the physical domain* $x \in [x_0, x_{max}]$.

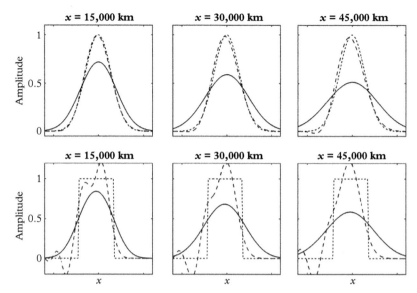

Fig. 8.10 *The finite-volume method, scalar advection. Simulation examples for the scalar advection problem, parameters given in Table 8.1.* **Top:** *Snapshots of an advected Gauss function (analytical solution, dotted line) are compared with the numerical solution of the first-order upwind method (solid line) and the second-order Lax–Wendroff scheme (dashed line) for increasing propagation distances.* **Bottom:** *The same for a boxcar function. In both cases the size of the window is 1.2 km.*

the increasing diffusion of the original signal in the case of the first-order up-wind method.[2] This disqualifies the upwind scheme for any realistic problem. The second-order Lax–Wendroff scheme does not show this strong diffusive behaviour; the peak amplitude is basically stable and the Gaussian waveform remains more or less unchanged. However, a slight shift of the original waveform with reference to the analytical solution develops with increasing propagation distance. This is also a well-known phenomenon that has to be taken into account when deciding on a final parameter set-up for a specific simulation problem. Fig. 8.10 (bottom row) illustrates the behaviour of the numerical solutions for a boxcar initial condition. Again, the upwind scheme diffuses the initial signal in a non-physical way. However, the second-order Lax–Wendroff scheme is not capable of advecting the boxcar function accurately. For seismic wave-propagation tasks this is usually no problem as we propagate smooth wavefields. For other physical problems, that contain discontinuities such as shock waves, more effort has to be undertaken to avoid this numerical behaviour, for example by using adaptive meshes refining during runtime at the discontinuities.

Concluding this introduction of the simplest finite-volume scheme we note that we developed a numerical scheme from first principles, advecting cell averages by analytically predicting the flow across cell boundaries (adopting the analytical solution to the advection problem). What remains to be shown is how this relates to the problem of elastic wave propagation. This will be the topic of the following sections.

[2] Note that this numerical diffusion can be derived analytically using Fourier analysis similar to the approach taken to understand numerical dispersion in the case of the finite-difference method. See Leveque (2002) for details.

8.5 Elastic waves in 1D

Let us look at the source-free version of the coupled first-order elastic wave equation in 1D that we encountered when we discussed staggered finite-difference schemes. Denoting $v = v_y(x, t)$ as the transverse velocity and $\sigma = \sigma_{xy}(x, t)$ as the shear stress component, we obtain

$$\partial_t \sigma - \mu \partial_x v = 0$$

$$\partial_t v - \frac{1}{\rho} \partial_x \sigma = 0. \tag{8.33}$$

where ρ is density and μ is the shear modulus. They can in general be space-dependent, but for the moment we assume they are constant. Here, we encounter for the first time a formulation of this equation that is fundamental for the treatment of coupled hyperbolic equations as linear systems. This applies in particular to the finite-volume method and the discontinuous Galerkin discussed in Chapter 9.

We proceed by writing this equation in matrix–vector notation

$$\partial_t \mathbf{Q} + \mathbf{A} \partial_x \mathbf{Q} = 0, \tag{8.34}$$

where $\mathbf{Q} = (\sigma, v)_T$ is the vector of unknowns and matrix $\mathbf{A}$ contains the parameters

$$\mathbf{A} = \begin{pmatrix} 0 & -\mu \\ -1/\rho & 0 \end{pmatrix}. \tag{8.35}$$

The above matrix equation is formally analogous to the simple advection equation $\partial_t f - a \partial_x f = 0$, which is descriptive of many physical phenomena. But it is coupled. To be able to apply the previously developed numerical tools to this equation we have to find a way to decouple it. This is the main purpose of this section. It will also allow us to come up with an analytical solution.

What needs to be done is to demonstrate the *hyperbolicity* of the wave equation in this form; that is, to show that $\mathbf{A}$ is diagonalizable. If we succeed, it will suffice to discuss the scalar form of this equation that will then be extendible to the vector–matrix form in a straightforward way. Mathematically, it means we have to diagonalize the matrix $\mathbf{A}$, which will allow us to decouple the two equations.

In the case of a quadratic matrix $\mathbf{A}$ with shape $m \times m$ ($m = 2$ in our case), this obviously leads to an eigenvalue problem. If we are able to obtain eigenvalues λ_p such that

$$\mathbf{A} \mathbf{x}_p = \lambda_p \mathbf{x}_p, \quad p = 1, ..., m, \tag{8.36}$$

we get a diagonal matrix of eigenvalues

$$\mathbf{\Lambda} = \begin{pmatrix} \lambda_1 & & \\ & \ddots & \\ & & \lambda_m \end{pmatrix} \tag{8.37}$$

and the corresponding matrix $\mathbf{R}$ containing the eigenvectors $\mathbf{x}_p$ in each column:

$$\mathbf{R} = \left(\mathbf{x}_1 \,|\, \mathbf{x}_2 \,|\, \ldots \,|\, \mathbf{x}_p \right). \tag{8.38}$$

The matrix $\mathbf{A}$ can now be expressed with the definitions

$$\mathbf{A} = \mathbf{R}\mathbf{\Lambda}\mathbf{R}^{-1}$$
$$\mathbf{\Lambda} = \mathbf{R}^{-1}\mathbf{A}\mathbf{R}. \tag{8.39}$$

Applying these definitions to Eq. 8.34 we obtain

$$\mathbf{R}^{-1}\partial_t\mathbf{Q} + \mathbf{R}^{-1}\mathbf{R}\mathbf{\Lambda}\mathbf{R}^{-1}\partial_x\mathbf{Q} = 0, \tag{8.40}$$

and introducing the solution vector $\mathbf{W} = \mathbf{R}^{-1}\mathbf{Q}$ results in

$$\partial_t\mathbf{W} + \mathbf{\Lambda}\partial_x\mathbf{W} = 0. \tag{8.41}$$

Bingo! As $\mathbf{\Lambda}$ is diagonal we now have two decoupled advection equations. Note that this means we have—in the eigensystem—two entirely decoupled scalar advection equations.

What remains to be shown is that in our specific case $\mathbf{A}$ has real eigenvalues. These are easily determined as $\lambda_{1,2} = \mp\sqrt{\mu/\rho} = \mp c$, corresponding to the shear velocity c. For the eigenvectors[3] we obtain

$$\mathbf{r}_1 = \begin{pmatrix} \rho c \\ 1 \end{pmatrix}, \ \mathbf{r}_2 = \begin{pmatrix} -\rho c \\ 1 \end{pmatrix}, \tag{8.42}$$

which, interestingly enough, contain as first elements values of the *seismic impedance* $Z = \rho c$ relevant for the reflection behaviour of seismic waves. Thus, the matrix $\mathbf{R}$ and its inverse are

$$\mathbf{R} = \begin{pmatrix} Z & -Z \\ 1 & 1 \end{pmatrix}, \quad \mathbf{R}^{-1} = \frac{1}{2Z}\begin{pmatrix} 1 & Z \\ -1 & Z \end{pmatrix}. \tag{8.43}$$

While all this seems rather theoretical, it has important practical consequences. The decoupling of the matrix–vector equation initially stated implies that we can proceed by employing the same numerical solution methods developed for the scalar linear advection equation. The eigenvalues take the meaning of the transport velocity. It is instructive to discuss the analytical solution to the homogeneous equation (for the initial value problem) as it can later be used as a benchmark.

[3] Watch out, the standard procedure leads to a truism. One of the components is set to 1 which constrains the other component.

The wave equation in the rotated eigensystem can be stated as

$$\partial_t \begin{pmatrix} w_1 \\ w_2 \end{pmatrix} + \begin{pmatrix} -c & 0 \\ 0 & c \end{pmatrix} \partial_x \begin{pmatrix} w_1 \\ w_2 \end{pmatrix} = 0, \tag{8.44}$$

with the simple general solution $w_{1,2} = w_{1,2}^{(0)}(x \pm ct)$, where the upper index 0 stands for the initial condition (i.e. waveform that is advected). The initial condition also fulfills $\mathbf{W}^{(0)} = \mathbf{R}^{-1}\mathbf{Q}^{(0)}$. We can therefore relate the so-called characteristic variables $w_{1,2}$ to the initial conditions of the physical variables as

$$
\begin{aligned}
w_1(x, t) &= \frac{1}{2Z}(\sigma^{(0)}(x+ct) + Zv^{(0)}(x+ct)) \\
w_2(x, t) &= \frac{1}{2Z}(-\sigma^{(0)}(x-ct) + Zv^{(0)}(x-ct))
\end{aligned}
\tag{8.45}
$$

to obtain the final analytical solution for velocity v and stress σ using $\mathbf{Q} = \mathbf{RW}$ as

$$
\begin{aligned}
\sigma(x, t) &= \frac{1}{2}(\sigma^{(0)}(x+ct) + \sigma^{(0)}(x-ct)) \\
&\quad + \frac{Z}{2}(v^{(0)}(x+ct) - v^{(0)}(x-ct)) \\
v(x, t) &= \frac{1}{2Z}(\sigma^{(0)}(x+ct) - \sigma^{(0)}(x-ct)) \\
&\quad + \frac{1}{2}(v^{(0)}(x+ct) + v^{(0)}(x-ct)).
\end{aligned}
\tag{8.46}
$$

In physical terms one can see that any initial condition in either stress $\sigma^{(0)}$ or velocity $v^{(0)}$ is coupled with the other variable and advected in both $\pm x$ directions with velocity c. In compact form, with the above definitions, this solution can be expressed as

$$\mathbf{Q}(x, t) = \sum_{p=1}^{m} \mathbf{w}_p(x, t)\mathbf{r}_p, \tag{8.47}$$

meaning that any solution is a sum over weighted eigenvectors, a superposition of m waves, each advected without change in shape. The pth wave has the shape $w_p^{(0)}\mathbf{r}_p$ and propagates with a velocity corresponding to eigenvalue λ_p.

Given the hyperbolicity of our linear system of equations descriptive of elastic wave propagation, we can directly apply the numerical schemes developed for the scalar advection case with only slight modifications. To illustrate some fundamental approaches to the flux calculations, we start with the case of homogeneous material.

8.5.1   Homogeneous case

To develop the numerical scheme for elastic wave propagation, we encounter a fundamental problem, briefly mentioned above: the Riemann[4] problem. This problem consists of a single discontinuity as initial condition to an advection problem (or any hyperbolic partial differential equation). This problem is illustrated

[4] Bernhard Riemann (1826–1866) is mostly known for his ground-breaking work in analysis and differential geometry. He died young, of tuberculosis.

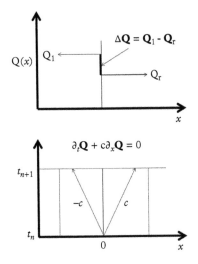

Fig. 8.11 *Riemann problem, homogeneous case.* **Top:** *A discontinuity* ΔQ *is located at* $x = 0$ *as initial condition to the advection equation (e.g. as initial stress discontinuity).* **Bottom:** *The discontinuity propagates along characteristic curves in the space–time domain. The figure illustrates adjacent cells and two time levels* t_n *and* t_{n+1}. *Two waves propagate in opposite directions modifying the values in the cells adjacent to* $x = 0$.

in Fig. 8.11. Why do we need to deal with this? We do not assume continuity of our solution field at the cell boundaries (as was necessary in the finite- or spectral-element method). Thus we have to calculate at each time step how much of the discontinuous field is flowing through the boundary in both directions. Riemann provided a general solution to this problem.

The Riemann problem

To make the link to our problem of elastic wave propagation note that, according to Eq. 8.47, the solution to our problem is a superposition of weighted eigenvectors $\mathbf{r}_p$, in our case $p = 1, 2$. We do not know how they are partitioned. Therefore, we can decompose the discontinuity jump into these eigenvectors with fractions α_1 and α_2, which we can determine according to

$$\Delta Q = Q_r - Q_l = \alpha_1 \mathbf{r}_1 + \alpha_2 \mathbf{r}_2$$
$$\mathbf{R}\alpha = \Delta Q \tag{8.48}$$
$$\alpha = \mathbf{R}^{-1}\Delta Q,$$

where $\mathbf{R}$ is the matrix of eigenvectors as defined above. As expected—for example for a stress discontinuity $\Delta Q = [1, 0]$—we obtain two waves propagating in opposite directions with equal weights $\alpha = [0.5, 0.5]$ (see exercises). The central concept of the flux calculations for finite-volume and discontinuous Galerkin methods is simply realizing that such a discontinuity propagates $c\, dt$ into adjacent cells, therefore changing the cell value by an amount $c\, dt/dx\, \Delta Q$. This is the approach taken in the scalar case. What makes the problem for linear systems more complicated is the fact that we need, first, to decompose the problem into its eigenvectors.

An elegant way of describing this for the elastic problem is to decompose the solution into positive (right-propagating) and negative (left-propagating) eigenvalues:

$$\mathbf{\Lambda}^- = \begin{pmatrix} -c & 0 \\ 0 & 0 \end{pmatrix}, \mathbf{\Lambda}^+ = \begin{pmatrix} 0 & 0 \\ 0 & c \end{pmatrix}. \tag{8.49}$$

Then we can derive matrices $\mathbf{A}^{\pm}$, corresponding to the advection velocity in the scalar case

$$\mathbf{A}^+ = \mathbf{R}\mathbf{\Lambda}^+\mathbf{R}^{-1}$$
$$\mathbf{A}^- = \mathbf{R}\mathbf{\Lambda}^-\mathbf{R}^{-1}, \tag{8.50}$$

allowing us to calculate analytically the fluxes in a finite-volume cell. This is graphically illustrated in Fig. 8.12 for one of the vector components of $\mathbf{Q}$. For cell i we have to propagate the solution fields $\mathbf{Q}$ from adjacent cells in both directions from the corresponding left and right boundaries. Taking the view of cell i we have:

$$-\mathbf{A}^{+}\mathbf{Q}_{i} \quad \rightarrow \text{ right outgoing}$$
$$+\mathbf{A}^{+}\mathbf{Q}_{i-1} \rightarrow \text{ left incoming}$$
$$+\mathbf{A}^{-}\mathbf{Q}_{i+1} \rightarrow \text{ right incoming} \tag{8.51}$$
$$-\mathbf{A}^{-}\mathbf{Q}_{i} \quad \rightarrow \text{ left outgoing.}$$

Adding these contributions we obtain

$$\mathbf{A}^{+}(\mathbf{Q}_{i-1}-\mathbf{Q}_{i})+\mathbf{A}^{-}(\mathbf{Q}_{i+1}-\mathbf{Q}_{i}). \tag{8.52}$$

Defining the cell differences as

$$\begin{aligned} \Delta\mathbf{Q}_{l}&=\mathbf{Q}_{i}-\mathbf{Q}_{i-1}\\ \Delta\mathbf{Q}_{r}&=\mathbf{Q}_{i+1}-\mathbf{Q}_{i} \end{aligned} \tag{8.53}$$

we can formulate an upwind finite-volume scheme for any linear hyperbolic system as

$$\mathbf{Q}_{i}^{n+1}=\mathbf{Q}_{i}^{n}-\frac{dt}{dx}(\mathbf{A}^{+}\Delta\mathbf{Q}_{l}+\mathbf{A}^{-}\Delta\mathbf{Q}_{r}). \tag{8.54}$$

Please note that we recover the signs of the flux contributions to cell i in Eq. 8.51 by recognizing that the terms with $\mathbf{A}^{-}$ are negative (in the scalar case corresponding to the term $-c$). We can relate this formulation to the basic flux concept of Eq. 8.21 with the definitions

$$\begin{aligned} \mathbf{F}_{l}&=\mathbf{A}^{+}\Delta\mathbf{Q}_{l}\\ \mathbf{F}_{r}&=\mathbf{A}^{-}\Delta\mathbf{Q}_{r}. \end{aligned} \tag{8.55}$$

As was already discussed in connection with the scalar advection problem, the first-order upwind solution is of no practical use because of its strong diffusive behaviour. Therefore we present the second-order Lax–Wendroff scheme analogous to the scalar advection case. Interestingly, the high-order scheme does not necessitate the separation into eigenvectors and the matrix $\mathbf{A}$ can be used in its original form. The extrapolation scheme reads

$$\begin{aligned} \mathbf{Q}_{i}^{n+1}=\mathbf{Q}_{i}^{n}&-\frac{dt}{2dx}\mathbf{A}(\mathbf{Q}_{i+1}^{n}-\mathbf{Q}_{i-1}^{n})\\ &+\frac{1}{2}\frac{dt^{2}}{dx^{2}}\mathbf{A}^{2}(\mathbf{Q}_{i-1}^{n}-2\mathbf{Q}_{i}^{n}+\mathbf{Q}_{i+1}^{n}). \end{aligned} \tag{8.56}$$

Let us illustrate the solution using the Lax–Wendroff scheme. The parameters for the simulation are given in Table 8.2. The 1D medium is divided into volume cells of equal size. The time-extrapolation scheme can be implemented as presented in the following code snippet:

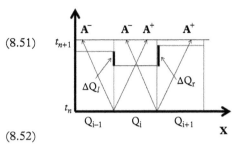

Fig. 8.12 *Finite-volume elastic wave propagation, homogeneous case. The constant average cell values of Q are illustrated for three adjacent cells $i-1, i, i+1$. The eigenvector decomposition leads to wave A^{+} propagating from the left boundary with velocity c into cell i and wave A^{-} propgating with velocity $-c$ into cell i from the right boundary. This determines the flux of discontinuities $\Delta\mathbf{Q}_{l,r}$ into cell i by the amount $c\,dt/dx\Delta\mathbf{Q}_{l,r}$.*

Table 8.2 *Simulation parameters for 1D elastic wave propagation. Homogeneous case.*

Parameter	Value
x_{max}	10,000 m
nx	800
c	2,500 m/s
ρ	2,500 kg/m^3
dt	0.025 s
dx	12.5 m
ϵ	0.5
σ (Gauss)	200 m
x_0	5,000 m

```
# [...]
# Specifications
Q = zeros((2,nx))
Qnew = zeros((2,nx))
# Initial condition
Qnew[0,:] = exp(-1/sig**2 * (x - x0)**2)
# Time extrapolation
for j in range(nt):
    # Lax Wendroff scheme
    Q = Qnew
    for i in range(1,nx-1):
        dQ1 = Q[:,i+1] - Q[:,i-1]
        dQ2 = Q[:,i-1] - 2*Q[:,i] + Q[:,i+1]

        Qnew[:, i] = Q[:,i] - dt/(2*dx) * A @ dQ1 + \
                        0.5 * (dt/dx)**2 * (A @ A) @ dQ2
    # Absorbing boundary conditions
    Qnew[:,0] = Qnew[:,1]
    Qnew[:,nx-1] = Qnew[:,nx-2]
# [...]
 end
```

Note here that the solution vector has the shape $Q(2, nx)$ containing stress values in $Q(1, :)$ and velocity values in $Q(2, :)$ for each volume cell of constant size dx. Note that the size of the grid cells can easily be modified provided that the Courant criterion is satisfied through the global time step.

Results of the simulations are shown in Fig. 8.13 for both solution fields stress (top) and velocity (bottom). The initial condition is a Gaussian-shaped function in stress that is advected in both positive and negative directions with velocity c, corresponding to the modulus of the eigenvalues of matrix **A**. The numerical solution is shown along with the analytical solution. With this simulation set-up they are indistinguishable. An extensive investigation of the accuracy of this approach is left for the computational exercises.

8.5.2   Heterogeneous case

Fortunately, the extension to the heterogeneous case is straightforward. The coefficient matrix **A** is allowed to vary for each element i as

$$\mathbf{A}_i = \begin{pmatrix} 0 & -\mu_i \\ -1/\rho_i & 0 \end{pmatrix}. \tag{8.57}$$

We use again the concept of separating the left- and right-propagating wavefields, defining for each element matrices

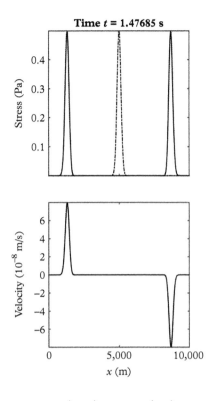

Fig. 8.13 *Finite-volume solution for elastic wave propagation: homogeneous case. The stress–velocity system is advected for an initial condition of Gaussian shape (top, dashed line, scaled by factor 1/2).* **Top:** *Stress snapshot at time t = 1.5 s.* **Bottom:** *Velocity snapshot at the same time. In both cases analytical solutions are superimposed.*

$$\Lambda_i^- = \begin{pmatrix} -c_i & 0 \\ 0 & 0 \end{pmatrix} \ , \ \Lambda_i^+ = \begin{pmatrix} 0 & 0 \\ 0 & c_i \end{pmatrix} \tag{8.58}$$

and using the definitions

$$\mathbf{R} = \begin{pmatrix} Z_i & -Z_i \\ 1 & 1 \end{pmatrix} \tag{8.59}$$

for the matrix with eigenvectors describing the solutions inside element i and $Z_i = \rho_i c_i$. We can determine the corresponding advection terms as

$$\begin{aligned} \mathbf{A}_i^+ &= \mathbf{R}\Lambda_i^+\mathbf{R}^{-1} \\ \mathbf{A}_i^- &= \mathbf{R}\Lambda_i^-\mathbf{R}^{-1}. \end{aligned} \tag{8.60}$$

We can now write down the Euler scheme for the elastic wave equation in the heterogeneous case to obtain

$$\begin{aligned} \Delta\mathbf{Q}_l &= \mathbf{Q}_i - \mathbf{Q}_{i-1} \\ \Delta\mathbf{Q}_r &= \mathbf{Q}_{i+1} - \mathbf{Q}_i \\ \mathbf{Q}_i^{n+1} &= \mathbf{Q}_i^n - \frac{dt}{dx}(\mathbf{A}_i^+\Delta\mathbf{Q}_l + \mathbf{A}_i^-\Delta\mathbf{Q}_r). \end{aligned} \tag{8.61}$$

This numerical solution is again too dispersive to be of any use in practice. However, it contains the flux scheme that is also used in the high-order discontinuous Galerkin method introduced in the next chapter.

As was the case in homogeneous media, a second-order scheme can be obtained with a Lax–Wendroff scheme that does not require separating the wavefields (see Leveque, 2002, for a full derivation). Using the above definition for $\mathbf{A}_i$ the Lax–Wendroff finite-volume scheme can be written as

$$\Delta \mathbf{Q}_l = \mathbf{Q}_i - \mathbf{Q}_{i-1}$$
$$\Delta \mathbf{Q}_r = \mathbf{Q}_{i+1} - \mathbf{Q}_i$$
$$\mathbf{Q}_i^{n+1} = \mathbf{Q}_i^n - \frac{dt}{2dx} \mathbf{A}_i \left[\Delta \mathbf{Q}_l + \Delta \mathbf{Q}_r\right] \qquad (8.62)$$
$$+ \frac{1}{2} \left(\frac{dt}{dx}\right)^2 \mathbf{A}_i^2 \left[\Delta \mathbf{Q}_r - \Delta \mathbf{Q}_l\right].$$

This scheme is implemented in the following Python code fragment with the co-efficient matrices of shape $\mathbf{A}(2, 2, nx)$ defined for each nx finite-volume cells.[5] Note here that changing the cell size is trivial and would only make the dx in the above equation space-dependent (see computer exercises).

[5] The flux scheme used for the homogeneous and heterogeneous cases is called the Godunov upwind flux. Exactly the same flux scheme is applied to the discontinuous Galerkin method. It is important to note that this flux scheme requires constant parameters inside the cells/elements. Other schemes (like so-called *fluctuations*) have to be used for schemes with parameters varying inside the cells/elements. See Leveque (2002) for details.

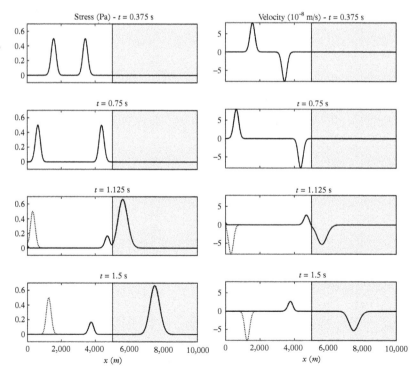

Fig. 8.14 *The finite-volume method, elastic waves in a heterogeneous medium. The two solution fields (stresses Q(1,:) left column, velocities (Q(2,:) right column, both solid lines) are shown for various time steps for an initial stress condition (dashed line, left column). We see stress waves propagating away from the source point and antisymmetric velocity components. At the interface the right-propagating wave speeds up, and a reflection propagates back into the left material. The staggered-grid finite-difference solution (without absorbing boundary) is superimposed (dotted line).*

```
# [...]
# Time extrapolation
for j in range(nt):
    for i in range(1,nx-1):
        dQl = Q[:,i] - Q[:,i-1]
        dQr = Q[:,i+1] - Q[:,i]
        Qnew[:,i] = Q[:,i]-dt/(2*dx)* A[:,:,i]
                    @ (dQl + dQr) + \
        0.5*(dt/dx)**2*A[:,:,i] @ A[:,:,i] @ (dQr - dQl)
    # Absorbing boundary conditions
    Qnew[:, 0] = Qnew[:, 1]
    Qnew[:, nx-1] = Qnew[:, nx-2]
# [...]
```

In our simulation example two media are separated by an interface. The left medium is the same as described in Table 8.2. The right medium has a velocity twice as fast (the shear modulus is 1/4 the value on the left side). The time step is adapted accordingly. The results of the simulation are shown in Fig. 8.14 and compared with the finite-difference method using a staggered-grid solution for the velocity–stress elastic wave equation using identical parameters.

Both solutions are indistinguishable except that the finite-difference solution (dotted line) is reflected from the domain boundaries as no absorbing condition is implemented. The solution shows the expected transmission and reflection behaviour at an elastic material interface. For the parameters used in this simulation the shape of the initial condition remains unchanged and no numerical dispersion is visible.

We can conclude that we have provided a complete description of the elastic wave-propagation problem in 1D, with the assumption that (1) the solution fields (stress and velocities) as well as (2) the elastic parameters μ and λ and thus the impedances $Z = \rho c$ are constant inside the finite-volume cells. Note that these concepts can be applied to grid cells of any shape with straight boundary edges and this is the strength of the finite-volume method.

8.5.3 The Riemann problem: heterogeneous case

Despite the fact that we were able to make use of the theoretical developments for the homogeneous case directly for the heterogeneous case in the previous section, it is instructive to present the theory of the Riemann problem for a material interface. We will closely follow the notation given in Leveque (2002). In fact, it allows us to develop fundamental results for seismic wave propagation, and the reflection and transmission coefficients for perpendicular incidence. The problem is illustrated in Fig. 8.15. At an interface the quantity **Q** to be advected is discontinuous. In addition, the advection velocities on both sides of the interface differ.

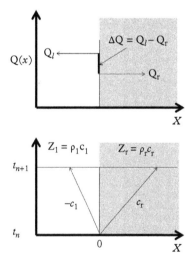

Fig. 8.15 *Riemann problem, heterogeneous case.* **Top:** *Single discontinuity separating two regions with different properties.* **Bottom:** *Velocities and impedances on both sides of the discontinuity. The Riemann problem solves the problem of how waves on both sides are partitioned.*

Mathematically the solution still has to consist of a weighted sum over eigenvectors that now describe solutions in the left and right parts. Following the developments above, the eigenvectors for this problem are

$$\Delta \mathbf{Q} = a_1 \begin{pmatrix} -Z_l \\ 1 \end{pmatrix} + a_2 \begin{pmatrix} Z_r \\ 1 \end{pmatrix} \tag{8.63}$$

for some unknown scalar values $a_{1,2}$. The first term corresponds to the left domain, and the second term to the right domain. This can be written as a linear system of the form

$$\mathbf{R}_{lr}\mathbf{a} = \Delta\mathbf{Q}, \tag{8.64}$$

where $\mathbf{a}$ is a vector and the matrix with eigenvector $\mathbf{R}_{lr}$ is

$$\mathbf{R}_{lr} = \begin{pmatrix} -Z_l & Z_r \\ 1 & 1 \end{pmatrix}, \tag{8.65}$$

with the inverse

$$\mathbf{R}_{lr}^{-1} = \frac{1}{Z_l + Z_r} \begin{pmatrix} -1 & Z_r \\ 1 & Z_l \end{pmatrix}. \tag{8.66}$$

We want to know how these eigenvectors are partitioned (left and right propagating waves with different velocities) given the field discontinuity $\Delta\mathbf{Q}$. This discontinuity cannot be arbitrary. For example, it could correspond to an incident wave from the left. In this case the discontinuity to partition would correspond to

$$\Delta\mathbf{Q} = \begin{pmatrix} Z_l \\ 1 \end{pmatrix}, \tag{8.67}$$

which could be arbitrarily scaled. Consequently, we ask what this implies for the waves propagating in the right domain and, after a possible reflection, in the left domain. With the discontinuous material parameters in matrix $\mathbf{R}_{lr}$ we seek $\mathbf{a}$ such that

$$\begin{aligned} \mathbf{a} &= \mathbf{R}_{lr}^{-1} \Delta\mathbf{Q} \\ &= \frac{1}{Z_l + Z_r} \begin{pmatrix} -1 & Z_r \\ 1 & Z_l \end{pmatrix} \begin{pmatrix} Z_l \\ 1 \end{pmatrix} \\ \begin{pmatrix} a_1 \\ a_2 \end{pmatrix} &= \begin{pmatrix} \frac{Z_r - Z_l}{Z_l + Z_r} \\ \frac{2Z_l}{Z_l + Z_r} \end{pmatrix} = \begin{pmatrix} R \\ T \end{pmatrix}. \end{aligned} \tag{8.68}$$

and obtain—suprise, surprise—the well-known transmission (T) and reflection (R) coefficients for perpendicular incidence at a material discontinuity (see Fig. 8.16). It is worth noting that this result is obtained from first principles, which simply require the conservation of a quantity (here, the vector of stress and velocity) and knowledge of the solution to an advection problem (assuming locally constant parameters).

In the next section we present an alternative view of the finite-volume method.

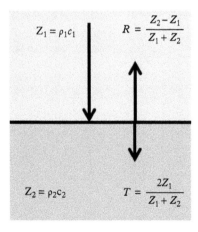

Fig. 8.16 *Reflection and transmission coefficients. Seismic waves incident perpendicular to a material discontinuity are reflected and transmitted according to coefficients R and T. These coefficients can be derived via the Riemann problem used to develop flux schemes for finite-volume methods.*

8.6 Derivation via Gauss's theorem

This view is based on discretizing Gauss's theorem (also known as the divergence theorem), and follows the results of Dormy and Tarantola (1995). Gauss's theorem states that the outward flux of a vector field $Q_i(x, t)$ through a closed surface S is equal to the volume integral of the divergence over the volume V inside the surface at some time t. Mathematically this can be expressed as

$$\int_V \partial_i Q_i dV = \int_S n_i Q_i dS, \tag{8.69}$$

where n_i are the components of the local surface-normal vector. First of all, note that this formulation is not restricted to vector fields—it can generalize to any tensor field $Q_{i,j...}$, and of course also applies to scalar fields $Q(x,t)$. We will use the above relation to estimate partial derivatives of scalar fields (which could represent one component of a vector field).

Assuming the gradient of the solution field is smooth enough and can be assumed constant inside volume V, we can take it out of the integral and obtain an expression for the derivative as a function of an integral over a surface S with segments dS in 3D or a line with segments dL surrounding a surface S in 2D:

$$\partial_i Q \int_V dV = \int_S n_i Q dS$$

$$\partial_i Q_{3D} = \frac{1}{V} \int_S n_i Q dS \tag{8.70}$$

$$\partial_i Q_{2D} = \frac{1}{S} \int_L n_i Q dL.$$

An example in 2D with a surface consisting of linear segments (a polygon) is shown on Fig. 8.17. Once a discretization of surfaces or surrounding lines is found we can develop a discrete scheme replacing the integrals in the above equations by sums to obtain

$$\partial_i Q_{3D} \approx \frac{1}{S} \sum_\alpha n_i^\alpha dS^\alpha Q^\alpha$$

$$\partial_i Q_{2D} \approx \frac{1}{L} \sum_\alpha n_i^\alpha dL^\alpha Q^\alpha. \tag{8.71}$$

The importance here is that this description is entirely independent of the shape of a particular volume. Provided that the space can be filled entirely with polygons, we have a numerical scheme that can be applied to solve partial differential equations.

Let us take a simple example illustrated in Fig. 8.18. A rhombus-shaped finite-volume cell is described by four edge points:

$$P_1 = (-\Delta_1, 0), \ P_2 = (+\Delta_1, 0),$$
$$P_3 = (0, -\Delta_2), \ P_4 = (0, +\Delta_2) \tag{8.72}$$

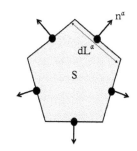

Fig. 8.17 *Concept of numerical gradient calculation using Gauss's theorem, illustrated in 2D. The constant gradient of a scalar field Q inside the finite volume S is approximated as $\partial_i Q = 1/S \sum_\alpha n_i^\alpha dL^\alpha Q^\alpha$. The polygon can have any shape.*

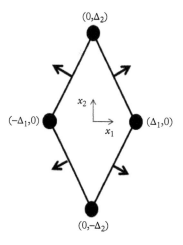

Fig. 8.18 *Rhombus-shaped finite-volume cell used to illustrate the finite-volume approach based on Gauss's theorem.*

with the length of the sides given by $\ell = \sqrt{\Delta_1^2 + \Delta_2^2}$ and the surface $S = 2\Delta_1\Delta_2$. The four normal vectors are defined by

$$n_{2,4} = \begin{pmatrix} \Delta_2 \\ \Delta_1 \end{pmatrix}; \quad n_{4,1} = \begin{pmatrix} -\Delta_2 \\ \Delta_1 \end{pmatrix};$$

$$n_{1,3} = \begin{pmatrix} -\Delta_2 \\ -\Delta_1 \end{pmatrix}; \quad n_{3,2} = \begin{pmatrix} \Delta_2 \\ -\Delta_1 \end{pmatrix}. \tag{8.73}$$

We now have all components necessary to apply Eq. 8.71 in the 2D case. We integrate along the paths suggested in the figure to obtain

$$\begin{aligned}
\partial_1 Q &= \frac{1}{S}\left(\frac{\ell}{2}Q_1\left(-\frac{\Delta_2}{\ell} - \frac{\Delta_2}{\ell}\right) + \frac{\ell}{2}Q_2\left(\frac{\Delta_2}{\ell} + \frac{\Delta_2}{\ell}\right)\right. \\
&\quad \left. + \frac{\ell}{2}Q_3\left(-\frac{\Delta_2}{\ell} + \frac{\Delta_2}{\ell}\right) + \frac{\ell}{2}Q_4\left(\frac{\Delta_2}{\ell} - \frac{\Delta_2}{\ell}\right)\right) \\
&= \frac{1}{S}(-Q_1\Delta_2 + Q_2\Delta_2) \\
&= \frac{Q_2 - Q_1}{2\Delta_1}
\end{aligned} \tag{8.74}$$

for the first derivative of Q with reference to x.

Maybe not so surprisingly—but again with an entirely different approach—we recover the well-known finite-difference formula for the first derivative with a first-order accuracy. While this is an elegant approach, unfortunately, the method is of low-order accuracy and can not in this form easily be extended to high-order accuracy. However, the applicability to arbitrarily shaped volumes is attractive. This allows solution to problems on entirely unstructured grids that can (in 2D) be triangulated and described with Voronoi cells.

Examples of operators so defined and applications to elastic wave-propagation problems can be found in Käser et al. (2001) and Käser and Igel (2001) (see Fig. 8.19).

The strong desire to solve elastic wave-propagation problems on unstructured grids, and frustration with the low accuracy of the simple finite-volume approach and other low-order techniques, eventually led to the search for other more accurate approaches and eventually to the adaptation of the discontinuous Galerkin method to seismic wave propagation (see Käser and Dumbser, 2006).

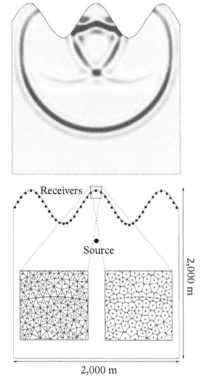

Fig. 8.19 *Example of elastic wave simulations in 2D using difference operators based on the finite-volume approach (Käser and Igel, 2001). In this example an unstructured grid follows a free-surface topography with ghost cells outside the surface to implement stress-free boundary conditions. Reprinted with permission.*

8.7 The road to 3D

In many publications in which numerical methods are compared, it is stated that finite-volume methods are disadvantageous because they are of low-order accuracy. In the developments illustrated above we restricted ourselves to the situation where the solution fields inside the cells are constant (or linear functions, one way of developing a second-order scheme like the Lax–Wendroff method). However, the restriction to low-order representations of the solution fields no longer applies: Dumbser et al. (2007*a*) presented an arbitrary high-order finite-volume scheme in space *and* time that allows the solution of viscoelastic wave propagation on unstructured meshes (see Fig. 8.20). They make use of the so-called ADER scheme (Arbitray high-order DERivative) developed by the group of Prof. Toro at the University of Trento (Titarev and Toro, 2002).

Dumbser et al. (2007*a*) discuss in detail the differences of their high-order finite-volume approach compared with the discontinuous Galerkin method the same group developed shortly before (Käser and Dumbser, 2006). They note that the finite-volume method has advantages concerning the number of degrees of freedom and the overall computation time. A disadvantage is the overhead when reconstructing the high-order representation inside the volume cells. Even in the high-order finite-volume approach it is still the cell average that is updated. As will be seen in Chapter 9 on the discontinuous Galerkin method, the fields inside elements are described by Lagrange polynomials and all polynomial coefficients are updated. Other developments, called spectral finite-volume methods, were pursued by Wang (2002).

In my view the potential of the finite-volume approach for elastic wave-propagation and rupture problems has not yet been fully explored, and there is room for further studies and applications in Earth sciences.

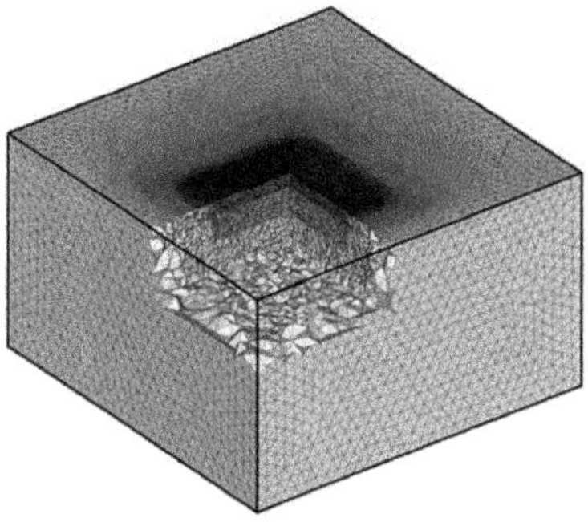

Fig. 8.20 *Computational mesh for finite-volume solution of wave propagation. The mesh based on tetrahedra is refined towards the centre of the model, where simulations are required to be highly accurate. From Dumbser et al. (2007a).*

Chapter summary

- The finite-volume method naturally follows from discretizing conservation equations considering fluxes between finite-volume cells of averaged solution fields.

- The fluxes across boundaries during an extrapolation step are estimated using solutions to the Riemann problem.

- The Riemann problem considers the advection of a single-jump discontinuity, taking into account the analytical solution of the homogeneous problem. It allows an analytical prediction of how much of the material (energy, stress, etc.) enters into or leaves a cell.

- The lowest-order finite-volume solution to the advection equation leads to a finite-difference algorithm with a forward (or backward) spatial differencing scheme, depending on the advection direction. This is called an upwind scheme.

- First-order finite-volume schemes are highly dispersive and are not appropriate for the solution of wave-propagation problems. The second-order Lax–Wendroff scheme does a much better job.

- The problem of elastic wave propagation can thus be formally cast as a first-order hyperbolic problem. Thus, and with only slight modifications, the fundamental schemes developed for the scalar advection problem can be applied to elastic wave propagation.
- In the finite-volume method the problem of estimating partial derivatives (finite differences) is replaced by the requirement to accurately calculate fluxes across cell boundaries.
- A major advantage of the finite-volume method is the fact that the scheme can be easily applied to volume cells of any shape.
- Finite-volume schemes for arbitrary high-order reconstructions inside the cells and high-order time-extrapolation schemes have been developed but not used extensively in seismology.

FURTHER READING

- Leveque (2002) provides an extensive discussion of the finite-volume method for the advection and acoustic/elastic wave-propagation problem. Supplementary electronic material for some of the numerical solutions is also available. The text provides an in-depth discussion of the Riemann problem and its impact for different physical problems.
- The most general formulation of the elastic wave equation for arbitrary high-order accuracy can be found in Dumbser et al. (2007a). Given the thorough discussion of the benefits of the finite-volume method over other schemes in some cases, it is surprising that this approach has so far not been used for research applications.

EXERCISES

Comprehension questions

(8.1) What is the connection between finite-volume methods and conservation equations?

(8.2) What is meant by a finite *volume*; is there any difference between this and a finite *element*?

(8.3) If you look at the upwind approach to the scalar advection problem (Eq. 8.29), why is the finite-volume method so closely linked to staggered-grid finite-difference schemes? Explain.

(8.4) What are the main advantages of finite-volume methods compared with finite-difference methods?

(8.5) Explain the Riemann problem and illustrate why it is so essential for finite-volume schemes.

(8.6) In what areas of natural sciences are finite-volume schemes mostly used? Explore the literature and try to give reasons.

(8.7) What is numerical diffusion? Why is it relevant for finite-volume methods?

(8.8) What is the connection between reflection/transmission coefficients of seismic waves and the finite-volume method?

(8.9) The finite-volume method extrapolates cell averages. What strategies do you see to extend the method to high-order accuracy?

Theoretical problems

(8.10) Show that Eq. 8.6 is a finite-difference solution to the equation $\partial_t Q - a\partial_x Q = 0$ using a forward difference in space.

(8.11) Derive the upwind scheme Eq. 8.17 starting with the scalar advection equation.

(8.12) The stability criterion for the finite-volume method is $cdt/dx \leq 1$. Starting with Fig. 8.7, derive this stability criterion from first principles.

(8.13) Starting with the advection equation $\partial_t Q - a\partial_x Q = 0$ derive the second-order wave equation by applying the so-called Cauchy–Kowalseski procedure (see text).

(8.14) Following the finite-volume approach based on the divergence theorem, calculate the spatial derivative operator for the hexagonal cell shown in Fig. 8.21 and functional values defined at three points P_i.

(8.15) The linear system for elastic wave propagation in 1D (transverse motion) is given in Eq. 8.33. The wave equation can also be formulated for compressional waves using the compressibility K as elastic constant. Reformulate the linear system for acoustic wave propagation and calculate the eigenvalues of the resulting matrix $\mathbf{A}$.

(8.16) For either an elastic or an acoustic linear system derive the eigenvectors of matrix $\mathbf{A}$; the matrix of eigenvectors and its inverse.

(8.17) Show that the superposition of left- and right-propagating stress and velocity waves (Eq. 8.46) are solutions to the linear system of equations (Eq. 8.33) for elastic wave propagation.

(8.18) Show that a discontinuity of the form $\Delta \mathbf{Q} = [1, 0]$ leads to an equi-partitioning of two seismic waves propagating in opposite directions. Start with the Riemann problem formulated for the homogeneous case (Eq. 8.48).

(8.19) Derive reflection and transmission coefficients for seismic waves with vertical incidence by considering the Riemann problem for material discontinuity (Eq. 8.68).

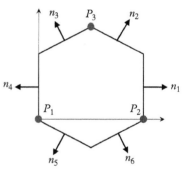

Fig. 8.21 *Hexagonal grid cell with functional values defined at three points.*

(8.20)   Show that the derivation of the reflection and transmission coefficients (Eq. 8.68) is also possible assuming a left-propagating wave with
eigenvector $\Delta \mathbf{Q} = [-Z_r, 1]$.

Programming exercises

(8.21)   Write a finite-volume algorithm for the scalar wave equation from scratch,
implementing both the Euler upwind and the Lax–Wendroff schemes.
Implement the scheme such that you can easily change between the two
approaches. Compare the solution behaviour and discuss the results. To
start, use the parameters given in Table 8.1. Code the analytical solution
and compare the results with the numerical solution.

(8.22)   Determine the stability limit of the Euler and Lax–Wendroff schemes for
the scalar advection equation.

(8.23)   Create a highly unstructured 1D mesh and investigate the accuracy of the
finite-volume method (Lax–Wendroff) for the scalar advection problem.

(8.24)   Investigate the concept of trapped elastic waves by inserting an initial condition in a low-velocity region. Use the Lax–Wendroff algorithm
in 1D.

(8.25)   Implement circular boundary conditions in the 1D elastic Lax–Wendroff
solution. Initiate a sinusoidal function $f(x) = sin(kx)$ that is advected in
one direction. Investigate the accuracy of the finite-volume scheme as a
function of wavelength and propagation distance by comparing with the
analytical solution.

(8.26)   The finite-volume method is supposed to conserve energy in the homogeneous case. Use the computer programs for scalar advection, set
up an example, and calculate the total energy in the system for each
time step. Check whether it is conserved. Explore this problem for the
heterogeneous case.

(8.27)   Scalar advection problem: Advect a Gaussian-shaped waveform for as
long as you can and extract the travel time difference with the analytical
solution in an automated way using cross-correlation. Plot the time error
as a function of propagation distance and your simulation parameters
(e.g. grid points per wavelength, Courant criterion).

The Discontinuous Galerkin Method

'Why yet another method?' you may ask. With the spectral-element method we seem to have given answers to the main simulation requirements in seismology. Yet the devil is in the detail, and there are still some cases where even the spectral-element method based on hexahedral meshes is not the optimal choice. Before we motivate the application of the discontinuous Galerkin method (sometimes also called the *discontinuous Galerkin finite-element method* but I find that too long) to the problem of seismic wave propagation, let us recall the pros and cons of the methods we have encountered so far.

The finite-difference method, still a major workhorse, was presented as a simple, quite flexible low-order method which, however, suffers from the difficulties encountered in implementing boundary conditions with high accuracy for complex shapes. In addition, because of the regular-grid discretization, adaptation to models with strong heterogeneities is difficult. While the pseudospectral method allowed the improvement of the spatial accuracy, this was only possible at the expense of substantially more operations per time step. Boundary conditions were even more difficult to implement efficiently (though this was fixed with the Chebyshev approach). For 3D wave propagation the method was more or less abandoned in its classic form, because of the required global communication scheme making parallelization inefficient.

With the finite-element method the extension to high-order approximations inside the elements was possible and hexahedral or tetrahedral grids allowed meshing of geometrically complex models. One of the main advantages of finite-element-type methods is the implicit accurate modelling of the free-surface (stress-free) boundary condition. However, the drawback is the implicit scheme requiring the inversion of huge system matrices. For hexahedral meshes and a clever combination of interpolation and integration schemes this problem can be fixed, leading to an explicit extrapolation scheme—the spectral element method. The finite-volume method can be considered another attempt to become more flexible in the choice of model geometry.

It turns out that computational meshes with hexahedral elements, even with curved edges, are difficult to generate when boundaries (e.g. surface topography, internal interfaces, faults with complex shapes) need to be honoured. On the other hand—as is well known in the engineering community—tetrahedral meshes are easily generated for arbitrary shapes (see Fig. 9.1) once the surfaces are known

Computational Seismology. First Edition. Heiner Igel.
© Heiner Igel 2017. Published in 2017 by Oxford University Press.

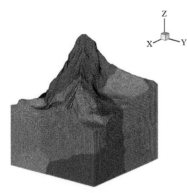

Fig. 9.1 *Tetrahedral grids. Mesh of the famous Matterhorn mountain at the border between Switzerland and Italy. The colours indicate mesh partitioning to different processors. Figure courtesy of Martin Käser.*

[1] A very peculiar remote location, near a giant extinct volcano 2,300 m above sea level in New Mexico. LANL was created during the Second World War for the development of the first atomic bombs. Amongst other things, today it has a strong focus on supercomputing in physics. And great bike riding!

[2] At the time, Martin Käser and Michael Dumbser shared an office at Trento University; their collaboration, together with Josep de la Puente in Munich, led to an impressive, high-speed development of the discontinuous Galerkin method for seismic wave propagation.

in parametric form. Therefore, the question arises of which numerical scheme is capable of efficiently solving the elastic wave equation on tetrahedral (or generally unstructured) grids.

This is the key motivation that led to the transference of the discontinuous Galerkin method to seismology. As the story unfolded, several other beneficial aspects of the method became apparent (e.g. the efficient implementation of local time stepping, high accuracy of frictional boundary conditions for dynamic rupture problems) that now constitute some of its most attractive features.

But let's go back to the beginning. We first review the history of the discontinuous Galerkin method and then describe the method in a nutshell. This will be followed by a discussion of the ingredients of the method and properties in comparison with the other methods encountered so far. Finally, we present the numerical solution to the elastic wave-propagation problem.

9.1 History

The discontinuous Galerkin method was developed in the Los Alamos National Laboratories (LANL)[1] for the problem of neutron transport by Reed and Hill (1973), formulated on triangular meshes. Starting in the late eighties, B. Cockburn and co-workers provided a theoretical framework for the discontinuous Galerkin method in connection with high-order Runge–Kutta-type time integration schemes, summarized in Cockburn et al. (2000). Until then, numerical methods for seismic wave propagation problems had primarily been based on regular grid methods.

The desire to solve wave-propagation problems on unstructured grids appeared when the first solvers for global wave propagation were developed in spherical coordinates (Igel and Weber, 1995; Igel and Weber, 1996). It was obvious that for 3D global wave propagation regular grid methods in spherical coordinates would not work. Therefore, finite-difference-type operators on unstructured grids were considered that could be applied to arbitrary point clouds with point densities following the seismic velocity models, keeping the number of grid points per wavelength constant in the whole domain (Käser et al., 2001; Käser and Igel, 2001).

At the time that the first European training network in computational seismology (SPICE) took off in 2004, with a strong focus on seismic forward modelling, an efficient solution for wave propagation on unstructured tetrahedral grids was still lacking. During a meeting on numerical methods organized by the finite-volume expert E. Toro in Trento, Italy, the idea to apply the discontinuous Galerkin method to seismic wave-propagation problems came up. A quick first glance at the mathematical structure revealed that a transfer from the arbitrary high-order approach developed for the aero-acoustic problem (later published by Dumbser and Munz (2005a and 2005b)) should be straightforward.[2]

The first application to elastic wave propagation was published by Käser and Dumbser (2006) for the 2D case (see Fig. 9.2), later extended to 3D by

Käser et al. (2007*b*). Further rheological models were incorporated, such as viscoelasticity (Käser et al., 2007*a*), anisotropy (de la Puente et al., 2007), and poroelasticity (de la Puente et al., 2008), extending substantially the domains of application.

Käser et al. (2008) carried out a detailed analysis of the convergence properties of the discontinuous Galerkin method. Substantial progress could be made by introducing the concept of local time stepping. Being able to arbitrarily change the mesh density by using tetrahedra is a major advantage. However, when Earth models are simulated with very strong velocity variations the global time step depends on the smallest grid cell and the largest velocity. This implies that often large parts of the models are oversampled. The local time-stepping approach by Dumbser et al. (2007*b*) circumvents this problem, thereby reducing the overall computations.

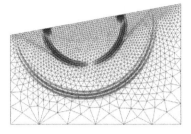

Fig. 9.2 *Wave propagation on triangular meshes. Simulation example with the discontinuous Galerkin method for a vertical force at an inclined free surface (Lamb's problem). Note the flexibility to vary the triangle density throughout the model. Figure from Käser and Dumbser (2006).*

The geometrical flexibility of the discontinuous Galerkin method is an attractive feature for kinematic and dynamic rupture simulation problems. Kinematic rupture scenarios with curved faults were first presented by Käser et al. (2007*b*). de la Puente et al. (2009*a*) implemented frictional boundary conditions that allow the simulation of dynamic rupture. These were later extended to 3D by Pelties et al. (2012). Gallovic et al. (2010) used these concepts to study strong ground motion in the presence of geometrically complex faults and topography.

For elastic wave-propagation problems without strong geometrical complexity or material heterogeneities the discontinuous Galerkin method is most likely not the method of choice, as finite-difference or spectral-element methods provide more efficient solutions. However, for dynamic rupture problems, the discontinuous Galerkin method is currently the most accurate solver, in particlar for complicated fault models (see Chapter 10 on applications).

Further developments of the discontinuous Galerkin method were carried out by the Austin group (Wilcox et al., 2010) using *nodal*[3] basis functions on hexahedral grids with applications to global wave propagation using the cubed-sphere approach (Ronchi et al., 1996). Hermann et al. (2011) introduced an algorithm that allowed combining tetrahedral and hexahedral grids. Etienne et al. (2010) introduced a discontinuous Galerkin method for tetrahedral meshes using nodal basis functions. This methodology was later extended to the problem of dynamic rupture by Tago et al. (2012).

The discontinuous Galerkin method has tremendous flexibility in terms of the variation of element sizes (also called *h*-adaptivity, where *h* stands for a representative element size) across the computational mesh, as well as the option to vary the polynomial order arbitrarily in each cell (called *p*-adaptivity). As any realistic simulation requires the parallelization of the numerical algorithm, it is obvious that *h*-adaptivity, *p*-adaptivity, and an unstructured mesh, combined with local time stepping, create a tremendous challenge for load-balancing a large-scale simulation on current standard parallel computer architectures.

In de la Puente et al. (2009*b*) a first analysis of scaling and synchronization of the discontinuous Galerkin method was undertaken. The challenging

[3] At this point we need to introduce the distinction between *nodal* and *modal* approaches to approximate functions. The nodal approach is what we have used so far, allowing exact interpolation at some set of points. Modal basis functions may be orthogonal but do not have this interpolation property. In this chapter we only discuss the nodal approach.

optimization task piqued the interest of the computational science community in Munich, initiating a re-engineering that eventually led to the *SeisSol* code becoming one of the fastest application codes with close to 50% peak performance (exceeding 1 PFlop in 2014). The code *SeisSol* became a finalist in the prestigious Gordon Bell Competition in 2014 (Breuer et al., 2014).

At the time of writing, the discontinuous Galerkin method is primarily used for dynamic-rupture and strong ground-motion problems, as well as wave-propagation problems with significant geometrical complexity or velocity variations. In the next section we will present a snapshot of the method, before discussing its details.

9.2 The discontinuous Galerkin method in a nutshell

In this section we illustrate qualitatively the most important features of the discontinuous Galerkin method. Basically all concepts that feature as components of this method have been discussed previously. These include (1) the finite-difference extrapolation using, for example, the Euler scheme; (2) the calculation of element-based stiffness and mass matrices; (3) flux calculations at the element boundaries as encountered in the finite-volume method; (4) exact nodal interpolation based on Lagrange polynomials as used in the spectral-element method; and (5) numerical integration schemes using collocation points.

We start with the wave equation in first-order form as used in Chapter 8 on the finite-volume method. With v as velocity, $\sigma = \sigma_{xy} = \sigma_{yx}$ representing the only non-zero stress component, and implicitly assuming space–time dependencies, the wave equation as a coupled system of two first-order partial differential equations reads

$$
\begin{aligned}
\partial_t \sigma &= \mu \partial_x v \\
\rho \partial_t v &= \partial_x \sigma + f.
\end{aligned}
\tag{9.1}
$$

This coupled system can be expressed in matrix–vector form:

$$
\partial_t \mathbf{Q} + \mathbf{A} \partial_x \mathbf{Q} = \mathbf{f},
\tag{9.2}
$$

where $\mathbf{Q} = (\sigma, v)^T$ is the vector of unknowns and $\mathbf{A}$ contains the coefficients of the equation given by

$$
\mathbf{A} = \begin{pmatrix} 0 & -\mu \\ -1/\rho & 0 \end{pmatrix}.
\tag{9.3}
$$

It turns out this is a linear hyperbolic system, with the same form as the classic advection equation. The basic strategy for the discontinuous Galerkin method is the same as for the finite-element method; multiplying the equation by an arbitrary test function combined with describing the unknown fields with the same set of basis functions (the Galerkin principle).

The main differences come with the freedom to allow the unknown fields to be discontinuous at the element boundaries. Obviously, the elements need to communicate information across the boundaries and this is achieved through a flux scheme based on solutions of the Riemann problem which we encountered in Chapter 8. What makes the discontinuous Galerkin method so powerful is the fact that the formulation leads to an entirely local scheme even for high-order extensions.

This principle is illustrated in Fig. 9.3. The wavefield inside each element is described by Lagrange polynomials exactly interpolating at appropriate collocation points. At each time step, a flux term F has to be evaluated at all element boundaries. The extrapolation scheme of the discontinuous Galerkin method can be expressed as

$$\partial_t \mathbf{Q}^k(t) \ = \ (\mathbf{M}^k)^{-1}(\mathbf{K}^k\mathbf{Q}^k(t) - \mathbf{F}^k(\mathbf{Q}^k(t))), \tag{9.4}$$

where $\mathbf{Q}^k(t)$ is the vector of unknowns, $\mathbf{M}^k$ and $\mathbf{K}^k$ are the elemental mass and stiffness matrices respectively, and $\mathbf{F}^k(\cdot)$ is the vector containing the flux terms at the left and right boundaries. The upper index k denotes the element (source term is omitted).

The fact that the elements are only connected through the boundary fluxes, and there is no global system of equations to solve, has important implications: (1) We obtain a fully explicit scheme which lends itself to element-based parallelization; (2) the choice of element size is arbitrary and has no impact on the solution algorithm (h-adaptivity); (3) the polynomial order in each element can be arbitrarily chosen and again has no impact on the algorithm (p-adaptivity); (4) the fact that we have to consider the boundary points twice to calculate the fluxes (see Fig. 9.3) implies an increase in the number of degrees of freedom that of course gets worse with increasing dimensionality.

From the above we can appreciate that the discontinuous Galerkin method in the nodal form is very close to the spectral-element method, with the only difference being the flux terms. There are many choices for how these fluxes can be evaluated. We will proceed with a presentation of the discretization scheme, which leads to the complete algorithmic implementation of the method for the elastic wave equation.

9.3   Scalar advection equation

As already discussed in the chapter on the finite-volume method we can treat the 1(2,3)D wave equation just like the classic advection equation as a hyperbolic partial differential equation. We proceed by seeking a solution of the scalar linear

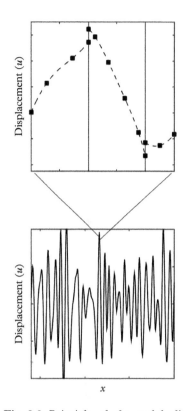

Fig. 9.3 *Principle of the nodal discontinuous Galerkin method.* **Bottom:** *Displacement Snapshot of 1D wave propagation through a heterogeneous medium.* **Top:** *Schematic (exaggerated) representation of the displacement field in three adjacent cells. The squares denote the Lagrange collocation points at which the fields are exactly interpolated. Note the varying size of the elements (h-adaptivity) as well as the different interpolation orders (p-adaptivity). The most important aspect is the discontinuous behaviour at the element egdes that is treated with a flux scheme.*

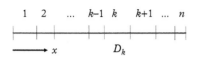

Fig. 9.4 *Indexing scheme for discontinuous Galerkin discretization scheme. The physical domain of each element is denoted by D_k with left and right boundaries x_l^k and x_r^k, respectively. Note the varying element sizes (h-adaptivity).*

advection equation using the discontinuous Galerkin approach. This will be followed by a generalization to the vector–matrix problem through the eigensystem analysis introduced in the previous chapter.

The solution vector for the coupled elastic wave equation was denoted by $\mathbf{Q}(x, t)$. In this section we use $q(x, t)$ for the unknown *scalar* solution and a for the given (possibly space-dependent, here positive) advection (wave) velocity to obtain the source-free advection equation

$$\partial_t q(x, t) + a\, \partial_x q(x, t) = 0, \tag{9.5}$$

which we proceed to solve using the discontinuous Galerkin approach. The discretization scheme in 1D is illustrated in Figure 9.4. The space domain is divided into n elements, each element having the physical domain $x \in D_k$ limited by left and right boundaries $[x_l^k, x_r^k]$ with element size $h^k = x_r^k - x_l^k$. Like in other finite-element-type schemes, the element size h^k for every D_k is variable (*h*-adaptivity). As indicated in the introduction to this chapter there are two approaches to formulating the approximation of the unknown field $q(x, t)$ inside each element: (1) the nodal, and (2) the modal approach. The nodal approach is slightly simpler in its formulation and allows a direct comparison with the spectral-element approach.

The solution field inside our element will be formulated using the same basis functions we encountered in the spectral-element method: Lagrange polynomials. To understand the spatial discretization scheme for the moment it is sufficient to know that for polynomials of order N each element has $N_p = N + 1$ collocation points. The two end points coincide with the boundaries. We do not assume continuity at the boundaries between elements. Therefore, we have two values (in 1D one defined in the left element, one defined in the right element). While we see the theoretical consequences later, this also implies that the number of degrees of freedom is larger. We illustrate this by presenting the shape of the solution vector and the vector of collocation points in the case of the scalar advection equation as it appears later in the implementation. The scalar solution matrix in 1D q_j^k is given as

$$q_j^k = \begin{pmatrix} q_1^1 & q_1^2 & \cdots & q_1^n \\ q_2^1 & q_2^2 & \cdots & q_2^n \\ \vdots & \vdots & \ddots & \vdots \\ q_{N_p}^1 & q_{N_p}^2 & \cdots & q_{N_p}^n \end{pmatrix}, \tag{9.6}$$

where $j = 1, \ldots, N_p$ is the number of points per element and $k = 1, \ldots, n$ the element number. The collocation points at which the solution is calculated are GLL points stored as

$$x_j^k = \begin{pmatrix} x_1^1 & x_1^2 = x_{N_p}^1 & \cdots & x_1^n \\ x_2^1 & x_2^2 & \cdots & x_2^n \\ \vdots & \vdots & \ddots & \vdots \\ x_{N_p}^1 = x_1^2 & x_{N_p}^2 & \cdots & x_{N_p}^n \end{pmatrix}, \tag{9.7}$$

with the same size as q_j^k. The last point of the first element and the first point of the second element coincide, etc. Again it is important to note that the $q(\cdot)$ values defined at these points do not have the same values (see Fig 9.3). In fact this is the *discontinuity* in our method's name. We now proceed with the derivation of the weak form following the discontinuous Galerkin approach.

9.3.1   Weak formulation

To obtain the weak formulation of the scalar advection equation we multiply

$$\partial_t q(x, t) + a\, \partial_x q(x, t) \; = \; 0 \tag{9.8}$$

with a general test function $\phi_j(x)$, integrate over the kth element domain D_k to obtain

$$\int_{D_k} \partial_t q(x, t)\phi_j(x)\,dx + \int_{D_k} a\partial_x q(x, t)\phi_j(x)\,dx = \; 0, \tag{9.9}$$

and integrate by parts[4] replacing the right term containing the space derivative by

$$\int_{D_k} a\partial_x q(x, t)\phi_j(x)\,dx \; = \; [aq(x, t)\phi_j(x)]_{x_l}^{x_r}$$
$$-\int_{D_k} aq(x, t)\partial_x\phi_j(x)\,dx, \tag{9.10}$$

where x_r and x_l are the right and left boundaries of element k, respectively. We assume constant velocity a inside the element.

In the derivation of the finite- (spectral-) element formulation of the second-order wave equation the term in square brackets contains the gradient of the wavefield. Assuming stress-free boundary conditions at the edges of the physical domain this term vanishes, leading to the implicit implementation of the free-surface boundary condition. In that sense, this is the important point of departure from the classic finite-element concept onto which we want to shed a little more light. What does the integration by parts look like with more than one dimension? The definition is known as

$$\int_\Omega \partial_{x_i} uv\,d\Omega = \int_\Gamma uvn_i\,d\Gamma$$
$$-\int_\Omega u\partial_{x_i} v\,d\Omega, \tag{9.11}$$

where u, v are arbitrary space-dependent functions, Ω denotes the entire volume, Γ its boundary, and n_i is a vector normal to the boundary. Note that by setting the function $u = 1$ in the above equation we recover Gauss's theorem by equating the volume integral of the divergence of a vector field with the surface integral

---

[4] Integration by parts is defined as: $\int qv' = [qv] - \int q'v$.

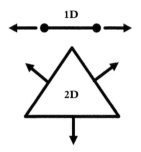

Fig. 9.5 *Integration by parts and Gauss's theorem. In higher dimensions > 1D the partial integration rule leads to Gauss's theorem (see text), illustrating that fluxes at the boundaries need to be considered for the discontinuous Galerkin ansatz.*

over the field itself at the closed boundary. The physical significance of this is important for our numerical method. The surface integral gives the net outflow of the field across the boundary. This justifies calling it the flux term, which will occupy us later for some time (see also Fig. 9.5).

Putting this result back into the advection equation we obtain in general for element k

$$\int_{D_k} \partial_t q(x,t)\phi_j(x)\,dx - \int_{D_k} aq(x,t)\partial_x \phi_j(x)\,dx \\ = -[aq(x,t)\phi_j(x)]_{x_l}^{x_r}, \tag{9.12}$$

where the right-hand-side term is evaluated at the left and right element boundaries x_l and x_r.

So far, our unknown field $q(x,t)$ is a continuous function. In the next step–identical to the procedure for the spectral-element method–we replace $q(x,t)$ by a finite polynomial representation in terms of a weighted sum over Lagrange polynomials[5] inside each element k of order N denoted as $\ell_i(x), i = 1,\ldots,N+1$ defined in the interval $x \in [-1,1]$. The exact interpolation property of the Lagrange polynomials at the GLL collocation points x_i (as presented in detail in Chapter 7 on spectral elements) means that the approximation $\hat{y}(x)$ of function $y(x)$ implies $\hat{y}(x_i) = y(x_i)$, where

$$\hat{y}_i = \sum_{j=1}^{N_p} y(x_j)\ell_j(x_i) = \sum_{j=1}^{N_p} y(x_j)\delta_{ij}. \tag{9.14}$$

For element k we obtain

$$q(x,t)\big|_{x=x_i} = \sum_{i=1}^{N_p} q_i(t)\,\ell_i(x), \tag{9.15}$$

where in each element the polynomial order may vary, and $x = x_i$ are the collocation points. To make notation easier we keep the same symbol $q(x,t)$ from now on for the discrete polynomial representation of the original continuous $q_{\text{exact}}(x,t)$. $N_p = N + 1$ denotes here in general the upper index required for polynomial order N. Again, there is a fundamental difference to the other approaches. This so-called *p-adaptivity* (i.e. varying the polynomial order in different elements) is in general also possible in higher dimensions and is one of the most attractive features of the discontinuous Galerkin approach.

In addition to replacing the unknown field by Lagrange polynomials, we also use them as test functions. This is the well-known *Galerkin* approach,[6] giving the method its name in combination with the *discontinuous* behaviour at the element boundaries.

Combining Eq. 9.15 and the left-hand side of Eq. 9.12 we obtain

$$\int_{D_k} \partial_t \sum_{i=1}^{N_p} q_i(t)\ell_i(x)\ell_j(x)\,dx - \int_{D_k} a\sum_{i=1}^{N_p} q_i(t)\ell_i(x)\partial_x\ell_j(x)\,dx, \tag{9.16}$$

---

[5] Remember the Lagrange polynomials are defined as

$$l_j(x) = \prod_{\substack{1 \le m \le N_p \\ m \ne j}} \frac{x - x_m}{x_j - x_m}. \tag{9.13}$$

See Fig. 7.8 for an illustration. For polynomial order N summation goes to $N_p = N + 1$.

[6] Named after Boris Galerkin (1871–1945), a Russian mathematician and engineer, who, however, referred to Swiss mathematician Walther Ritz (1878–1909) as having discovered the method.

and after re-ordering

$$\sum_{i=1}^{N_p} \left[\left[\int_{D_k} \ell_i(x)\ell_j(x)dx \right] \partial_t q_i(t) - \left[\int_{D_k} a\ell_i(x)\partial_x\ell_j(x)dx \right] q_i(t) \right]. \qquad (9.17)$$

We recognize the familiar ingredients of this equation; the elemental mass M_{ij} and stiffness K_{ij} matrices with similar form as in the spectral element method. Assuming implicit matrix–vector operations we obtain

$$\mathbf{M} \, \partial_t q(t) - \mathbf{K}^T q(t), \qquad (9.18)$$

where the matrices are given by

$$M_{ij} = \int_{D_k} \ell_i(x)\ell_j(x)dx$$
$$K_{ij} = a \int_{D_k} \ell_i(x)\partial_x\ell_j(x)dx, \qquad (9.19)$$

assuming constant velocity a inside element k. Note that the lower index D_k indicates that we are still in physical space and we have to map to the local coordinate system. How are the elements of mass M_{ij} and stiffness K_{ij} matrices calculated in detail?

9.3.2 Elemental mass and stiffness matrices

The local matrices (nodal or modal approach) for arbitrary test functions $\phi_i(x)$ are defined by

$$M_{ij} = \int_{D_k} \phi_i(x)\phi_j(x) \, dx$$
$$K_{ij} = a \int_{D_k} \phi_i(x)\partial_x\phi_j(x) \, dx \qquad (9.20)$$

containing integrals over (derivatives of) the test functions $\phi(x)$. These integrals can in general not be calculated analytically and we have to employ a numerical integration scheme. We proceed with the same approach as in the spectral-element method and replace the integrals by a weighted sum over the function values $f(x_i)$ at carefully chosen points x_i inside the elements

$$\int_\Omega f(x) \, dx \approx \sum_{i=1}^N w_i f(x_i). \qquad (9.21)$$

In the nodal case with Lagrange polynomials, obviously the best choice is to use the GLL collocation points, leading to a diagonal mass matrix as we will see below.

As in any finite-element type method, we need to map our physical coordinates into an element-based system. In 1D this is quite simple using ξ as local variable and transforming via

$$x^k(\xi) = x_l^k + \frac{(1+\xi)}{2}h^k, \quad \xi \in [-1, 1], \tag{9.22}$$

where x_l^k and x_r^k are the left and right physical boundaries of element k, respectively, and h^k is the element size. In general the mapping of the differential used to evaluate integrals is called the Jacobian, which is defined for element k as

$$\mathcal{J}^k = \frac{dx}{d\xi}, \quad \mathcal{J}^k = \frac{x_r^k - x_l^k}{1 - (-1)} = \frac{h^k}{2}, \tag{9.23}$$

where h^k is the size of element k. For the elemental matrices we obtain for arbitrary test functions

$$M_{ij}^k = \int_{D_k} \phi_i(\xi)\phi_j(\xi)\mathcal{J}^k \, d\xi$$

$$K_{ij}^k = a \int_{D_k} \phi_i(\xi)\partial_\xi((\mathcal{J}^k)^{-1})\mathcal{J}^k\phi_j(\xi) \, d\xi \tag{9.24}$$

$$= a \int_{D_k} \phi_i(\xi)\partial_\xi\phi_j(\xi) \, d\xi$$

and we note that for the calculation of the stiffness matrix the Jacobians cancel out. Finally, we can replace the test function with the Lagrange polynomials of order N leading to the definition of the mass and stiffness matrices:

$$M_{ij}^k = \int_{-1}^{1} \ell_i(\xi)\ell_j(\xi) \, \mathcal{J}^k \, d\xi = \sum_{m=1}^{N_p} w_m \, \ell_i(x_m)\ell_j(x_m) \, \mathcal{J}^k$$

$$= \sum_{m=1}^{N_p} w_m \delta_{im} \, \delta_{jm} \, \mathcal{J}^k$$

$$= \begin{cases} w_i \, \mathcal{J}^k & \text{if } i = j \\ 0 & \text{if } i \neq j \end{cases}$$

$$K_{ij}^k = \int_{-1}^{1} a\ell_i(\xi)\partial_x\ell_j(\xi) \, d\xi = \sum_{m=1}^{N_p} a \, w_m \, \ell_i(x_m)\partial_x\ell_j(x_m) \tag{9.25}$$

$$= \sum_{m=1}^{N_p} a \, w_m \delta_{im}\partial_x\ell_j(x_m)$$

$$= a \, w_i \partial_x\ell_j(x_i).$$

Note, as previously obtained for the spectral-element method, the beautifully simple structure of these matrices, in particular the diagonal mass matrix. Again, this was obtained by combining the Lagrange interpolation with the GLL integration scheme. The fundamental difference from the classic finite- (or spectral-) element method is that we *only* determine these matrices at an elemental level. We do not need to assemble them to a global system matrix.

Even though we aim at having each numerical method presented independently at this point we refer to the earlier Section 7.4.3 for the calculation of the derivatives of the Lagrange polynomials at the collocation points $\partial_x l_j(x_i)$. The same routine developed for the spectral-element method can be used. The integration weights required are given in Table 7.1.

We are not ready yet. At this point we have formulations for our unknown field inside the elements but no information is transferred from one element to another. Essentially, the values at the element boundaries are defined twice and it is not clear how to calculate and connect them (see Fig. 9.6 for a sketch). Therefore, we introduce the concept of fluxes, to solve the remaining term in Eq. 9.12.

9.3.3 The flux scheme

As the flux concept is the key difference between the classic Galerkin methods (finite or spectral elements) and the discontinuous Galerkin concept, it deserves some special attention. Let us start with the right-hand side of Eq. 9.12 stated in the original integral form that also holds for higher dimensions (see discussion above and Eq. 9.11):

$$\int_{\partial D_k} a\, q(x,t)\phi_j(x)\, \mathbf{n}\, dx, \qquad (9.26)$$

where $\mathbf{n} = \pm 1$ denotes the vector normal to the boundary, in 1D taking the values $n = -1$ and $n = 1$ at the left and right boundaries, respectively. We proceed systematically as above and replace the space-dependent part of $q(x,t)$ by a sum over Lagrange polynomials (Eq. 9.15) to obtain

$$\sum_{i=1}^{N_p} \int_{\partial D_k} \ell_j(x)\, \ell_i(x)\, a\, q_i(t)\, \mathbf{n}\, dx, \qquad (9.27)$$

also replacing the arbitrary test function with Lagrange polynomials. The orthogonality of the Lagrange polynomials and the fact that we are integrating over the surface ∂D_k (which in 1D consists of the left and right boundaries) leads to

$$\sum_{i=1}^{N_p}(\ell_i(x_r^k)\, \ell_j(x_r^k)(a\, q(x_r^k,t))^* - \ell_i(x_l^k)\, \ell_j(x_l^k)(a\, q(x_l^k,t))^*)$$

$$= \ell_j(x_r^k)(a\, q(x_r^k,t))^* - \ell_j(x_l^k)(a\, q(x_l^k,t))^* \qquad (9.28)$$

where we introduced the starred terms $(a\, q(x_{r,l}^k))^*$ that are the presently undefined values at the element boundaries (see Fig. 9.7). Note that in the general integral formulation above the surface normal $\mathbf{n}$ leads to the signs of the two terms in the bottom line of Eq. 9.28. The expression above corresponds to a flux vector in $j = 1, \dots, N_p$ with only the boundary values x_l^k, x_r^k being non-zero, and we denote it as

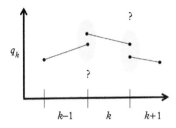

Fig. 9.6 *The discontinuity at the element boundaries. The key task in the discontinuous Galerkin scheme concerns the question of what values to allocate to the points at the element boundaries. This involves the use of flux schemes originating from finite-volume techniques.*

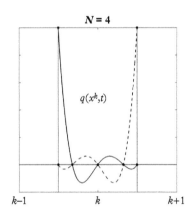

Fig. 9.7 *Illustration of the interpolation scheme for element boundaries. In Eq. 9.28 the terms $\ell_j(x_{r,l}^k)$ single out the boundary points. The values of the solution field $q(x,t)$ at a certain time level is determined by the flux scheme. Lagrange polynomials $\ell_{1,N}$ are plotted with GLL collocation points for $N = 4$.*

$$\mathbf{F} = \begin{pmatrix} F_1 \\ 0 \\ \vdots \\ 0 \\ F_{N_p} \end{pmatrix}. \tag{9.29}$$

The key question is how to determine the specific flux values F_1 and F_{N_p} for each element. As discussed in Chapter 8 on finite volumes, this is called the *Riemann problem* and deals with the question on how to transport a discontinuity (knowing that we are solving an advection problem).

A natural choice seems to be to take the average of the values on both sides of the boundaries, which is called the *central flux* F^c and can be expressed as

$$
\begin{aligned}
F_1^c &= \frac{1}{2}a(q(x_r^{k-1}, t) + q(x_l^k, t)) \\
F_{N_p}^c &= \frac{1}{2}a(q(x_r^k, t) + q(x_l^{k+1}, t))
\end{aligned} \tag{9.30}
$$

for left and right element boundaries, and a is the velocity in element k. The more stable choice is the so-called *upwind flux* F^{up} that basically uses the information from the direction it is coming from. In the scalar advection problem there is only one speed a. In this case we obtain

$$
\begin{aligned}
F_1^{up} &= \begin{cases} a\,q(x_l^k) & \text{if } a \leq 0 \quad (1) \\ a\,q(x_r^{k-1}) & \text{if } a > 0 \quad (2) \end{cases} \\
F_{N_p}^{up} &= \begin{cases} a\,q(x_l^{k+1}) & \text{if } a \leq 0 \quad (3) \\ a\,q(x_r^k) & \text{if } a > 0 \quad (4), \end{cases}
\end{aligned} \tag{9.31}
$$

where we basically use the boundary value of the neighbouring (or current) element depending on the sign of the advection velocity (see Fig. 9.8). Both central and upwind fluxes can be formulated in a compact way, convenient for coding. This formulation reads (implicit time t)

$$
\begin{aligned}
F_1 &= -a\frac{1}{2}(q(x_l^k) + q(x_r^{k-1})) - \frac{|a|}{2}(1-\alpha)(q(x_r^{k-1}) - q(x_l^k)) \\
F_{N_p} &= a\frac{1}{2}(q(x_r^k) + q(x_l^{k+1})) + \frac{|a|}{2}(1-\alpha)(q(x_r^k) - q(x_l^{k+1})),
\end{aligned} \tag{9.32}
$$

where $\alpha = 0$ corresponds to the upwind flux and $\alpha = 1$ to the centred flux scheme.

At this point we have all the ingredients (except the straightforward time extrapolation) to allow us to write our first discontinuous Galerkin solver and investigate its properties.

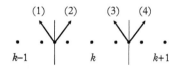

Fig. 9.8 *Illustration of fluxes in the scalar case that will modify the values of element k. The unevenly spaced Gauss–Lobatto–Legendre points are indicated by dots. Refer to Eq. 9.32.*

9.3.4   Scalar advection in action

In this section we will turn the algorithm developed so far into a computer program, and investigate its performance. In matrix notation we obtained for one element

$$\mathbf{M}\partial_t q(t) - \mathbf{K}^T q(t) = -\mathbf{F}(a, q(t)), \tag{9.33}$$

requiring an extrapolation scheme of the form

$$\partial_t q(t) = \mathbf{M}^{-1}(\mathbf{K}^T q(t) - \mathbf{F}(a, q(t))), \tag{9.34}$$

where $F(a, q(t))$ is the flux vector as defined above. We seek to extrapolate the system from some initial condition and obtain for each element, using the simple Euler method,

$$q(t_{n+1}) \approx q(t_n) + dt \left[\mathbf{M}^{-1}(\mathbf{K}^T q(t) - \mathbf{F}(a, q(t))) \right], \tag{9.35}$$

where for the flux scheme $\mathbf{F}(\cdot)$ we use the upwind approach. Note that this is a local (i.e. usually tiny) elemental system of equations. Communication to the outside world happens only through the flux vector $\mathbf{F}(\cdot)$ according to Eq. 9.32. It will turn out that the low-order Euler scheme (which actually works fairly well for the velocity–stress formulation using the finite-difference scheme) is pretty useless for the discontinuous Galerkin formulation as it becomes unstable even at low Courant values. Therefore, we employ a high-order extrapolation procedure known as the *predictor–corrector* method (or Heun's method, or two-stage Runge–Kutta method).[7] Now we are in a position to put everything together and write our first discontinuous Galerkin solver (at least) for the 1D advection problem that we will, with very little modification, later extend to the problem of elastic wave propagation.

The most important initialization step is the calculation of the elemental matrices, mass $\mathbf{M}$ and stiffness $\mathbf{K}$. The following Python code part illustrates a possible implementation looping through all elements *ne* and calculating these matrices as a function of the Jacobian $\mathcal{J}(k) = h^k/2$ where h^k is the size of element k:

```
# Initialize vectors, matrices
Minv = zeros([N+1,N+1,ne])
K = zeros([N+1,N+1,ne])
q = zeros([N+1,ne])
# [...]
for k in range(ne):
    for i in range(N+1):
        Minv[i,i,k] = 1. / w[i] * J[k]
# [...]
for k in range(ne):
    for i in range(N+1):
        for j in range(N+1):
            K[i,j,k] = a[k] * w[i] * l1d[j, i]
# [...]
```

[7] The general formulation for the predictor–corrector method given the problem $\partial_t y = f(t, y)$ is the following. At time step t_i using time increment dt

$$k_1 = f(t_i, y_i)$$
$$k_2 = f(t_i + dt, y_i + dt k_1)$$
$$y_{i+1} = y_i + \frac{1}{2} dt(k_1 + k_2),$$

where in our case $f(\cdot)$ corresponds to the right-hand side of Eq. 9.34. Have a look at the code example to see how this works in practice.

Fig. 9.9 *The matrix–vector form of the discontinuous Galerkin method. The system of questions at an elemental level is illustrated by plotting the absolute matrix/vector values. The corresponding equation is given at the bottom. Only the calculation of the flux vector F necessitates communication with adjacent elements.*

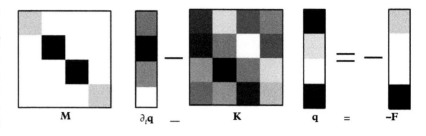

Note the dimensions of the solution vector q and the matrices, depending on the highest degree N of the polynomial approximation leading to a scheme accurate to order $N + 1$ and having $N + 1$ degrees of freedom, that is, points per element. The diagonal mass matrix is kept as a matrix for illustrative purposes, but of course can be stored as a vector. The subroutine $l1d()$ is the same as in the spectral-element method and provides the derivatives of the Lagrange polynomials at the collocation points. Here, we assume that the polynomial order is constant; however this can be easily modified (see exercises). An illustration of the shape and structure of these matrices is given in Fig. 9.9 for an example with $N = 3$, $N_p = N + 1 = 4$ and values taken from a simulation. The mass matrix $\mathbf{M}$ is diagonal, the stiffness matrix $\mathbf{K}$ is full, and the flux $\mathbf{F}$ vector non-zero only in first and last element.

Inside the time loop the flux vector is responsible for the spatial interaction (i.e. transport). The flux vector has to be calculated anew for each time step (or intermediate step when using high-order extrapolation schemes) as it depends on the current values of the wavefield q. An implementation following the definition of Eq. 9.32 is

```
# Flux calculation
F = zeros([N+1,ne])
# [...]
for k in range(ne):
    F[0,k]   =  -0.5 * a *(q[0, k] + q[N+1,k-1]) \
                -0.5*abs(a)*(1-alpha)*(q[N+1,k-1] + q[0,k])
    F[N+1,k] = 0.5*a*(q[N+1,k] + q[0,k+1]) \
                -0.5*abs(a)*(1-alpha)*(q[N+1,k] + q[0,k+1])
# [...]
```

where the variable *alpha* can be used to change the flux scheme from central (*alpha* = 1) to upwind (*alpha* = 0). Note that in the nodal form of the discontinuous Galerkin method the flux vector has only non-zero values at the first and last elements.

The element-wise system of equations can be extrapolated by the Euler scheme as

```
# Extrapolation for every element
for it in range(nt):
    for k in range(ne):
        q[:,k] = q[:,k]+dt*(Minv[:,k]@(K[:,k].T@q[:,k]-
        F[:,k]))
# [...]
```

where *nt* is the overall number of time steps, *dt* is the global time increment, and *ne* is the number of elements. Note that we can directly update the solution vector *q* without intermediate storage at different time level(s). A high-order extrapolation scheme like the predictor–corrector method can be implemented as

```
for it in range(nt):
    # [...]
    # Predictor corrector scheme
    # Initialize flux vectors F for all k
    # [...]
    # First step (predictor)
    for k in range(ne):
        k1[:,k] = Minv[:,k]@(K[:,k].T@q[:,k]-F[:,k])
    # [...]
    # Initialize flux vectors F for q + dt*k1
    # [...]
    # Second step corrector
    for k in range(ne):
        k2[:,k] = Minv[:,k]@(K[:,k].T@(q[:,k]+dt*k1[:,k])
                  -F[:,k])
    # [...]
    # Update
    for k in range(ne):
        q[:,k] = q[:,k] + 0.5*dt*(k1[:,k] + k2[:,k])
# [...]
```

We can see that the price for a high-order extrapolation is basically another solution of the forward problem.

Let us take a concrete example and compare the discontinuous Galerkin approach to other methods. The simulation parameters for our example are given in Table 9.1. Note that, to keep it simple, we initiate the simulation with a spatial initial condition using a Gaussian function $e^{-1/\sigma^2 (x-x_0)^2}$. A source term can be added to the system of equations in a straightforward way (see also the spectral-element method).

The results of the simulations are illustrated in Fig. 9.10. In this figure we compare numerical solutions for the same linear advection problem using four different algorithmic implementations: (1) the simplest upwind finite-volume scheme (equivalent to the finite-difference method); (2) the Lax–Wendroff

Table 9.1 *Simulation parameters for 1D discontinuous Galerkin advection*

Parameter	Value	Meaning
ne	200	elements
N	3	order
a	2,500 m/s	velocity
x_{max}	10,000 m	x-domain
dx_{min}	13.82 m	increment
dt	4.4×10^{-4} s	time step
eps	0.08	Courant
σ	300 m	Gauss width
x_0	1,000 m	source x
t_{max}	3 s	duration

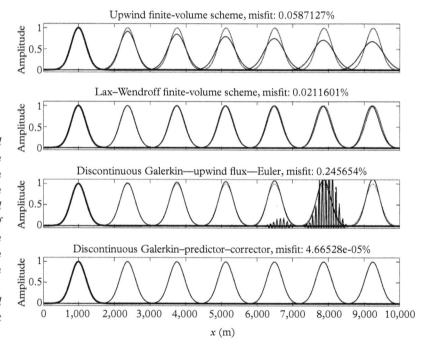

Fig. 9.10 *Comparison of numerical schemes for the advection equation. The parameters of the simulations are given in Table 9.1 and in the text. The title of each plot indicates the method and the relative misfit energy. Snapshots of the advected Gaussian waveform are superimposed for seven different time steps (solid lines). Propagation is from left to right. Initial condition in bold. The analytical solution is superimposed at each time frame (dashed line). See text for details.*

finite-volume scheme (accurate to second order); and the discontinuous Galerkin scheme using (3) the Euler method and (4) the predictor–corrector scheme.

The advecting field is superimposed at seven different times, equal for all four different methods. This presentation allows us to visually appreciate the stability (or lack thereof) of the original waveform given at the left of the spatial domain in bold. Remember that the analytical solution of the advection problem predicts that the initial waveform should be advected without any change in shape at all. This clearly does not seem to be the case in the topmost example, where the upwind finite-volume approach is shown. This numerical solution is characterized by strong numerical diffusion (see Chapter 8 on finite volumes) and is thus of little use for high-accuracy calculations.

A much improved finite-volume solution is obtained by the Lax–Wendroff approach. However, if one looks carefully, an increasing time shift of the numerical solution compared to the analytical solution can be seen. The Euler-based discontinuous Galerkin scheme shows unstable behaviour in the wake of the advected waveform. Keeping the Courant criterion the same and implementing the high-order predictor–corrector method leads to a much improved solution, easily the most accurate of the four presented schemes.

At this point it is instructive to use this simple discontinuous Galerkin scheme to exploit the options concerning the p- and h-adaptivity (see exercises). We aim at providing a solution to the elastic wave equation in 1D. We have shown in the chapter on the finite volume method that the coupled velocity–stress system of equations is formally equivalent to the advection problem. In the next sections we will develop the solution for this case.

9.4 Elastic waves in 1D

We recall the source-free elastic wave equation in 1D with unknown velocities $v(x, t)$ and stresses $\sigma(x, t)$, and the elastic model defined with shear modulus $\mu(x)$ and density $\rho(x)$. With implicit space–time dependencies the coupled system reads

$$\begin{aligned} \partial_t \sigma &= \mu \partial_x v \\ \rho \partial_t v &= \partial_x \sigma, \end{aligned} \tag{9.36}$$

which can be written in matrix–vector form as

$$\partial_t \mathbf{Q} + \mathbf{A} \partial_x \mathbf{Q} = 0, \tag{9.37}$$

where $\mathbf{Q}(t) = (\sigma(t), v(t))$ is the vector of unknowns and $\mathbf{A}(x)$ contains the coefficients defined as

$$\mathbf{A} = \begin{pmatrix} 0 & -1/\rho(x) \\ -\mu(x) & 0 \end{pmatrix}. \tag{9.38}$$

Vectors and matrices are given in bold letters.

We follow the same procedure as above, developing the weak form of this equation assuming an arbitrary test function ϕ_j. The system of equations is multiplied with the test function ϕ_j and integrated over the physical domain D_k representing a 1D element (and **not** the entire physical domain like in the finite element derivation):

$$\int_{D_k} \partial_t \mathbf{Q}(x, t) \phi_j(x) \, dx + \int_{D_k} \mathbf{A} \, \partial_x \mathbf{Q}(x, t) \phi_j(x) \, dx = 0. \tag{9.39}$$

Integration by parts of the second term, omitting most space–time dependencies, leads to

$$\int_{D_k} \partial_t \mathbf{Q} \phi_j(x)\, dx \;-\; \int_{D_k} \mathbf{A}\, \mathbf{Q} \partial_x \phi_j(x)\, dx \;+\; \int_{\partial D_k} \mathbf{A} \mathbf{Q} \phi_j(x)\mathbf{n}\, dx = 0, \qquad (9.40)$$

where $\mathbf{n}$ denotes a normal vector at the element boundaries. This continuous notation is still quite general. For multi-faceted element shapes the boundary integral over ∂D_k has to be evaluated separately taking into account local normal vectors. The first two terms are well known from any Galerkin-type methods we have already encountered. In our 1D system the right term in Eq. 9.40 is an antiderivative and corresponds to the evaluation of $\mathbf{A} \mathbf{Q}$ at both element boundaries. These are the flux terms that will be detailed in the next section.

To develop the full discrete scheme we have to introduce the basis functions for our solution field inside the elements. We follow the same approach as in the scalar advection case (and the spectral-element method) and introduce a nodal representation for the two-element solution vector $\mathbf{Q}(x, t)$. We replace the unknown continuous field $\mathbf{Q}(x, t)$ by a sum over the same nodal basis functions as we employed as test functions. Again we use Lagrange polynomials of order N. Following notation of earlier chapters this leads to N_p collocation points (GLL) for $\xi \in [-1, 1]$. With Nth order Lagrange polynomials $\ell_i(\xi)$ we obtain

$$\mathbf{Q}(\xi, t) = \sum_{i=1}^{N_p} \mathbf{Q}(\xi_i, t)\ell_i(\xi). \qquad (9.41)$$

Following the convention of earlier chapters we use the same letter $\mathbf{Q}$ for the coefficients of the polynomials.[8]

Note that despite the fact that we are solving a 1D physical problem $\mathbf{Q}$ is a 3D matrix with size $[ne, N_p, 2]$ where ne is the number of elements, and N_p the number of GLL collocation points inside an element. The third dimension corresponds to stress and velocity values at the collocation points. For example, element k has solution values

$$\mathbf{Q}_k(\xi_i, t) = \begin{pmatrix} \sigma(\xi_1, t) & \sigma(\xi_2, t) & \dots & \sigma(\xi_{N_p}, t) \\ v(\xi_1, t) & v(\xi_2, t) & \dots & v(\xi_{N_p}, t) \end{pmatrix} \qquad (9.42)$$

at the discrete time level t.

We are ready to assemble our full system of equations by inserting the discrete approximation of $\mathbf{Q}$ into the weak form of the wave equation. Defining a matrix **Flux** with the same shape as our solution field $\mathbf{Q}$ as

$$\mathbf{Flux} \;=\; \int_{\partial D_k} \mathbf{A} \mathbf{Q} \ell_j(\xi)\mathbf{n}\, d\xi, \qquad (9.43)$$

which we will detail later, we re-write Eq. 9.40 as

---

[8] Because of the interpolation properties of the Lagrange polynomials the coefficients correspond directly to the values at the GLL points ξ_i.

$$\sum_i (\partial_t \mathbf{Q}(\xi_i, t) \int_{D_k} \ell_i(\xi) \ell_j(\xi) \mathcal{J} \, d\xi$$

$$-\mathbf{A} \, \mathbf{Q}(\xi_i, t) \int_{D_k} \ell_i(\xi) \partial_\xi \ell_j(\xi) \, d\xi \; = \; - \mathbf{Flux},$$

(9.44)

where $\mathcal{J}$ is the Jacobian $\mathcal{J} = dx/d\xi$. We recognize the well-known mass matrix $\mathbf{M}$ and stiffness matrix $\mathbf{K}$

$$\mathbf{M} = \int_{D_k} \ell_i(\xi) \ell_j(\xi) \mathcal{J} \, d\xi$$

$$\mathbf{K} = \int_{D_k} \ell_i(\xi) \partial_\xi \ell_j(\xi) \, d\xi$$

(9.45)

which are calculated in the same way as in the scalar case. Finally, the semi-discrete scheme can be written in matrix–vector form as

$$\mathbf{M} \partial_t \mathbf{Q} = \mathbf{A} \mathbf{K} \mathbf{Q} - \mathbf{Flux}.$$

(9.46)

By applying a standard first-order finite-difference approximation to the time derivative (Euler scheme) we obtain

$$\mathbf{Q}^{n+1} \approx \mathbf{Q}^n + dt \mathbf{M}^{-1} (\mathbf{A} \mathbf{K} \mathbf{Q} - \mathbf{Flux}),$$

(9.47)

which is the scheme that we will implement in our code. The predictor–corrector scheme can be implemented by analogy with the scalar case (see supplementary electronic material).

9.4.1 Fluxes in the elastic case

We have not completely finished yet; we must draw one more time on results from the finite-volume method. To make notation light, let us consider the situation for one element (avoiding subscripts k everywhere). We assume the coefficients of matrix $\mathbf{A}$ to be constant inside the element. To recall,[9] $\mathbf{A}$ is defined as

$$\mathbf{A} = \begin{pmatrix} 0 & -\mu \\ -\frac{1}{\rho} & 0 \end{pmatrix}$$

(9.48)

and it can be diagonalized by

$$\mathbf{A} = \mathbf{R} \, \Lambda \, \mathbf{R}^{-1},$$

(9.49)

where

$$\mathbf{R} = \begin{pmatrix} Z & -Z \\ 1 & 1 \end{pmatrix}$$

(9.50)

[9] The following relations are described in much more detail in the previous chapter on the finite-volume method, elastic case.

with impedance $Z = \rho c = \rho\sqrt{\mu/\rho}$, $(Z,1)^T$ and $(-Z,1)^T$ representing the eigenvectors. The diagonal matrix Λ

$$\Lambda = \begin{pmatrix} -c & 0 \\ 0 & c \end{pmatrix} \tag{9.51}$$

contains the eigenvalues of matrix $\mathbf{A}$ with $\mp c = \mp\sqrt{\frac{\mu}{\rho}}$. We define a matrix $|\mathbf{A}|$ such that

$$|\mathbf{A}| = \mathbf{R}|\Lambda|\mathbf{R}^{-1} = \begin{pmatrix} c & 0 \\ 0 & c \end{pmatrix} = \begin{pmatrix} \sqrt{\mu/\rho} & 0 \\ 0 & \sqrt{\mu/\rho} \end{pmatrix}, \tag{9.52}$$

and separate Λ into positive (right-) and negative (left-propagating) eigenvalues to obtain the definition of $\mathbf{A}^{\pm}$:

$$\mathbf{A}^- = \mathbf{R}\,\Lambda^-\,\mathbf{R}^{-1} = \frac{1}{2}\begin{pmatrix} -c & -cZ \\ -\frac{c}{Z} & -c \end{pmatrix}$$

$$\mathbf{A}^+ = \mathbf{R}\,\Lambda^+\,\mathbf{R}^{-1} = \frac{1}{2}\begin{pmatrix} c & -cZ \\ -\frac{c}{Z} & c \end{pmatrix}. \tag{9.53}$$

The $\mathbf{A}^{\pm}$ take the meaning of advection velocities in the scalar case previously described. With

$$\mathbf{A} = \mathbf{A}^+ + \mathbf{A}^- \tag{9.54}$$

we arrive at the definitions that are commonly used in discontinuous Galerkin and finite-volume flux formulations:

$$\mathbf{A}^+ = \frac{1}{2}(\mathbf{A} + |\mathbf{A}|)$$

$$\mathbf{A}^- = \frac{1}{2}(\mathbf{A} - |\mathbf{A}|). \tag{9.55}$$

How does the flux scheme work for element k? The term we previously called **Flux** (Eq. 9.43) leads to four flux contributions for left and right sides of the elements by analogy with the scalar case and the finite-volume elastic case, and we obtain

$$\int_{\partial D_k} \mathbf{A}\mathbf{Q}\ell_j(\xi)\mathbf{n}d\xi \;=$$

$$-\mathbf{A}^-\mathbf{Q}_l^k \int_{\partial D_k}^l \ell_i(\xi)\ell_j(\xi)\,d\xi$$

$$+\mathbf{A}^+\mathbf{Q}_r^k \int_{\partial D_k}^r \ell_i(\xi)\ell_j(\xi)\,d\xi$$

$$-\mathbf{A}^+\mathbf{Q}_r^{k-1} \int_{\partial D_k}^r \ell_i(\xi)\ell_j(\xi)\,d\xi$$

$$+\mathbf{A}^-\mathbf{Q}_l^{k+1} \int_{\partial D_k}^l \ell_i(\xi)\ell_j(\xi)\,d\xi, \tag{9.56}$$

where the integral superscripts denote the point (in general boundary) at which the integral has to be evaluated. The latter four integrals in the system of equations correspond to matrices $\mathbf{F}^{l,r}$ defined as

$$
\begin{aligned}
\mathbf{F}^l &= \int_{\partial D_k}^{1} \ell_i(\xi)\ell_j(\xi) \, d\xi \\
\mathbf{F}^r &= \int_{\partial D_k}^{r} \ell_i(\xi)\ell_j(\xi) \, d\xi.
\end{aligned}
\tag{9.57}
$$

In the nodal case $\mathbf{F}^{l,r}$ are of shape $N_p \times N_p$ and basically single out the points at which the fluxes are evaluated (boundaries). Due to the definition of Lagrange polynomials $\ell_1(\xi) = \ell_1(-1) = 1$ or $\ell_{N_p}(\xi) = \ell_{N_p}(1) = 1$. Thus we obtain

$$
\mathbf{F}^l = \begin{pmatrix} 1 & 0 & \dots & 0 \\ 0 & 0 & \dots & 0 \\ & & \ddots & \\ 0 & 0 & 0 & 0 \end{pmatrix}, \quad \mathbf{F}^r = \begin{pmatrix} 0 & 0 & \dots & 0 \\ 0 & 0 & \dots & 0 \\ & & \ddots & \\ 0 & 0 & 0 & 1 \end{pmatrix}.
\tag{9.58}
$$

With these definitions we can specify the **Flux** matrix that will be calculated in the final algorithm at each time step. We obtain

$$
\mathbf{Flux} = -\mathbf{A}_k^- \mathbf{Q}_l^k \mathbf{F}^l + \mathbf{A}_k^+ \mathbf{Q}_r^k \mathbf{F}^r - \mathbf{A}_k^+ \mathbf{Q}_r^{k-1} \mathbf{F}^l + \mathbf{A}_k^- \mathbf{Q}_l^{k+1} \mathbf{F}^r,
\tag{9.59}
$$

where we indicate with subscript k that the coefficient matrix $\mathbf{A}$ may vary for each element, allowing heterogeneous media to be simulated.

We can further illustrate the meaning of the indices l, r in the flux formulation of Eq. 9.59 in the context of an *upwind* flux (see Fig. 9.11). When we are at the left boundary of an element, the *upwind* scheme implies that for a right-propagating wave $\mathbf{A}^+$ we have to consider the element to the left $\mathbf{Q}^{k-1}$ (inflow), and for a left-propagating wave $\mathbf{A}^-$ we use the value inside the current element $\mathbf{Q}^k$ (outflow), and accordingly for the right boundary.

In the code implementation, the flux matrix has the same size as the solution field $\mathbf{Q}$ with $ne \times N_p \times 2$ elements, where ne is the number of elements, N_p the number of collocation points for a scheme of order N, and the last dimension contains the values of stress and velocity at each collocation point.

In a pseudo-code form the flux matrix for element k, calculated at each time step, looks like

$$
\begin{aligned}
\mathrm{Flux}(k,1,:) &= -\mathbf{A}_k^- \mathbf{Q}_l^k - \mathbf{A}_k^+ \mathbf{Q}_r^{k-1} \\
\mathrm{Flux}(k,j,:) &= 0, \quad j = 2, \dots, N_{p-1} \\
\mathrm{Flux}(k,N_p,:) &= \mathbf{A}_k^+ \mathbf{Q}_r^k + \mathbf{A}_k^- \mathbf{Q}_l^{k+1},
\end{aligned}
\tag{9.60}
$$

with implicit matrix–vector multiplication. We will further illustrate this below in the simulation example.

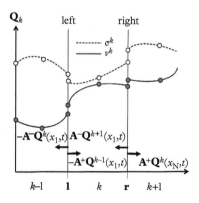

Fig. 9.11 *Flux scheme in the elastic case. For two possible propagation directions fluxes have to be calculated at each element boundary for the two solution fields σ and v. That makes 8 flux terms in total in the 1D case for each element.*

9.4.2 Simulation examples

Finally, let us show how the algorithm works in practice. The beauty is that the discontinuous Galerkin method can basically be assembled using modules from the spectral-element method (Lagrange polynomials, their derivatives, numerical integration) and the finite-volume method (flux scheme).

The algorithm for the extrapolation of the solution matrix $\mathbf{Q}$ (Euler scheme) is

$$\mathbf{Q}^{t+1} \approx \mathbf{Q}^t + dt\mathbf{M}^{-1}(\mathbf{AKQ} - \mathbf{Flux}), \qquad (9.61)$$

with the flux terms given in the previous section. In the following we show several code snippets in Python that illustrate the code structure. The initialization of mass and stiffness matrices is omitted as these are identical to those used in the spectral-element method. Please refer to the supplementary material for details.

The solution field $\mathbf{Q}$ and the relevant coefficient matrices $\mathbf{A}p$ and $\mathbf{A}m$ for positive and negative directions respectively, are initialized with the following shapes:

```
# [...]
# Initalize solution vectors
Q    = zeros((ne, N+1, 2))
Qnew = zeros((ne, N+1, 2))
# initialize heterogeneous A
Ap = zeros((ne,2,2))
Am = zeros((ne,2,2))
R  = zeros((ne,2,2))
L  = zeros((ne,2,2))
```

where *ne* is the number of elements and N the order of the solution scheme. Once the element-dependent impedances $Z_i = \rho_i c_i$ are initialized the system matrix $\mathbf{A}$ can be decomposed into positive and negative parts as coded below:

```
# Initialize flux matrices
for i in range(1,ne-1):
    # Z[i]=rho[i]*sqrt(mu/rho[i])
    # Left side positive direction
    R = array([[Z[i], -Z[i]], [1, 1]])
    L = array([[0, 0], [0, c[i]]])
    Ap[i,:,:] = R @ L @ linalg.inv(R)
    # Right side negative direction
    R = array([[Z[i], -Z[i]], [1, 1]])
    L = array([[-c[i], 0 ], [0, 0]])
    Am[i,:,:] = R @ L @ linalg.inv(R)
```

The matrices $\mathbf{A}m$ and $\mathbf{A}p$ enter the function *flux* that returns the overall fluxes with the same shape as the solution field $\mathbf{Q}$. This routine is called at each time step.

Further variables that are passed are the current solution matrix **Q**, the number of elements *ne* and the order of the scheme *N*.

```
def flux(Q, N, ne, Ap, Am):
# [...]
    # for every element we have 2 faces
                      # to other elements (left and right)
    out = np.zeros((ne,N+1,2))
    # Calculate Fluxes inside domain
    for i in range(1, ne-1):
        out[i,0,:] = Ap[i,:,:]@(-Q[i-1,N,:])+Am[i,:,:]
                     @(-Q[i,0,:])
        out[i,N,:] = Ap[i,:,:]@  Q[i,N,:]    +Am[i,:,:]
                     @Q[i+1,0,:]
    # Boundaries
    # [...]
    return out
```

With this function definition, precalculated mass matrix **M** (and its inverse **M***inv*), and stiffness matrix **K**, we can extrapolate the system for a given initial condition (we employ a Gaussian stress field **Q0** at the first time step) using

```
# [...]
# Euler extrapolation scheme
for it in range(nt):
    # Calculate Fluxes
    # Extrapolate each element using flux F
    Flux = flux(Q, N, ne, Ap, Am)
                  # Loop through all elements
    for i in range(1,ne-1):
        Qnew[i,:,0] = dt*Minv @ \
        (   -mu[i]*K@Q[i,:,1].T-Flux[i,:,0].T)+Q[i,:,0].T
        Qnew[i,:,1] = dt*Minv @ \
        (-1/rho[i]*K@Q[i,:,0].T-Flux[i,:,1].T)+Q[i,:,1].T
# [...]
```

where the stresses **Q***new*[*i*,:, 0] and velocities **Q***new*[*i*,:, 1] are extrapolated separately. As was indicated in the scalar case the Euler implementation is useless from a practical point of view because of its dispersive properties. Therefore, we implement a predictor–corrector scheme as in the scalar case (see supplementary electronic material). The parameters for a homogeneous simulation are given in Table 9.2.

The results are shown in Fig. 9.12. The initial Gaussian-shaped stress distribution leads to stress and velocity waves propagating in both directions away from the source. Visually there is no difference between the analytical solution

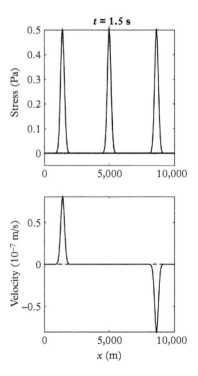

Fig. 9.12 *1D Elastic Case.* **Top:** *Stress waves propagating to the left and right from the initial distribution (dotted line).* **Bottom:** *Velocities propagating with reversed polarity in opposite directions. In both graphs the analytical solution is superimposed. The difference is shown as a dashed line.*

Table 9.2 *Simulation parameters for 1D elastic case.*

Parameter	Value	Meaning
ne	200	elements
N	4	order
c	2,500 m/s	velocity
ρ	2,500 kg/m^3	density
x_{max}	10,000 m	x-domain
dx_{min}	8 m	increment
dt	7×10^{-4} s	time step
eps	0.2	Courant
σ	200 m	Gauss width
x_0	5,000 m	source x

and the numerical solution. Careful analysis of this scheme and a comparison with the other methods encountered so far is left as an exercise. From an algorithmic point of view it is important to note that for Lagrange polynomials of order $N = 0$ (constant function) we recover the classic finite-volume method as introduced in the previous chapter.

Finally, we present a simulation of the discontinuous Galerkin method for a heterogeneous model and compare with a solution of the 1D velocity–stress staggered-grid solution introduced in Chapter 4 on the finite-difference method. We use the same simulation set-up as presented in Table 9.2 except that the density ρ is decreased by a factor of 4 in the right half of the model ($x > 5,000$ m). This leads to a velocity increase by a factor of 2. The stress initial condition is located at $x = 4,000$ m. The results are shown in Fig. 9.13. There is an excellent fit between the two numerical solutions (finite-difference method vs. discontinuous Galerkin method). Both methods capture well the behaviour at the discontinuity.

In the introduction to this chapter we motivated the method by pointing out its flexibility with respect to complex geometries. Of course this cannot be exploited in 1D. However, interesting aspects to investigate are: (1) allowing **A** to vary at each collocation point (possible also with the spectral-element method), and (2) varying the polynomial order (i.e. p-adaptivity) in each element (not advisable for the spectral-element method in higher dimensions). These options can be explored with the supplementary electronic material.

9.5 The road to 3D

The key challenge for 2D or 3D implementations of the discontinuous Galerkin method is the fact that, in practice, you probably want to deal with unstructured meshes. For regular meshes with planar element faces the implementation

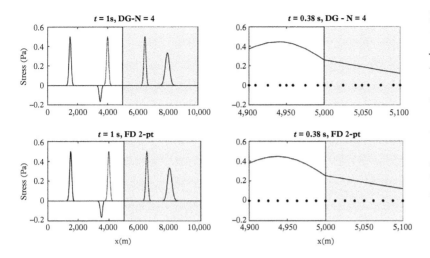

Fig. 9.13 *Simulation in a heterogeneous medium, comparison with the finite-difference method (bottom row). A Gaussian-shaped stress initial condition (dotted line) is propagating in both directions. In the middle of the domain the velocity increases by a factor of 2 leading to both transmission and reflection.* **Top left:** *Discontinuous Galerkin simulation with order* $N = 4$ *(solid line), initial condition (dotted line) and analytical solution for homogeneous case (dashed line).* **Bottom left:** *Same as top left but with finite-difference method.* **Top right:** *Detail around interface, discontinuous Galerkin method.* **Bottom right:** *Same with finite-difference method.*

is straightforward. For triangular, tetrahedral, or curved hexahedral meshes the difficulty lies in the implementation of the flux integrals over all boundaries.

It is important to note that there are two different implementation schemes. The *nodal* approach is what he have used in this demonstration, following the concepts of spectral elements. In this case the solution field is evaluated at the collocation points. An alternative is the *modal* approach in which basis functions are used that do not have these interpolation properties.

The modal approach was used in the 2D isotropic case by Käser and Dumbser (2006), extended to the viscoelastic Käser et al. (2007*a*) and anisotropic de la Puente et al. (2007) cases. Further articles with algorithms in 2D and 3D following this approach are Dumbser et al. (2007*a*) and Dumbser et al. (2007*b*). The latter paper introduced local time-stepping. In that case in principle each element can have its own time step. While a difficult task for load-balancing, this is an attractive feature to reduce overall computational costs. Further analysis of the tetrahedral, modal implementation in 3D was given by Pelties et al. (2012).

Etienne et al. (2010) introduced a 3D nodal scheme using Lagrange basis functions on tetrahedral grids. Mazzieri et al. (2013) published a 3D nodal implementation. Both these algorithms were used to simulate earthquake scenarios.

Because of its intrinsic local character, the natural allowance for arbitrary high-order schemes, and the possibility to vary order in space, there has been an increasing interest in the discontinuous Galerkin method for wave and rupture problems. This is partly due to the fact that in order to be able to use the rapidly increasing number of processors in supercomputers, scaling is a key issue. Further applications of the discontinuous Galerkin method and pointers to community codes are given in Part III and the Appendix.

Chapter summary

- The discontinuous Galerkin method is a finite-element-type method. The main difference to standard finite-element methods is that the solution fields are not continuous at the element boundaries.

- The elemental mass and stiffness matrices are formulated very similarly to how they are in the classic finite-element schemes. However, they are never assembled to a global system of equations. Therefore no large system matrix needs to be inverted.

- Elements are linked by a flux scheme, similar to the finite volume method. This scheme leads to an entirely local algorithm in the sense that all calculations are carried out at an elemental level. Communication happens only to direct neighbours.

- The discontinuous Galerkin scheme can easily be extended to higher orders, keeping the local nature of the solution scheme. This leads to high efficiency when parallelizing.

- The solution fields can be expanded using nodal and modal approaches.

- The discontinuous behaviour at the element boundaries and the associated discretization of the element boundaries increases the number of degrees of freedom compared to other methods.

- The flexibility with polynomial order, element size, and local time stepping leads to a formidable problem when parallelizing a discontinuous Galerkin method. Solution to this problem requires close cooperation with computational scientists.

- In seismology, the discontinuous Galerkin method is useful for problems with highly complex geometries (by using tetrahedral meshes) and for problems with non-linear internal boundary conditions (e.g. dynamic rupture problems).

FURTHER READING

- The book by Hesthaven and Warburton (2008) provides an extensive discussion of the history of and motivation for the introduction of the Discontinuous Galerkin method. The focus is on the nodal version.

- Leveque (2002) discusses in detail the concept of fluxes in connection with the finite-volume method. The same flux schemes are used in the discontinuous Galerkin method.

. . . .

EXERCISES

Comprehension questions

(9.1) List the key points that led to the development of the discontinuous Galerkin method in seismology. Discuss the pros and cons of
the method compared to finite-element-type methods and the finite-
difference method.

(9.2) Explain qualitatively the difference between nodal and modal approaches.

(9.3) Explain why the discontinuous Galerkin method lends itself to parallel
implementation on supercomputer hardware.

(9.4) What are p- and h-adaptivity? Why is it straightforward to have this adaptivity with the discontinuous Galerkin method and not with others? Give
examples in seismology where this adaptivity can be exploited and why.

(9.5) What is local time-stepping? For what classes of Earth models and/or
problems in seismology might it be useful?

(9.6) What is the problem that arises on computers when using algorithms with
h-/p-adaptivity and local time-stepping?

(9.7) Compare the spectral-element and the discontinuous Galerkin methods as described in this volume. Point out their strong similarities
and their differences. Based on this discussion, formulate domains of
application.

Theoretical problems

(9.8) Show that the advection problem $\partial_t q + a \partial_x q = 0$ has a hyperbolic form.

(9.9) The coupled 1D wave equation for longitudinal velocity v and pressure p
can be formulated with compressibility K and density ρ as

$$\partial_t p + K \partial_x v = 0$$
$$\partial_t v + \frac{1}{\rho} \partial_x p = 0. \tag{9.62}$$

Formulate the coefficient matrix $\mathbf{A}$ of the coupled system of equations.
Calculate its eigenvalues and eigenvectors. Compare with the solutions
developed in this chapter for transversely polarized waves.

(9.10) Show that the rule of integration by parts corresponds to Gauss's theorem
in higher dimensions (assuming one of the functions under the integral
to be unity). Explain the relevance of this for the discontinuous Galerkin
method.

(9.11) Show that setting $\alpha = 0$ in Eq. 9.32 leads to the upwind flux scheme.

(9.12) Discuss the size of all matrices and vectors for the 1D solution presented
in Eq. 9.47.

(9.13) Search in the literature for the *classical* four-term Runge–Kutta method. Formulate a pseudo-code for the scalar advection problem for this extrapolation scheme.

Programming exercises

(9.14) Apply the 1D discontinuous Galerkin solution for the scalar advection problem and find numerically the stability limit for the Euler scheme and the Lax–Wendroff scheme. Vary the polynomial order and investigate whether the stability limit changes. Compare with the stability behaviour of the finite-difference method for similar grid-point density.

(9.15) How *discontinuous* is the discontinuous Galerkin method? For the example problems given in the supplementary material, extract the field values at the element boundaries from the adjacent elements and calculate the relative amount of the field discontinuity. How do the discontinuities compare with the flux values?

(9.16) Formulate an upwind finite-difference scheme for the scalar advection problem and write a computer program. Discuss the diffusive behaviour. Compare with the results of the scalar discontinuous Galerkin implementation.

(9.17) Modify the sample code such that each element can have its own polynomial order (*p*-adaptivity) and size (*h*-adaptivity). (Suggestion: Initialize the size of the solution matrices using the maximum number of degrees of freedom N^p_{max}).

(9.18) Extend the sample code for the scalar advection problem to the four-term Runge–Kutta method. Compare the accuracy of the method with the lower-order extrapolation schemes as a function of spatial order N inside the elements.

(9.19) Formulate the analytical solution to the advection problem (see Chapter 8 on the finite-volume method) and plot it along with the numerical solution in each time you visualize during extrapolation. Formulate an error between analytical and numerical results. Analyse the solution error as a function of propagation distance for the Euler scheme and the predictor–corrector scheme.

(9.20) Explore the *p*- and *h*-adaptivity of the discontinuous Galerkin method in the following way. Using an appropriate Gaussian function defined on the entire physical domain, decrease the element size by a factor of 5 towards the centre of the domain. Find an appropriate variation of the order inside the elements to obtain a reasonable computational scheme (in the sense that the grid-point distance does not vary too much). Hint: Use high-order schemes at the edges of the physical domain and low(est)-order schemes at the centre of the domain.

Part III

Applications

Applications in Earth Sciences

How can we make best use of the technologies described in this volume for real Earth science problems? Unfortunately, as indicated many times in the previous chapters, there is no single numerical method that works best in all situations. It is also important to note that the material we have covered here is really just scratching the surface. However, it should allow you to proceed with the next step, studying the literature which has solutions for problems in 3D, often with further improvements of the numerical schemes to make them more accurate.

The goal of this chapter is (1) to provide some fundamental questions you should ask yourself before choosing a specific solver, and (2) to point to some milestone papers and recent examples where research questions were addressed with (2.5D or) 3D algorithms. It is important to note that the following sections are *not* reviews of the specific research fields. This is beyond the scope of this volume. The stress is on *recent* applications of the methods discussed in the previous chapters, mostly in a 3D context. For more information on the developments in the research fields discussed, the reader is referred to the *Further reading* list at the end of Chapter 1.

What are key issues for seismic wave-propagation simulation tasks? Certainly, the list below is non-exhaustive, but it should help you in getting started.

- What is the geometry of your problem? Does it suffice to work with Cartesian coordinates? Do you have spherical or cylindrical geometry?

- Can your problem be reduced to a 2.5D problem (substantially reducing the required resources)?

- What is your desired wavefield frequency range in relation to the size of your model? What is roughly the expected memory requirement for your Earth model and wavefields?

- What is the source–receiver geometry of your problem? Can you make use of reciprocity (interchange sources and receivers) to reduce costs?

- What rheology do you need in order to solve your problem (e.g. elastic, anisotropic, viscoelastic, poroelastic)?

Computational Seismology. First Edition. Heiner Igel.
© Heiner Igel 2017. Published in 2017 by Oxford University Press.

- Which wave types do you want to simulate (body waves, surface waves, complete wavefield)? How important is a highly accurate implementation of the free-surface boundary condition?

- What is the level of geometric complexity of your Earth model (surface topography, internal boundaries such as faults or interfaces?)

- What is the degree of heterogeneity in your geophysical parameters? How much do they change? Is the specific numerical method capable of coping with these parameter changes?

- How would you describe the properties of your Earth model (smooth, layered, etc)? Do you need to honour internal interfaces?

- Can your problem be handled with regular (structured) meshes or do you require irregular (unstructured) meshes?

- Does the problem (or the solver available to you) require computational mesh generation? Are there meshes available for the specific region you want to model or do you have to generate them yourself?

- Does your problem require parallel resources? Is your program adapted to these resources in terms of software (e.g. MPI implementation) or hardware (e.g. CPU or GPU, or both)?

- What is your strategy to check the accuracy of your results?

- What are the data volumes you will create? Can you (or your processing tools) handle these data volumes (data transfer, available disk space, available core memory)? Alternatively, will you have to consider co-processing (i.e. analysing or visualizing results during runtime)?

- Does your problem require specific boundary conditions to be applied (e.g. absorbing boundaries, free-surface boundary)?

- Are you targeting dynamic rupture problems (special internal frictional boundaries apply to pre-defined fault surfaces)?

Once you have given (rough) answers to these questions, there might already be a tendency to favour one or the other numerical method or piece of community software available today. In the following sections we will show some examples from a variety of geo-scientific domains in which the numerical methods discussed in this volume were used to solve scientific problems. The focus here is deliberately not on studies that present technical developments (such references are given in Part II) but on applications to science problems (in 2.5D or 3D) often involving comparison with observations.

As routine (and sufficiently large numbers of) 3D simulations have only become possible in the past few years and continue to require sometimes substantial parallel resources, the number of relevant large-scale research projects is still quite small; but it is increasing rapidly. The examples given in what follows are by no means exhaustive and shall merely reflect the thought processes that go into choosing appropriate solution strategies.[1]

10.1   Geophysical exploration

Several of the applications of numerical methods to the seismic wave-propagation problem were pioneered and developed within the exploration domain. This is not surprising. Exploring the subsurface for resources and monitoring their extraction over time requires the analysis of seismic waves. Without doubt the more realistic the forward modelling, the better the images are. Also, no other field has so much control over source–receiver geometries, and the current 3D marine or terrestrial experiments produce breathtaking data sets. Therefore, it is also fair to say that many of the technical developments and extensions today are done in the research labs of large geophysical exploration companies.[2] Thus, many of these developments are reported in the annual meetings of professional societies such as the European Association of Geoscientists and Engineers (EAGE) and the Society of Exploration Geophysicists (SEG).[3]

What is the character of exploration-type models? Let us list some of the most important features

- The physical domains are such that the spherical nature of the Earth does not need to be taken into account.

- You are simulating in limited domains. Thus, adequately performing absorbing boundaries are important.

- For marine simulation scenarios, free-surface boundary conditions are trivial; however, you have to make sure the fluid–solid boundaries (sea bottom) are reflecting and transmitting correctly.

- When comparing with observations, anisotropy and viscoelasticity have to be taken into account.

- Depending on the target frequency range, the geometry of strong discontinuities might have to be honoured.

- In most cases, body waves are the target seismic phases. In this case low-order implementations of the-free surface boundary condition might suffice.

Exploration geophysicists developed benchmarking projects for forward modelling and inversion with which the emerging technologies could be tested. A famous example is the *Marmousi* model, that was introduced in the late eighties primarily to test migration schemes. The 2D model is shown in Fig. 10.1. There are some interesting features that indicate the challenges of modelling (and imaging) reservoir wave propagation.

First, note that the P-wave velocities range from 1.5 km/s to 5.5 km/s. For an elastic model with corresponding S-velocities this means that wave velocities vary by almost an order of magnitude. From a computational point of view this implies

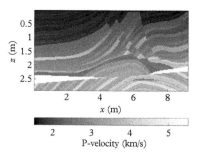

Fig. 10.1 *Marmousi velocity model. A typical exploration-type seismic velocity model used to benchmark solvers and inverse problems. It contains many features that one finds in (marine) sedimentary structures.*

[1] The authors of the projects I have missed out here may forgive me!

[2] Here is a message for students. If you want a job in the exploration industry, do yourself a favour by studying and using simulation technology. The exploration domain job profiles often refer to computational skills in particular. Having said that, many other applied branches outside the geo-domain are making more and more use of simulation technologies.

[3] Note that the level of research in the exploration field is well expressed by the expanded abstracts that are published with these meetings. It is not easy to find large-scale simulation problems with real data in the exploration domain published in journals, though, as much of it is confidential.

that, while in general possible, the simulation of waves through such models with regular grids will necessarily lead to oversampling in parts of the model. Second, the model contains inclined layers of complex discontinuous shape. This raises the question of whether for such models meshes have to be generated that follow these interfaces and honour the shape of faults and discontinuities. As indicated in the introduction, this is the topic of ongoing research. *Homogenization* (see next chapter) would in principle allow replacing the discontinuous structures with a smooth version offering more flexibility concerning the mesh and the numerical method to be used.

Another interesting feature in the Marmousi model is the presence of the high-velocity inclusion (white horizontal feature at the bottom of Fig. 10.1 mimicking a salt body intrusion). There is a drastic velocity jump at the boundaries. It is not obvious that the numerical solvers can handle these jumps correctly (something that would need to be tested before performing final simulations). In terms of imaging, such reflectors are so dominant that they hide the structure underneath.

The efforts to test and verify numerical modelling and imaging methods are currently continued in the SEG-SEAM project (SEG Advanced Modeling Corporation) with substantially more sophisticated test models; see Fehler and Keliher (2011) for an account of the first phase, and subsequent reports in the *Leading Edge* journal.

Recommendations as to which numerical method works best in this situation are difficult. The most common approaches are the finite-difference method, and the finite/spectral-element methods. As indicated above, regular grid methods would require very fine meshes to accurately account for model complexities. On the other hand, honouring interfaces involves the generation of meshes, a process that can be very time consuming.

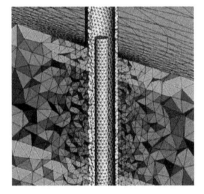

Fig. 10.2 *Waves around boreholes. Unstructured tetrahedral mesh of a cylindrical borehole including a sensor inside the borehole and the medium around it. From Käser et al. (2010). Reprinted with permission.*

Certain classes of exploration-type models are extremely difficult to construct with regular and/or hexahedral grids. An example is shown in Fig. 10.2. To understand the waveforms recorded with high-frequency sources and sensors around boreholes, the cylindrical borehole and the sensor inside have to be meshed. Tetrahedral meshes are the method of choice for such complex geometries. Käser et al. (2010) applied the discontinuous Galerkin method (*SeisSol*) using tetrahedral meshes to exploration problems. The overhead is the substantially higher computation time compared to regular meshes with similar mesh density (that would not be able to properly honour geometry).

Recent examples of reservoir wave simulations using 3D finite-difference methods are: A tutorial on 3D acoustic wave propagation with reservoir applications by Etgen and O'Brien (2007); Regone (2007) simulating wide-angle survey for subsalt imaging; Rusmanugroho and McMechan (2012) modelling a vertical seismic Profiling (VSP) experiment. Spectral-element simulations were presented by Boxberg et al. (2015) for porous media with application to CO_2 monitoring, and Morency et al. (2011) for acoustic, elastic, and poroelastic simulations of CO_2 sequestration and crosswell monitoring.

General accounts of numerical wave propagation for reservoir problems with a variety of methods can be found in the excellent paper collection by Robertsson et al. (2012).

10.2 Regional wave propagation

Regional (or continental-scale) wave propagation stands for problems with O (1,000 km) dimensions. As nuclear tests have been banned[4] the source models are medium to large earthquakes (or other types of seismic sources such as ocean waves). Let us summarize the main requirements for regional wave propagation:

- For 3D Earth models with dimensions >1,000 km, the spherical (or elliptical) shape has to be taken into account.
- Seismic wavefields for these propagation distances are dominated by surface waves. Therefore, the accurate implementation of the free-surface boundary condition is important.
- Structures near the Earth's surface are characterized by low velocities (e.g. crustal velocities and/or ocean layers). This indicates the need to refine the computational meshes near the Earth's surface.
- For comparison with observations, the inclusion of viscoelastic and anisotropic effects is important.
- Wave propagation in spherical sections can be formulated using either Cartesian or spherical coordinates (details below).
- For regional wave-propagation problems the physical domain can often be limited to the crust and (upper) mantle.
- The limitation of regional models implies that efficient absorbing boundaries are important.

The first algorithms for 3D regional wave propagation were presented by Igel (1999) using the Chebyshev pseudospectral approach and a formulation of the wave equation in spherical coordinates. Later a high-order staggered-grid finite-difference approach was applied to the same mathematical formulation (Igel et al., 2002), with an application to wave propagation in subduction zones.

Note the problem of discretizing a (regular) mesh in spherical coordinates: For models that include one of the poles the mesh elements become very small and in addition the equations are not defined at the axis $\theta = 0$. These problems can be avoided by centring the mesh of a spherical section on the equator (i.e. rotating/shifting regions appropriately). Then the application of numerical methods to wave propagation in spherical coordinates has some beauty: For Chebyshev pseudospectral or finite-difference methods the implementation of free-surface boundary conditions is straightforward due to the orthogonality of local coordinates at the Earth's surface.

[4] The former Soviet Union has recorded so-called peaceful nuclear explosions (PNEs) on seismic profiles that were as long as 3,000 km. Analysis of these data were published in Fuchs (1997).

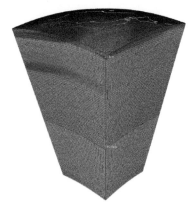

Fig. 10.3 *Cubed-sphere chunk. Computational mesh for regional spectral-element simulation based on the cubed-sphere approach. In this case one chunk is extended into the mantle. The element size is adapted to the increasing velocities at depth. (From* specfem3d *manual, Figure courtesy of M. Chen.)*

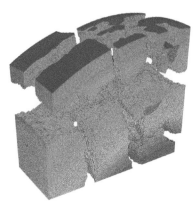

Fig. 10.4 *Tetrahedral mesh parts assembled for regional wave propagation. The mesh was used for simulations with the discontinuous Galerkin method. The mesh density follows the seismic velocity model keeping the number of points per wavelength approximately constant throughout the model. Figure courtesy of S. Wenk.*

An entirely different approach and a milestone for regional and global wave propagation was introduced through the *cubed-sphere* concept (Ronchi et al., 1996). Shortly after, the spectral-element method was adapted to wave-propagation problems such that the mass matrix was diagonal, allowing a fully explicit scheme, which is highly efficient for parallel implementations. These developments paved the way for the combination of spectral elements with the cubed-sphere approach based on a formulation of the wave equation in Cartesian coordinates and appropriate mapping of the elements to spherical geometry (Chaljub et al., 2003; Chaljub et al., 2007). For the problem of regional waves only one (or several) cubed-sphere chunk(s) are used (for an example see Fig. 10.3).

To make up for the decreasing element size with depth for regular cubed-sphere meshes (while having at the same time increasing seismic velocities) the element size increases by an integer factor of two at several depth levels. This approach is implemented in the specfem3d software and plays an important role today for the application of full-waveform inversion on regional scales (see Section 10.5).

At a time when full-waveform inversion using earthquake data on regional scales was around the corner Fichtner and Igel (2008) and Fichtner et al. (2009*b*) presented an alternative, formulating a spectral-element method in spherical coordinates on a regular mesh. As this obviously leads to problems with accurately discretizing the Earth's crust, a local homogenization scheme was introduced making sure that the surface waves are properly modelled. This leads to a very efficient scheme that was used for the first full-waveform inversion on a continental scale (Fichtner et al., 2009*a*) with an application to Australia. Another spectral-element implementation specifically designed for regional wave propagation (*regsem*, see appendix) was presented by Cupillard et al. (2012).

The need to adapt meshes to the low velocities near the Earth's surface led to the application of the discontinuous Galerkin method to the problem of regional wave propagation (Wenk et al., 2013). An example is shown in Fig. 10.4. Mesh generation in this case is straightforward. However, because the implementation is based on tetrahedra with planar faces (compared to curved hexahedra in the cubed-sphere approach), the convergence to the reference solutions is slow. In other words, extremely small tetrahedra have to be used and the high-order potential of the discontinuous Galerkin method (on large elements) cannot be exploited. A snapshot example is shown in Fig. 10.5.

In summary, because of the strong requirement for high-order implementation of the free-surface condition, the most widely used numerical method today for regional wave propagation is the spectral-element method with hexahedral meshes. Other alternatives that have not yet been fully explored include the discontinuous Galerkin method in nodal form on hexahedral grids. Given the current and future large-scale seismic array projects such as USArray or AlpArray, efficient numerical tools for regional wave propagation will be important in the coming years.

10.3 Global and planetary seismology

Let us list the requirements for 3D global or planetary seismic wave propagation:

- The shape of the body to be simulated is a complete sphere (or ellipsoid, or a deformed ellipsoid).
- Planets are naturally limited areas. Therefore we do not need absorbing boundaries (hurray!).
- The obvious way of describing physical problems on a sphere is by means of spherical coordinates. However, this leads to problems when applying numerical methods due to singularities.
- Seismic velocities inside the Earth span more than an order of magnitude. This indicates that the density of computational meshes should vary accordingly.
- Global (teleseismic) seismograms are (mostly) dominated by surface waves. Thus accurate implementation of the free-surface boundary condition is crucial.
- When comparing with observations, attenuation and anisotropy have to be taken into account.
- Unless you are targeting very long periods, waves are likely to propagate many wavelengths. Therefore the numerical scheme has to be extremely accurate.
- At least *our* planet has a substantial number of oceans. In general, their load (depth) and velocity structure have to be taken into account.
- Earth's crustal structure has a strong impact on seismic waveforms observed at the surface. Therefore, knowledge of this structure and proper implementation of it in a numerical scheme is crucial (still a hot topic today).

Fig. 10.5 *Regional wave propagation in Europe. Artistic view of seismic wave propagation based on the discontinuous Galerkin method. The edges of the tetrahedral grid are indicated by blue lines under the oceans. Figure courtesy of M. Meschede.*

It is interesting to note that one of the first applications of the finite-difference method to wave-propagation problems was targeting global wave propagation. The pioneering work by Alterman et al. (1970) was way ahead of its time, but due to the limitations of computational resources the simulations were only possible at very long periods. Parallel computing was established in the nineties and numerical schemes were developed in 2D and 3D Cartesian frameworks primarily for exploration problems or earthquake seismology. An obvious further domain of application was global seismology. However, because of the tremendously long propagation distances (in terms of number of wavelengths) the actual value of computational global wave simulation was questionable.

By that time global seismology was dominated by quasi-analytical methods, such as normal-mode solutions that are exact for spherically symmetric Earth

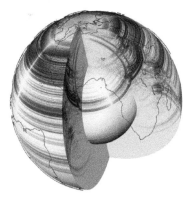

Fig. 10.6 *Axisymmetric modelling. Snapshot of global seismic wave propagation using a spectral-element method for the visco-elastic wave equation in cylindrical coordinates. Figure from van Driel et al. (2015a).*

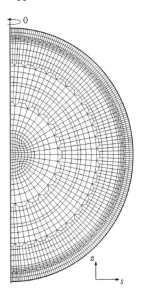

Fig. 10.7 *Spectral-element mesh for axisymmetric case. Element size is adapted to accommodate varying seismic velocities in the Earth's interior. Figure from van Driel et al. (2015a).*

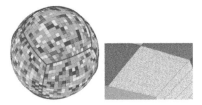

Fig. 10.8 *Cubed sphere. The cubed-sphere concept (Ronchi, 1996) allows discretizing a spherical object with deformed cubes. This opened the way for hexahedral-based spectral elements to be applied to global wave propagation. From Tsuboi et al. (2003). Reprinted with permission.*

---

[5] At the time I remember strong scepticism with the argument that the Earth is spherically symmetric to first order and the rest can be dealt with using perturbation methods.

models and the incorporation of 3D effects using perturbation methods; see the excellent book by Dahlen and Tromp (1998). Following the early work by Alterman et al. (1970) an attractive computational scheme is the reduction of the complete spherical domain to a hemisphere, assuming that all fields are invariant along lines of constant latitude. This so-called axisymmetric (or zonal) approach corresponds to a 2.5D scheme as described in Section 3.3.1.

First realizations of this approach using high-order staggered-grid finite-difference approximations to the wave equation in spherical coordinates were presented by Igel and Weber (1995), Igel and Weber (1996), and Chaljub and Tarantola (1997).[5] An example is illustrated in Fig. 10.6. It is important to note that the wave equation in spherical coordinates is singular at the axis $\theta = 0$ (just check the Laplace operator in spherical coordinates) and the implementation of general seismic sources is tricky.

However, while limited concerning direct comparison of synthetic seismograms with observations, axisymmetric methods proved useful to estimate effects of laterally heterogeneous structures in the mantle (Igel and Weber, 1996; Igel and Gudmundsson, 1997; Jahnke et al., 2008; Thorne et al., 2013b).

A major boost to this 2.5D approach came with the option to use arbitrary seismic sources (Toyokuni and Takenaka, 2006) with the finite-difference method or the spectral-element method on a mesh with depth-dependent element size (Fig. 10.7). Nissen-Meyer et al. (2007) introduced a scheme by which seismograms for arbitrary moment tensors could be obtained by summation. This approach was recently extended, allowing extremely fast calculation of high-frequency synthetic seismograms using the superposition principle and pre-calculated Green's functions (van Driel et al., 2015b).

What about synthetic seismograms for an entire sphere? While it may appear natural to stick to spherical coordinates, this is a no-go! The reason is, as indicated earlier, that the wave equation in spherical coordinates has singularities and regular discretization of spherical coordinates leads to decreasing grid spacing near the poles (certainly familiar from looking at a globe). From a stability point of view this is not acceptable. A solution to the problem came with the work of Ronchi et al. (1996) who introduced the cubed-sphere concept as already mentioned in Section 10.2.

A spherical object can be discretized by deforming cubes such that they fill an entire sphere (see Fig. 10.8). This opens the path to applying numerical methods like spectral elements to the problem of global wave propagation. The flexibility of Galerkin-type methods perfectly matches the requirement to describe the wavefield on curved hexahedral elements that make up an entire Earth model. The original work in the dissertation of Chaljub (2000) and Chaljub et al. (2003) was taken up and extended by Komatitsch and Tromp (2002a) and Komatitsch and Tromp (2002b). A special formulation for the coupling of fluid and solid parts of the Earth's interior was presented by (Chaljub et al., 2007). The subsequently developed community code (*specfem3dglobe*) is in my view one of the great success stories of computational seismology.

This solver allowed for the first time the exact (to possible accuracy) calculation of waves through 3D global Earth models, including ellipticity, anelasticity, surface topography, and anisotropy. The open-source philosophy and the distribution via the NSF-funded CIG project[6] allowed extensions and improvements by the seismological community. Today this approach is the method of choice for the simulation of global or planetary wave propagation.

A further milestone was the publication of synthetic seismograms for the M7.9 2002 Denali fault earthquake calculated on the *Earth Simulator* in Japan, a supercomputer that was installed in response to the devastating M7.2 Kobe earthquake in Japan in 1995. The comparison between observation and theory presented by Tsuboi et al. (2003) (see Fig. 10.9) illustrated the state of knowledge of the 3D structure of the Earth's interior (at the frequency range possible at the time).

It was the starting point for full wavefield modelling of global wave propagation, with the ultimate goal of applying adjoint inversion techniques on a global scale. Tromp et al. (2010) started an initiative by which synthetic seismograms and wavefield animations are automatically calculated after each sizeable earthquake. Further applications include the study of seismic waves following a meteorite impact (Meschede et al., 2011). An excellent review of the spectral-element method for both forward and inverse functionalities is presented in Peter et al. (2011).

There is a tremendous amount of research related to global (and planetary) wave propagation using the *specfem3dglobe* code and to list it all is impossible. Examples are 3D mantle structure forward modelling experiments (Schuberth et al., 2009; Schuberth et al., 2012; Schuberth et al., 2015), studies exploring the sensitivities to certain parameter classes (Sieminski et al., 2009; Sieminski et al., 2009), and improving global earthquake parameters using 3D planetary-scale simulations (Lentas et al., 2013).

An alternative approach was presented by Capdeville et al. (2003) who combined a spectral-element formulation in the Earth's mantle with a normal-mode solution in the core. This approach is particularly useful for applications with focus on mantle waves.

To my knowledge the only alternative to the spectral-element method for global wave propagation was presented by Wilcox et al. (2010), who introduced a nodal discontinuous Galerkin approach for a spherical mesh but only for simplified Earth models. An example of a tetrahedral mesh adapted to the velocity structure inside the Earth is shown in Fig. 10.10. While it is possible to use such meshes for global wave propagation in combination with the discontinuous Galerkin method, the computational effort far exceeds the one required when using spectral elements.

Computational global wave propagation is dominated by the spectral-element method both in the 3D case (cubed sphere) and in the axisymmetric case (where finite-difference approximations still have some attraction due to their simple algorithms).

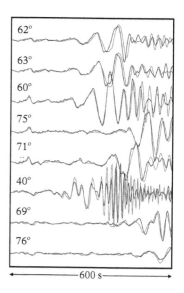

Fig. 10.9 *3D global simulation. Observed (black) and synthetic (red) transverse component seismograms for the M7.9 2002 Denali fault earthquake calculated on the* Earth Simulator *using the spectral-element method. The level of fit between observations (black) and simulation (red) reflects the knowledge about global 3D structure. From Tsuboi et al. (2003). Reprinted with permission.*

[6] CIG stands for computational infrastructure in geodynamics (<http://www.geodynamics.org>) and provides access to a number of geophysical simulation codes. See Appendix.

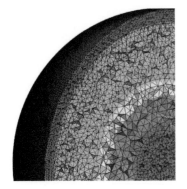

Fig. 10.10 *Global unstructured grid. Tetrahedral grids offer flexible adaptation to seismic velocity structures. However, computational requirements exceed those of alternative schemes. Figure courtesy of S. Wenk.*

10.4 Strong ground motion and dynamic rupture

Simulation codes for strong ground motion and dynamic rupture simulation in 3D have similar requirements, which is the reason why they are presented here in one section. In fact, many algorithms were first developed for modelling wavefields due to kinematic (i.e. predefined) sources and later extended to allow spontaneous (dynamic) rupture at predefined fault surfaces.

In the absence of any hope of predicting earthquakes in a deterministic way, the calculation of seismic wavefields for realistic earthquake scenarios has become one of the most important tasks in seismology and earthquake engineering. The term *strong ground motion* in this context takes the meaning that the earthquakes to be studied (and the associated ground motion) are sizeable (i.e. potentially lead to damage), implying usually that the region to be studied is not too far away from the source. Strong earthquake ground motions may well exceed $1g(\approx 10 \text{ m/s}^2)$ which would saturate standard broadband velocity sensors. Therefore, strong motion networks are predominantly based on accelerometers.

The scientific importance of earthquake simulations and their substantial societal relevance led to early applications of numerical methods to this problem, a vast number of publications, and continuous developments that are still ongoing. Therefore, only a few aspects of this field can be covered. An example of a 3D earthquake simulation involving a sedimentary basin structure is shown in Fig. 10.11.

Let us list the key features relevant for earthquake scenario simulations:

- Realistic earthquake scenario simulations only make sense in 3D, as strong ground motion is strongly affected by 3D structure, rupture behaviour, and radiation patterns.

- Except for mega-earthquakes (M9), a Cartesian framework is usually sufficient (e.g. Los Angeles Basin, San Francisco Basin).

- Accurate earthquake scenario calculations that can be compared with observations necessitate good structural subsurface models.

- To make civil engineers happy (i.e. to simulate wavefields relevant for structural damage) you need to achieve frequencies well above 1 Hz

- The necessity for high frequencies in turn implies good knowledge of the near-surface structure. (Difficult! One reason why stochastic earthquake simulations are so popular).

- Large earthquakes happen on finite faults. These have to be initialized using the superposition principle.

- Earthquake faults may be of complex shape.

- Rupture behaviour can be predefined (*kinematic*) or the result of a process initiated on a predefined fault with unknown outcome (*dynamic* or *spontaneous* rupture).

- In many cases the accurate modelling of complicated 3D sedimentary (i.e. low-velocity) structures is important.

- Strong ground motion simulations may require the incorporation of non-linear rheologies (e.g. plastic behaviour, damage rheology).

- Because of the high-frequency requirements, earthquake scenario calculations to date are computationally very demanding.

- To obtain a reasonable picture about shaking hazard in a seismically active region, a large number of 3D simulations might be necessary, accounting for uncertainties in 3D structure and earthquake characteristics.

- Recent extensions are the incorporation of structures (bridges, buildings) in the modelling (soil–structure interaction).

Before listing some of the key studies it is important to distinguish *kinematic* and *dynamic* (rupture) simulations. When the space–time behaviour (also called slip history) on a fault is predefined, then one speaks of a *kinematic* rupture. A more physical approach is to define a fault plane, introducing a frictional boundary condition that determines what happens to a fault patch when it breaks, and then initialize an earthquake at the desired hypocentre. This is called *dynamic* or *spontaneous* rupture modelling. At first, dynamic rupture simulations were performed primarily to understand the rupture itself. Only recently have they been used in connection with strong ground motion simulations and to model real observations.

Strong ground motion simulations

Early *kinematic* 3D calculations using a staggered-grid finite-difference method were presented by Graves (1993) and Graves (1995). The group of Jacobo Bielak applied the classic finite-element method with implementation on parallel hardware (Li et al., 1994) and many applications to strong ground motion problems (e.g. Bielak and Xu, 1999; Bielak et al., 1998). Bielak et al. (2005) applied the *octree* approach allowing local mesh refinement to regions with low velocities. Ground motion calculations can be used to extract for so-called *shaking hazard maps* quantifying the maximum expected shaking at the Earth's surface (an example is shown in Fig. 10.12).

Wang et al. (2008) used the finite-difference method to calculate a database with subfault Green's functions allowing subsequent synthesis of arbitrary slip histories and associated ground motion scenarios. Yin et al. (2011) simulated earthquakes with the finite-element method (*GeoFEM*) specifically developed for the *Earth Simulator* in Japan at the time.

Recent applications of the spectral-element method to kinematic rupture problems include simulations of the l'Aquila earthquake (Smerzini and Villani, 2012; Magnoni et al., 2013) and earthquakes in Taiwan (Lee et al., 2009) incorporating and discussing effects of strong topography. Subduction earthquakes of the Cascadian area were simulated by Delorey et al. (2014) using a finite-difference approach. Kinematic rupture simulations were carried out with the discontinuous

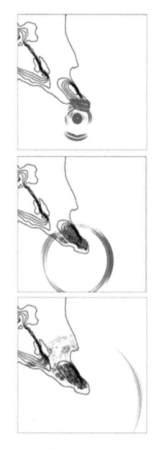

Fig. 10.11 *Earthquake scenario calculations. Surface snapshots of the horizontal component of ground velocity for an earthquake happening in the Cologne basin, Germany. The map corresponds to a 120 km × 120 km region. The epicentre is outside the sedimentary basin denoted by (depth) contour lines. The wavefield is trapped inside the basin, leading to prolonged shaking compared to bedrock sites. From Ewald et al. (2006).*

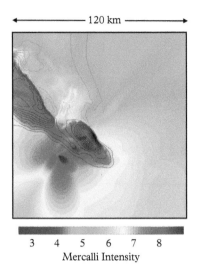

<--------- 120 km --------->

3   4   5   6   7   8
Mercalli Intensity

Fig. 10.12 *Shaking hazard maps. Mercalli intensity derived from simulations of earthquakes in the region discussed in Fig. 10.11. The sedimentary basin (contour lines) leads to increased shaking amplitudes and therefore higher Mercalli intensity. From Ewald et al. (2006).*

[7] In my view the organizers of these projects are **heroes**. It is extremely hard to motivate groups to participate as getting benchmarks right takes time. Also, researchers involved do not necessarily get the credit they deserve. However, these verification exercises are important steps towards credible and reproducible science in our field.

Galerkin approach (Käser et al., 2007*b*; Käser and Gallovic, 2008; Gallovic et al., 2010). An example is shown in Fig. 10.13. Mazzieri et al. (2013) developed an open-source nodal discontinuous Galerkin scheme (SPEED) for multi-scale problems that was subsequently applied to earthquake simulation problems. Further milestones include the work of Moczo et al. (2010*b*), Moczo et al. (2011), and Maufroy et al. (2015) comparing various methods with respect to their capability of modeling high v_p/v_s ratios. An extensive recent review of the state of the art of earthquake simulation including soil–structure interaction is given by Paolucci et al. (2014) and in the book by Moczo et al. (2014).

Dynamic rupture simulation

Following the pioneering work on 2D rupture problems using the finite-difference method by Andrews (1973), Andrews (1976*a*), Andrews (1976*b*), Madariaga (1976), the applications were extended to 3D by Day and Boatwright (1982) and Day (1982). The potential of this approach to understand strong ground motion observations and to quantify shaking hazard was recognized by Kim Olsen and co-workers (Olsen et al., 1995; Olsen and Archuleta, 1996; Olsen et al., 1997; Madariaga et al., 1998), leading to many realistic large-scale parallel earthquake scenario simulations (e.g. Olsen et al. (2008) within the framework of the *Terashake* project). This line of research continues with recent efforts to match simulations with geological observations (precariously balanced rocks, Lozos et al., 2015). The impact of various dynamic rupture scenarios for a bi-material interface was investigated by Brietzke et al. (2009) using a 3D finite-difference method. A recent hybrid approach combining boundary-element methods (rupture) with a 3D finite-difference method (wave propagation) was presented in Aochi and Ulrich (2015).

The spectral-element method is also widely used for dynamic rupture simulations (Festa and Vilotte, 2006; Kaneko et al., 2008; Kaneko et al., 2011; Galvez et al., 2014). Due to the specific description of wavefields in the discontinuous Galerkin method the application to rupture problems (with their natural discontinuous behaviour across the fault plane) seemed obvious. The 2D version (de la Puente et al., 2009*a*) was soon after extended to 3D (Pelties et al., 2014; Pelties et al., 2015). An attractive feature of this approach is the possibility of complex fault shapes (see Fig. 10.14). At the present time it appears that the discontinuous Galerkin method has considerable advantages over other numerical approaches in dealing with the nonlinear frictional boundary conditions.

Community projects have been set up that aim at comparing various numerical techniques for given 3D structures, as well as kinematic and/or dynamic rupture specifications. These projects are extremely important and should be supported.[7]

Recent examples are the finite-source scenarios for complex basins ((Chaljub et al., 2010*b*; Chaljub et al., 2015; Maufroy et al., 2015)), and the *Euroseistest* project (Chaljub et al., 2010*a*). Other initiatives are the *Shakeout* project (Bielak et al., 2010) and the SCEC–USGS rupture dynamic code comparison exercise

(Harris et al., 2010). Let me finish by quoting the conlusions of one of these validation exercises (Chaljub et al., 2010*b*):

> The main recommendation to obtain reliable numerical predictions of earthquake ground motion is to use at least two different but comparably accurate methods, for instance the present formulations and implementations of the finite-difference method, the spectral-element method, and the arbitrary high-order (ADER) discontinuous Galerkin method.

That says it all. In any case, I highly recommend careful study of the publications of these validation exercises before choosing a solution strategy for a specific problem.

10.5 Seismic tomography—waveform inversion

The imaging of Earth's interior structure using seismic data is so fundamental that it deserves some special attention. It is tightly linked to the history of computational seismology and today one of the most active and expanding fields in seismology. For decades, the Earth's interior structure was mapped using (more or less only) travel time information and ray-theoretical concepts for the forward problem. In the past few decades this has dramatically changed. With increasing computational power the goal is now to calculate entire waveforms for 3D models (e.g. using the methods described in this volume) and match them with observations as well as possible. What do we really mean by *waveform inversion*?

Let us start with an example. Fig. 10.15 illustrates seismic observations at station OUZ in New Zealand following an earthquake in Indonesia (solid black line). Synthetic seismograms for some initial 3D Earth structure (dashed line) do not match the observations well. Following an iterative scheme that progressively alters the Earth model thereby minimizing the misfit between theory and observations leads to the red line. The final fit is much better. This procedure is called *waveform inversion* or *waveform fitting*. In simple terms, the goal is to minimize a discrete functional like:

$$\text{misfit} = \sum_{\text{sources}} \sum_{\text{receivers}} \sum_{\text{components}} \| \mathbf{d} - \mathbf{g}(\mathbf{m}) \|. \tag{10.1}$$

Here **d** is a vector containing seismogram samples of one motion component at a receiver location. **g(m)** denotes the forward problem, that is, the solution of the wave equation for Earth model **m** calculated with the same sampling interval. The double bars denote the mathematical *norm* (e.g. the least-squares norm ℓ_2) defining distances in model spaces. Other comparative measures like cross-correlations (time shifts, coefficients) are also possible.

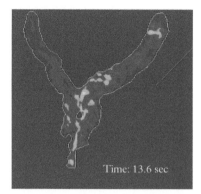

Time: 13.6 sec

Fig. 10.13 *Grenoble valley simulation using the discontinuous Galerkin method.* **Top:** *Snapshot of a finite-source earthquake originating on a fault (red line) outside a sedimentary basin. The wavefield continues to vibrate once the body waves have passed.* **Bottom:** *Tetrahedral meshes that make up the 3D model. Figures courtesy of M. Käser. Reprinted with permission.*

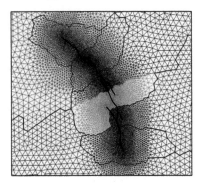

Fig. 10.14 *Fault system mesh. Triangular mesh of the Landers fault system. The mesh is severely densified towards the fault to allow accurate calculations of the rupture process. From de la Puente et al. (2009a).*

There are many inverse approaches that work with the waveform-fitting concept; from the pioneering work on surface wave inversion (Woodhouse and Dziewoński, 1984) using a ray-based approach to the recent finite-frequency waveform-matching approach using cross-correlation measurements of phase misfit, for example Sigloch et al. (2008), Tian et al. (2011). Perhaps the most holistic and most computationally expensive approach is the use of full 3D modelling schemes for forward (and usually also for inverse) calculations. In the following we will focus on these types of applications using numerical schemes as discussed in this volume. As before let us summarize some aspects of wave simulations in the context of waveform inversion:

- Formal waveform-fitting algorithms require *many* forward problems to be solved. 3D simulations are expensive anyway so this means potentially severe restrictions concerning the frequency range.

- The target is the modelling of real observations. Therefore, viscoelasticity and in many cases anisotropy have to be taken into account.

- Waveform inversion can be done at local (Cartesian), regional, or global scale (spherical geometry).

- Waveform-fitting solutions (nonlinear or linearized) can be built around standard forward simulation algorithms.

- Structural inversion requires good knowledge of source properties (e.g. source time function in exploration-type problems and source depth, moment tensors for earthquake data).

- Because of the large computational requirements, numerical solutions are preferred that do not require re-meshing with Earth model updates.

Ideally, we would like to explore all (or many) possible Earth models, compare synthetic seismograms with observations, and find the best one, or, even better, a collection of models that fit the data well.

Unfortunately, it is not that easy. Because the *forward* problem is so expensive, the way forward is to start with a good guess (*initial* model) that hopefully is close enough to the final solution such that the solution can be found by gradient search (akin to the Newton algorithm).

Fig. 10.15 *Full waveform inversion. The improvement of iterative adjoint waveform misfit. The graph shows observed data (solid black), synthetic seismograms for initial 3D model (dashed black), and final 3D model (red line). Figure courtesy of A. Fichtner.*

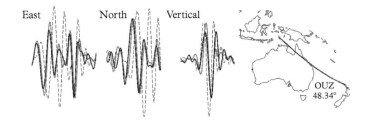

This approach was pioneered for the acoustic and elastic wave equations by Bamberger et al. (1982), Tarantola (1986), and Tarantola (1988). Tarantola and Valette (1982) and Tarantola (2005) embedded these concepts in a probabilistic inversion framework (Fig. 10.16.).

The theoretical concepts for waveform inversion led to a hype in exploration geophysics with industrial consortia (e.g. the G^TG Group in Paris, see Fig. 10.16, or the Stanford Exploration Project SEP) heavily promoting research and the development of (parallel) inversion software (e.g. Crase et al., 1990; Igel et al., 1996). But these applications, due to computational restrictions, were initially at most in 2D with limited applicability to the emerging 3D acquisition geometries.

Therefore, the hype came to a halt in the nineties, when 3D solutions were impossible. While preparing this volume, I wondered whether this intuitive notion can be supported by data, and I think it can (see Fig. 10.17). The number of publications on full waveform inversion increased quite slowly in the nineties until around 2005, when computational power had evolved such that 3D inversion seemed possible. Activity in this field then exploded, and the rate since then has not relented, as can be seen in the figure (the data show exploration problems only; the same holds for earthquake seismology).

In seismic exploration, alternative strategies based on numerical solutions in the frequency domain (Pratt et al., 1998) led to a number of applications in 2D (e.g. Bleibinhaus et al., 2007; Bleibinhaus et al., 2007) and in 3D (Ben Hadj Ali et al., 2009a, Ben Hadj Ali et al., 2009b, Krebs et al., 2009). The latter studies were based on the *source encoding* concept exploiting the superposition principle. Excellent reviews on forward and inverse modelling methodologies can be found in Virieux and Operto (2009) and Virieux et al. (2009).

Fig. 10.16 *Albert Tarantola (1949–2009). With his work on seismic waveform inversion, the probabilistic formulation of inverse problems, his vision of the role of computations, the leadership of the Geophysical Tomography Group in Paris, and his vibrant personality, he had a strong impact on computational seismology. Picture shows A.T. as referee during the soccer match at the 2006 SPICE workshop in Kinsale, Ireland.*

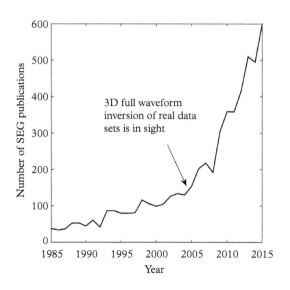

Fig. 10.17 *Publication history of waveform inversion. The number of publications in the field of seismic waveform inversion published by the Society of Exploration Geophysicists (SEG) including extended abstracts, special volumes, journals, and books.*

A fundamental milestone in applied waveform inversion is the work of Sirgue et al. (2010) and Sirgue et al. (2012) showing inversion results for the Valhall data set (3D acquisition geometry with 2,300 sea-bottom receivers and 50,000 sources) of a resolution that is reminiscent of medical tomography.[8] Further applications on this data set were carried out by Etienne et al. (2012) and Schiemenz and Igel (2013) using spectral elements (an example of a model gradient is shown in Fig. 10.18, indicating the high resolution potential). A strategy based on the finite-difference method was presented by Butzer et al. (2013) with an application to small-scale heterogeneities.

What about *earthquake seismology*? For a long time this field was dominated by ray-based inversion methods. It is remarkable how much we have learned about the structure of Earth's interior, enabling us to reduce a seismogram to a few bytes of information and throw away the rest. It is fair to say that a breakthrough for the application of waveform inversion tools to earthquake seismology came through the adoption of alternative misfit measures. Luo and Schuster (1991) suggested the use of cross-correlation functions as a misfit measure, allowing phase information to be extracted from waveform data.

In the seminal paper by Tromp et al. (2005) the concepts of adjoints, time reversal, and banana-doughnut kernels are merged, providing a framework for waveform inversion of earthquake data (for source and structure). Fichtner et al. (2008) introduced an alternative time–frequency domain misfit criterion based on the work of Kristekova et al. (2006). An inversion scheme using the finite-element method was presented by Askan and Bielak (2008).

These theoretical developments laid the foundations for some of the first applications to earthquake data. Chen et al. (2007) used the finite-difference method to invert for the velocity structure of the Los Angeles Basin. Tape et al. (2009) and Tape et al. (2010) applied the spectral-element method and an automated way of finding appropriate time-windows for matching synthetics with data (Maggi et al., 2009).

The first application on a regional (continental) scale was presented by Fichtner et al. (2009*a*) and Fichtner et al. (2010). They used a regular-grid spectral-element method in spherical coordinates to image the structure beneath the continent of Australia, providing proof that, despite the potential uncertainties in source parameters, waveform inversion on this scale is possible. By now, waveform inversion approaches are applied to all regions where station density and data quality is sufficient. Examples are the European continent (Zhu et al., 2012), the South Atlantic (Colli et al., 2013), North Anatolia (Fichtner et al., 2013*a*), the western Mediterranean region (Fichtner and Villasenor, 2015), and several regions in Asia (Chen et al., 2015*a*; Chen et al., 2015*b*). An illustration of continental-scale imaging is shown in Fig. 10.19. The emergence of waveform inversion-based images at various scales all over the place raises the question of how these models can be merged, combined, and reused. This is the topic of recent research (Fichtner et al., 2013*b*).

Fig. 10.18 *Full waveform inversion. Slice at depth 234 m through the model update of an acoustic full waveform inversion. The acoustic Valhall ocean cable data set has 2,300 receivers and 50,000 shots were fired. The image shows acoustic velocities (dark—low, bright—high). The white bands are paleo-rivers now embedded in the sediments. From Schiemenz and Igel (2013).*

[8] Let us be honest: we have always envied the medical tomographers for their controlled source–receiver geometries, their straight ray paths, and the fascinating 3D images they obtain. But the gap is narrowing!

An exciting current research front is the application of full waveform inversion for the whole planet using 3D simulations. As computational resources increase and our knowledge of Earth's interior structure improves, these techniques are paving the way to explaining more and more energy in the observed global seismic wavefield. Recent results based on spectral-element methods were presented in French and Romanowicz (2015). Work in progress is shown in Fig. 10.20 using the global version of *specfem* for the wave simulations. Capdeville et al. (2005) investigated theoretically the potential of source stacking for global waveform inversion, but with limited success.

While these studies indicate a major advance in matching theory with observations, we are still not doing very well at quantifying uncertainties in the resulting tomographic models. Despite recent developments (Fichtner and Trampert, 2011*b*) it appears that only a substantial level of random search allows proper quantification of model uncertainties. An attempt was made by Käufl et al. (2013) to try to test structural hypotheses with reduced dimensionality using Monte Carlo techniques (Sambridge, 1999*a*; Sambridge, 1999*b*).

In summary, full waveform inversion is an extremely active field. It is likely that in a few years it will become the standard technology to invert for structure on all scales. From experience, inverting real data is always hard. In terms of numerical techniques, finite-difference and spectral-element methods with regular or general hexahedral meshes have been used extensively. The computationally more expensive methods based on tetrahedral meshes provide little advantage.

Excellent presentations on the theory behind full waveform inversion can be found in Tromp et al. (2005), Peter et al. (2011), and Fichtner (2010), and references therein. A recent open-source Python-based framework (LASIF) for the entire inversion workflow from data recovery to final model was presented by Krischer et al. (2015*a*). A full waveform inversion implementation using spectral elements on GPUs can be found in Gokhberg and Fichtner (2015).

10.6   Volcanology

The seismic monitoring of an active volcano is key to understanding the state of its eruptive system. By their nature, volcanoes are usually difficult to access (see Fig. 10.21), hard to instrument, and observations of ground motions are extremely complex.[9] Let's summarize some requirements to model wave propagation in a volcano:

- Rough topography will influence the wavefield and has to be taken into account.
- Mesh generation based on digital elevation models (DEMs) may be necessary.

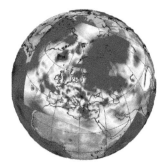

Fig. 10.19 *Regional waveform inversion under Europe. Seismic velocity model at 100 km depth (red—low velocities, blue—high velocities) obtained by full waveform inversion. Figure courtesy of A. Fichtner.*

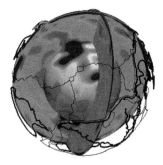

Fig. 10.20 *Global waveform inversion. Velocity perturbations in Pacific region after 15 iterations (red—low velocities, blue—high velocities), obtained by full waveform inversion using the spectral-element method. Figure courtesy of E. Bozdag.*

[9] Some say a volcano is a seismologist's nightmare—unknown internal structure, strongly scattering, rough topography, etc. But of course it is fun to face these multiple challenges.

Fig. 10.21 *Mount Fuji, Japan. Most (active) volcanoes are characterized by strong topography and severely scattering internal structures. Photo courtesy of M. Goll.*

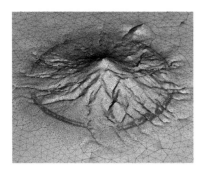

Fig. 10.22 *Volcano Merapi, Indonesia. Snapshot of waves generated by a source near the summit of the volcano. Wavefield was simulated with a discontinuous Galerkin method on a tetrahedral grid. Figure courtesy of A. Breuer.*

- The internal structures are often highly uncertain. Observations suggest strongly scattering material that might have to be modelled with a random media approach.

- There are many sources of seismic energy (volcano–tectonic events, tremor-like signals, rock falls, dome collapses, bubble explosions, etc.), some of which might be difficult to describe with standard moment tensors.

- Some seismic signals (e.g. bubble explosions) may require the modelling of interaction between atmosphere and solid volcano edifice.

Because the problem of seismic wave propagation in volcanoes is a truly 3D problem, the application of numerical methods to this problem started only when resources were sufficient to allow 3D calculations. At an early stage, methods allowing flexible mesh geometry at the surface were not available. Therefore, attempts were made to adapt regular-grid finite-difference methods to allow complicated surfaces. Ohminato and Chouet (1997) introduced a method with which regular finite-difference blocks with appropriate boundary conditions were adapted to real volcano topographies.

While it was in principle possible to use the approach to understand the topographic effects of volcanoes (e.g. Ripperger et al., 2003), it required an extremely large number of grid points per wavelength to get the waveforms right. An elegant alternative with an effort similar to finite-difference methods was presented by O'Brien and Bean (2004) and O'Brien and Bean (2009), who extended earlier work (Toomey and Bean, 2000) in which a method based on a discrete particle scheme for seismic wave propagation was introduced. This particle-based methodology was extensively used to study wave propagation and inverse problems on volcanoes (Lokmer et al., 2007; Davi et al., 2010; Métaxian et al., 2009).

The spectral-element method does allow sufficient flexibility with hexahedral meshes to model complex tomography. This approach was adopted by van Driel et al. (2012), Kremers et al. (2013), and van Driel et al. (2015c) who investigated strain–rotation coupling, moment tensor inversion, and tilt effects on moment tensor inversion in models including realistic volcano topography. Further recent examples can be found in Kim et al. (2014) who studied infrasound signals at volcanoes using the finite-difference method.

Last, but not least, the discontinuous Galerkin method on unstructured tetrahedral meshes lends itself to problems like wave propagation inside volcanoes. The meshing based in arbitrary topography models is straightforward. The modelling of wave propagation inside the Indonesian Merapi volcano was the chosen test case when the *SeisSol* code exceeded 1 PFlop performance (Breuer et al., 2014), see Fig. 10.22.

I am convinced that the simulation of seismic wave propagation inside volcanoes will continue to develop, and become a standard procedure for volcano monitoring. Methods allowing complex geometries (e.g. Galerkin-type methods, finite-volume methods) will play a more important role than finite-difference methods. However, because of the large computational requirements and the necessity to develop high-quality meshes, it will take time to make these tools available to scientists working in volcano observatories.[10]

10.7 Simulation of ambient noise

The study of permanently recorded ambient seismic noise (ocean/atmosphere-generated or anthropogenic) is one of the most vibrant fields in seismology today. In particular the study of ocean-generated noise, with the option of performing tomography in the absence of earthquake sources (Shapiro et al., 2005) and the possibility of studying the structure of Earth's interior as a function of time (Brenguier et al., 2008) is revolutionizing our field. Ambient noise studies are predominantly data-processing tasks. Simulating the required long time series using simulation techniques is a challenging problem. Nevertheless, in the quest to fully understanding the source mechanisms of the noise fields as well as their interaction with 3D structure, simulations are likely to play an important role.

The requirements are:

- Ambient noise is dominated by surface waves because sources usually act at the surface. Therefore, accurate implementation of the free-surface boundary condition is crucial.

- For the noise field to develop, many wavelengths need to be propagated. This is challenging in 3D, requiring large models and highly accurate time-extrapolation schemes.

- Depending on the desired Earth model complexity, surface topography (or bathymetry) needs to be taken into account (e.g. to explain Love waves in the ocean-generated noise).

- Ocean-generated noise is characterized by continental or global-scale propagation distances and thus requires spherical geometry.

- Cultural or anthropogenic noise can be simulated using a Cartesian framework.

- Because of the required long time series, efficient absorbing boundaries are important when performing limited-area calculations.

A seminal study using simulations in a spherically symmetric Earth model with normal modes (not discussed in this volume) was presented by Gualtieri et al. (2013). The noise sources are modelled using ocean wave information, along with bathymetry. The vertical component seismic noise spectra fit observed spectra. However, a discrepancy is found between the modelled and observed horizontal component spectra, which has been attributed to the existence of Love waves in the observed noise. At the time of writing, the origin of the Love waves in the ocean-generated noise is still not well understood. Answers are expected from 3D

[10] While this can be a dream job, often the observatories are understaffed and the maintenance of observational infrastructure takes up all the time.

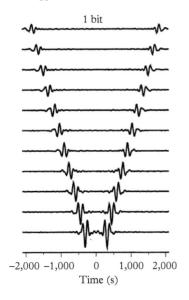

1 bit

-2,000 -1,000   0   1,000   2,000
Time (s)

Fig. 10.23 *Green's function from noise simulations. One-bit normalized cross-correlations between the vertical displacements in the case of a uniform distribution of noise sources. Due to the uniform distribution of noise, symmetric Green's functions emerge. From Cupillard and Capdeville (2010).*

simulations including bathymetry and internal 3D structures. In another synthetic normal-mode study Cupillard and Capdeville (2010) investigated the amplitude of surface waves reconstructed by noise correlation in the context of noise sources located on the surface of the Earth. An example of Green's functions emerging from synthetic noise calculations is shown in Fig. 10.23.

Noise simulations using a regional-scale spectral-element method can be found in Stehly et al. (2011). They study the effects of homogeneous distributions of random noise sources located at the free surface of an attenuating and spherically symmetric Earth model.

Noise studies at higher frequencies have been undertaken for quite some time to estimate local site effects studying the spectral ratio of horizontal components and vertical components of ground motion (H/V). A simulation example based on the finite-difference method (Moczo et al., 2002) can be found in Guillier et al. (2006). Several simple 2D and 3D structures are considered, and the locations of the H/V peaks in the spectra are investigated and discussed.

It is fair to say that both spectral-element and finite-difference methods (flat surfaces) seem to be well suited for the tasks of ambient noise modelling. With increasing frequencies and more and more domains of application, simulations will play an important role in supporting the results obtained from real data processing.

10.8 Elastic waves in random media

Everyone who has looked at a geological outcrop in detail accepts that the Earth's crust (and most likely the deep interior) is far from homogeneous, or even locally homogeneous. As discussed in the introductory chapter, whether the spatial scales of heterogeneities have to be taken into account for seismic wave propagation strongly depends on the (dominant) wavelengths of seismic waves. In the field of seismic scattering the effects of (random) spatial heterogeneities on wavefields are investigated. Strong effects are expected when the wavelengths of scatters and elastic waves are similar.

Some requirements for random media calculations are:

- The scale of the scattering medium has to be properly discretized by the computational mesh.
- For strong scattering media and coda studies the implementation of absorbing boundaries is crucial for limited-area calculations.
- In many cases the scattering of body waves is the target in which case free-surface boundary conditions are not relevant.
- The sampling requirements of heterogeneities favour low-order methods (e.g. finite-difference methods).
- Scattering problems require careful initialization of the elastic models with specified statistical properties.

Understanding waves in random media was one of the first domains of application for computational seismology. Frankel (1989) reviews the early applications of numerical methods to the problem of waves in random media, discussing finite-element and finite-difference methods. A problem that is still under discussion today is the partitioning of energy into P- and S-waves, already investigated by Dougherty and Stephen (1988) for the oceanic crust using 2D finite-difference simulations. A comprehensive way to generate random media in 2D using ellipsoidal autocorrelation functions and resulting wave effects simulated with the finite-difference methods was presented in Ikelle et al. (1993).

Early this century, the extension to 3D media as well as more complex rheologies became possible (Frenje and Juhlin, 2000; Martini et al., 2001; Bohlen, 2002). All these studies were carried out with finite-difference methods. The effect of random media on the analysis of travel times was investigated by Baig et al. (2003) and Baig and Dahlen (2004). Pham et al. (2009) used the 3D discontinuous Galerkin method to study crustal P-SH scattering, modelling rotational ground motions observed with a ring laser.

Further recent finite-difference applications include the analysis of random media characterized by von Karmann correlation functions (Imperatori and Mai, 2013) and the modelling of observed high-frequency P-wave fields in Japan (Takemura and Furumura, 2013) in the presence of irregular surface topography. A recent application of the spectral-element method to random media calculations can be found in Obermann et al. (2016) who investigate the depth sensitivity of time-dependent velocity changes observed using ambient seismic noise.

Finally, the question of the short-scale structure of the Earth's mantle is still under debate. Finite-difference simulations of random mantle structures using the axisymmetric approximation can be found in Igel and Gudmundsson (1997) for the upper mantle and in Jahnke et al. (2008) for the entire mantle. Large-scale global wavefield simulations in 3D random mantle models were performed by Meschede and Romanowicz (2015) (see Fig. 10.24).

Fig. 10.24 *Random model in Cartesian coordinates with a stationary exponential covariance function. The model was computed by filtering white noise to the power spectrum that is given by the Fourier transform of the covariance function descriptive of the random field. Figure from Meschede and Romanowicz (2015).*

Chapter summary

- Today, seismic wave simulation in 3D media is essential for the solution of many solid Earth science problems.

- There is no *one-method-fits-all* situation. The suitable simulation technology depends on the specific problem.

- While mature simulation codes for 3D wave propagation have existed for some time, their extensive use to model observations has only recently started.

- In the near future, computational resources should allow us to perform massive 3D calculations for parameter space studies, improved 3D imaging, the assessment of uncertainties, and our understanding of earthquakes and the associated strong ground motion.

. . . .

EXERCISES

(10.1) Create a matrix of application domains and numerical methods discussed in this volume. Discuss pros and cons of the methods for the various applications.

(10.2) Search the literature for current applications of 3D seismic wave simulations in your domain of interest. Discuss the computational set-up (mesh type, mesh size, regular vs. irregular) in connection with the specific numerical method employed. Is this the optimal method for the problem? How was the accuracy of the method verified? Would the results be fully reproducible?

Current Challenges in Computational Seismology

At least to some extent the motivation for writing this volume was the fact that now a large number of 3D simulation codes are in place. Most conceivable numerical methods have been applied to the wave-propagation problem. Some say *the forward problem for seismic wave propagation is solved*. This final chapter aims at briefly highlighting a few issues, showing that surprises might still be around the corner and there are many exciting challenges that will keep us busy for some time.

11.1 Community solutions

As indicated in the introduction, there is no way out of having to rely more and more on professionally engineered software solutions. This applies also to computational wave propagation, in particular when running on increasingly parallel supercomputer infrastructure. Projects like CIG (Computational Infrastructure in Geodynamics, <http://www.geodynamics.org>) provide parallelized software for a variety of problems in geophysics, ranging from mantle convection to crustal deformation to seismology on all scales. The software can be downloaded and must be installed by the researchers themselves. In many cases this works, but with increasing complexity of hardware, even this approach becomes more and more difficult. In addition, for the developing groups, raising funds to maintain the software is hard or impossible.

To decrease time needed for research by those Earth scientists requiring 3D simulation technology we must go beyond this mode of operation. Ideally, a few well-developed codes that cover most of the standard workflow parts in computational seismology should be installed on the supercomputer infrastructure as modules, readily compiled, and permanently benchmarked with regression tests. An attempt in this direction was made with the EU-funded VERCE project (<http://www.verce.eu>, Atkinson et al., 2015). The goal was to develop a Web-based platform through which 3D simulation tasks can be initialized. In the course of this, community software was installed as pre-compiled modules on the European supercomputer infrastructure PRACE (<http://www.prace-ri.eu>).

Does the community want that? What makes the acceptance of such models difficult is the desire to make (slight or substantial) modifications to existing

Computational Seismology. First Edition. Heiner Igel.

software solutions. Another issue is that many codes require the provision of computational meshes, and currently there are no real standards to do that (see Section 11.3).

In any case, funding bodies and the communities have to acknowledge that today software *is* infrastructure, and it requires a substantial amount of funding for maintenance. The benefit for research will be substantial. The hope is that initiatives like EPOS (European Plate Observing System, <http://www.epos-eu.org>) and CIG or EarthCube (both in the USA) can play a leading role in this direction.

11.2   Structured vs. unstructured: homogenization

With seismic observations we are looking at the Earth's interior with a severely band-limited wavefield. How (on Earth) can we expect to recover a model that contains infinite spatial frequencies (e.g. layer boundaries)? Let me rephrase that: You want to simulate waves through a layered model (i.e. with sharp discontinuities). Is there a smooth model that leads to the same wavefield (within some bounds)? This is the question at the heart of *homogenization*, a method that has been developed by Yann Capdeville and co-workers in the past few years (Capdeville and Marigo, 2007; Capdeville and Marigo, 2008; Capdeville et al., 2010*a*; Capdeville et al., 2010*b*; Capdeville et al., 2013; Capdeville et al., 2015), extending the work by Backus (1962) and others. The concept is illustrated in Fig. 11.1 with the structurally complex 2D Marmousi model containing many velocity discontinuities. To obtain accurate seismic wavefields for such models the most promising approach is currently to honour the discontinuities and develop

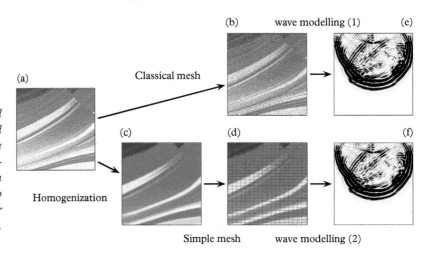

Fig. 11.1 *Homogenization. An initial discontinuous Earth model (a) is meshed (honouring discontinuities) (b) and the wavefield is simulated (e). A smooth version of the model (c) is simulated with a regular grid method (d) leading to the same wavefield (f) within some error bounds. From Capdeville et al. (2015). Reprinted with permission.*

a mesh that follows the layer boundaries. This is (1) extremely time-consuming, and (2) requires a solver that can handle irregular meshes.

Homogenization allows the original discontinuous model to be replaced by a *smooth* model that does not contain layer discontinuities. Within some (tiny) error bounds the wavefields are the same (Fig. 11.1f). However, the latter approach has tremendous advantages: a solver based on a regular mesh at lower resolution implies a tremendous speed-up. This comes at the cost of a preprocessing step, which converts the initial discontinuous model into a smooth version. The hope is that libraries will be provided to do this.

In my view, it is not only that homogenization could revolutionize forward modelling, leading to a revival of simple regular grid methods–the underlying theory is also important for the seismic inverse problem. In addition, recent developments (Al-Attar and Crawford, 2016) allow mapping Earth models with complex topography to a computational model with flat surface. In combination with homogenization, this may open a route to modelling schemes where time-consuming computational mesh generation is replaced by a clever preprocessing scheme. The future will tell.

11.3 Meshing

For many problems of seismic wave propagation (volcanoes, rupture on complex faults, regions with complex topography) there is no way out of generating a high-quality mesh either using hexahedra or tetrahedra. As indicated in the introductory chapters, hexahedral meshes are more efficient in terms of overall simulation time whereas meshing is much easier using tetrahedra.

The problem for seismology is the fact that there are no standardized workflows for geometry creation and meshing, even though efforts in connection with spectral-element solvers (e.g. Casarotti et al., 2008) have helped in this direction. My experience is that the necessity to generate meshes for interesting seismological problems has significantly slowed down projects and endangered final success. The reasons are that (1) seismologists are usually not trained in computational geometry, (2) there does not seem to be any *one-and-only* software that solves all problems, (3) getting into the software and creating meshes is very time-consuming, (4) even well-matured commercial solutions are sometimes not well adapted to the requirements of seismology, and (5) automation of the meshing process is close to impossible.

While the developments in the field of homogenization might release the pressure of meshing for some problems, we still need to come up with stable, standardized meshing solutions. In addition, community libraries should be set up that collect computational meshes (e.g. for volcanoes, sedimentary basins, and continents) for further use (potentially with automated remeshing to finer scale).

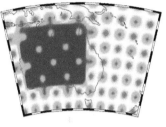

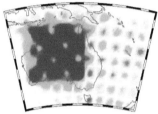

Fig. 11.2 *Chequerboard test. Tomographic resolution test with a chequerboard-like velocity perturbations (red and blue colours) on top of a layered background model und the Australian continent (true model at the top). For a given source-receiver geometry using full waveform inversion the model at the bottom is recovered. While this gives some qualitative estimation on how structures are recovered it does not replace a full (but computationally very expensive) quantitative uncertainty analysis. Figure from Fichtner et al. (2009a).*

11.4 Nonlinear inversion, uncertainties

In a recent international board meeting on supercomputing, a statement was made that the majority of projects seeking large computational resources are targeting problems related to uncertainties. This is not surprising, as (1) there is the expectation that Exaflops, not long after this volume is published, will open the way to increasing use of Monte Carlo approaches to inverse problems, and 2) in many fields the problem of properly estimating uncertainties (of any kind) is still basically unsolved.

Seismology is not an exception. Everyone knows that chequerboard tests (see Fig. 11.2) are not the right approach, as usually the infamous 'inversion crime' is committed.[1] Despite recent progress in connection with waveform inversion (Fichtner and Trampert, 2011*a*; Fichtner and Trampert, 2011*b*; Trampert et al., 2013; Zhang et al., 2013, to name but a few) I think it is fair to say that we are far from being able to appraise our final tomographic models in a proper quantitative way.

This has been a dangerous situation for many decades, as other fields, such as tectonics, geology, and geodynamics, base their research on results coming from seismic tomography. The fact that many of the velocity-perturbation models converge should not keep us from seeking a more quantitative approach to uncertainties. The efficient forward solvers that now exist, combined with increasing computer power which allows Monte Carlo model space searches, should point us in the right direction.

[1] If you calculate a *true* solution with a certain forward solver and then solve the inverse problem with the same solver, even if you add random noise, this is called an *inversion crime*. This is standard practice for chequerboard tests. Assessing uncertainties when fitting real data is far more difficult.

Appendix A
Community Software and Platforms in Seismology

In this appendix, some information is provided on current open-source software, which, while not exhaustive, provides a useful focus on 3D seismic simulation (and inversion), seismic processing, benchmarking, and general services useful for seismological research (certainly incomplete). In addition, a brief description of the *Jupyter Notebooks* is given with which the supplementary electronic material of this volume is provided.

A.1 Wave propagation and inversion

Finite-difference method

- SOFI3D (<https://git.scc.kit.edu/GPIAG-Software/SOFI3D>) stands for Seismic mOdeling with FInite differences, and is a 3D viscoelastic time domain massive parallel modelling code.

- SW4 (<http://www.geodynamics.org/cig/software/sw4>) implements substantial capabilities for 3D seismic modeling, with a free-surface condition on the top boundary, absorbing super-grid conditions on the far-field boundaries, and an arbitrary number of point force and/or point moment tensor source terms.

- FDSim (<http://www.nuquake.eu/fdsim>), Fortran95 computer codes for numerical simulations of seismic wave propagation and earthquake motion in structurally complex 3D heterogeneous viscoelastic media.

- SEISMIC_CPML (<https://geodynamics.org/cig/software/seismic_cpml>) is a set of eleven open-source Fortran90 programs to solve the two-dimensional or three-dimensional isotropic or anisotropic elastic, viscoelastic, or poroelastic wave equation using a finite-difference method with Convolutional or Auxiliary Perfectly Matched Layer (C-PML or ADE-PML) conditions.

- AWP-ODC (<http://hpgeoc.sdsc.edu/AWPODC>), anelastic wave propagation, independently simulates the dynamic rupture and wave propagation that occurs during an earthquake. Dynamic rupture produces friction, traction, slip, and slip rate information on the fault. The moment function is constructed from this fault data and used to initialize wave propagation.

- FDMAP   (<https://pangea.stanford.edu/~edunham/codes/codes.html>), dynamic ruptures and seismic wave propagation in complex geometries.

Finite/Spectral-element method

- SPCEFEM3D-CARTESIAN (<http://www.geodynamics.org/cig/software/specfem3d>). SPECFEM3D CARTESIAN simulates acoustic (fluid), elastic (solid), coupled acoustic/elastic, poroelastic, or seismic wave propagation in any type of conforming mesh of hexahedra (structured or unstructured).
- SPECFEM3D GLOBE (<https://geodynamics.org/cig/software/specfem3d_globe/>) simulates global and regional (continental-scale) seismic wave propagation. Effects due to lateral variations in compressional-wave speed, shear-wave speed, density, a 3D crustal model, ellipticity, topography and bathymetry, the oceans, rotation, and self-gravitation are all included.
- SES3D (<http://www.cos.ethz.ch/software/ses3d.html>) is a program package for the simulation of elastic wave propagation and waveform inversion in a spherical section. The package is based on a spectral-element discretization of the seismic wave equation combined with adjoint techniques.
- AXISEM (<http://www.geodynamics.org/cig/software/axisem>) is a parallel spectral-element method for 3D (an-)elastic, anisotropic, and acoustic wave propagation in spherical domains. It requires axisymmetric background models and runs within a 2D computational domain, thereby reaching all desired highest observable frequencies (up to 2Hz) in global seismology.
- REGSEM (<http://www.ipgp.fr/~paulcup/RegSEM.html>) is a versatile code based on the spectral-element method to compute seismic wave propagation at the regional scale.
- EFISPEC (<http://efispec.free.fr>) stands for *Element-FInis SPECtraux*. It solves the three-dimensional wave equations using a spectral-element method.
- SEM2DPACK   (<http://web.gps.caltech.edu/~ampuero/software.html>) for dynamic rupture simulations non-planar faults in heterogeneous or non-linear media with the spectral-element method.

Discontinuous galerkin method

- SEISSOL (<http://www.seissol.org>) is a modal discontinuous Galerkin method primarily for wave and rupture propagation on tetrahedral meshes.
- SPEED (<http://www.speed.mox.polimi.it/SPEED>) is a discontinuous Galerkin spectral-element code that incorporates the open-source libraries METIS and MPI for the parallel computation (mesh partitioning and

message passing). It has been designed with the aim of simulating large-scale seismic events, allowing the evaluation of the typically multi-scale wave-propagation problems in its complexity, from far-field to near-field and from near-field to soil-structure interaction effects.

- NEXD (<http://www.rub.de/nexd>) is a software package for high-order simulation of seismic waves using the nodal discontinuous Galerkin method

Other useful software

- MINEOS (<http://www.geodynamics.org/cig/software/mineos>) computes synthetic seismograms in a spherically symmetric non-rotating Earth by summing normal modes.
- GEMINI (www.geophysik.ruhr-uni-bochum.de/trac/gemini) is a program package to calculate Green's functions and surface wave modes of the elastic wave equation for one-dimensional depth-dependent media. Applications of the code range from high-frequency, small-scale wave-propagation problems like ultrasonic waves, seam waves, and shallow seismics to continental-scale seismic waves from earthquakes.
- LASIF (<http://www.lasif.net>) (LArge-scale Seismic Inversion Framework) is a data-driven end-to-end workflow tool to perform adjoint full seismic waveform inversions.
- ASKI (<http://www.rub.de/aski>) is a highly modularized program suite for sensitivity analysis and iterative full waveform inversion using waveform sensitivity kernels for 1D or 3D heterogeneous elastic background media, and makes use of different forward modelling codes.
- QSEIS (<http://www.gfz-potsdam.de>) is a Fortran code for calculating synthetic seismograms based on a layered viscoelastic half-space Earth model.
- QSSP (<http://www.gfz-potsdam.de>) is a code for calculating complete synthetic seismograms of a spherical Earth using normal mode theory.

A.2 Data processing, visualization, services

- SEISMO-LIVE (<http://www.seismo-live.org>) is an open-source library with Jupyter Notebooks for seismology running on a dedicated server. Notebooks can be run on any browser. This site contains the supplementary material of this volume ($\rightarrow$ *computational seismology*).
- ObsPy (<http://www.obspy.org>). ObsPy is an open-source project dedicated to providing a Python framework for processing seismological data. It provides parsers for common file formats, client access to data centres,

and seismological signal processing routines which allow the manipulation of seismological time series.

- ASDF (<http://www.seismic-data.org>). The Adaptable Seismic Data Format (ASDF) is a modern file format intended for researchers and analysts. It combines the capability to create comprehensive data sets including all necessary meta information with high-performance parallel I/O for the most demanding use cases.

- SAC (<http://www.ds.iris.edu/ds/nodes/dmc/software/downloads/sac>). The Seismic Analysis Code is a general purpose interactive program designed for the study of sequential signals, especially time series data. Emphasis has been placed on analysis tools used by research seismologists in the detailed study of seismic events.

- EPOS (<http://www.epos-eu.org>). The European Plate Observing System (EPOS) is a planned research infrastructure for European solid Earth science, integrating existing research infrastructures to enable innovative multidisciplinary research, recently prioritized by the European Strategy Forum on Research Infrastructures (ESFRI) for implementation.

- VERCE (<http://www.verce.eu>). VERCE developed a data-intensive e-science environment to enable innovative data analysis and data modelling methods that fully exploit the increasing wealth of open data generated by the observational and monitoring systems of the global seismology community.

- IRIS (<http://www.iris.edu>). The Incoporated Research Institutions in Seismology provide management of, and access to, observed and derived data for the global Earth science community.

- ORFEUS (<http://www.orfeus-eu.org>). Observatories and Research Facilities for European Seismology is the non-profit foundation that aims at coordinating and promoting digital, broadband (BB) seismology in the European–Mediterranean area.

- CMT (<http://www.globalcmt.org>) provides moment tensor estimates for globally observable earthquakes.

- INSTASEIS (<http://www.instaseis.net>). Instaseis calculates broadband seismograms from Green's function databases generated with AxiSEM and allows for near instantaneous (on the order of milliseconds) extraction of seismograms.

A.3 Benchmarking

- SISMOWINE (<http://www.sismowine.org>) is an interactive seismological web interface used for numerical modelling benchmarking. Participants calculate solutions for the defined models using their numerical or analytical computational method and compare the solutions with those submitted by other participants.

- NUQUAKE (<http://www.nuquake.eu>) maintains a number of forward solutions in 1–3D and also provides access to analytical solutions.

- QUEST (<http://www.quest-itn.org/library/software>) contains some simple numerical simulation codes and several analytical solutions (e.g. Lamb's problem).

- SCEC–USGS (<http://scecdata.usc.edu/cvws>) Spontaneous Rupture Code Verification Project. Comparison of various 3D methods to simulate (spontaneous) rupture dynamics.

- SIV (<http://equake-rc.info/SIV>). The SIV project aims at quantifying the uncertainty in earthquake source inversion through a series of verification & validation experiments.

- SEAM (<http://www.seg.org/resources/research/seam>). The SEG Advanced Modelling Program (SEAM) is a partnership between industry and SEG designed to advance geophysical science and technology through the construction of subsurface models and generation of synthetic data sets.

A.4   Jupyter Notebooks

The material presented in this volume is complemented by a substantial amount of *Jupyter Notebooks* (previously IPython Notebooks). The Jupyter Notebooks are based on an interactive computational environment, in which you can combine code execution, rich text, mathematics, plots, and rich media. The choice of Python over other programming languages is primarily due to its independence from commercial software for computer practicals. In addition, the Jupyter Notebooks offer a fascinating new tool to exchange software or practicals in a platform-independent way. The potential of this approach was recently recognized an article in *Nature* (Shen, 2014) providing access to an online example. A snapshot is shown in Fig. A.1. All Jupyter Notebooks will be provided in a novel openly accessible library of notebooks for seismology with online execution options (www.seismo-live.org).

A.5   Supplementary material

Details, instructions on how to install the Python environment, and access to the computer exercises are given on the website maintaining the electronic material:

<http://www.computational-seismology.org>

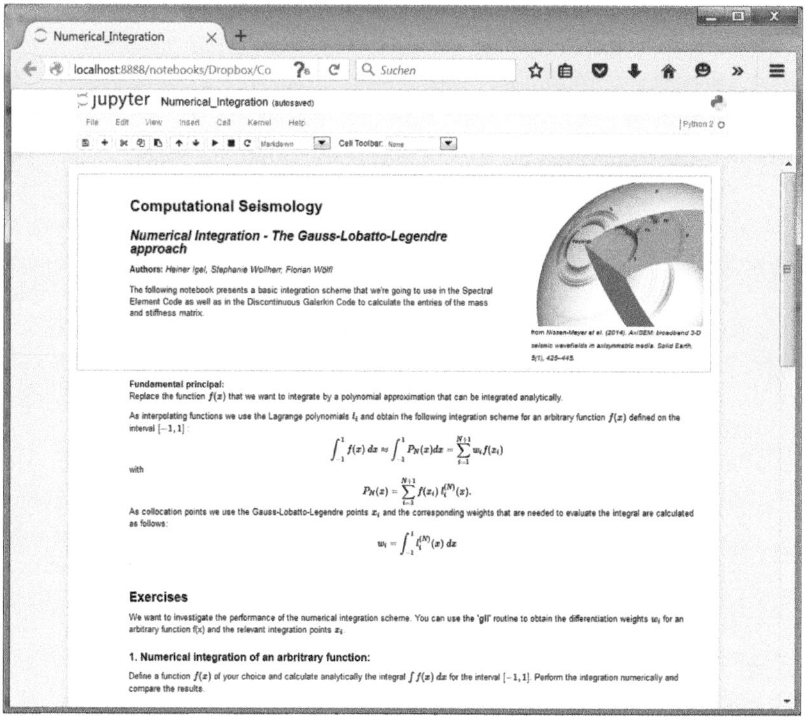

Fig. A.1 *Jupyter Notebook. The notebooks can be run by any web browser. Text, graphs, and equations can be inserted using the mark-up language and Latex scripting. Code blocks can be run in dedicated windows, return data, and graphs inside the notebook.*

Below is a non-exhaustive list of programs and tools available online. Many of these programs are also available in Matlab®. Complete codes are given as well as solutions to the computer exercises. You are advised to write as many codes from scratch as you can and compare them with the available solutions. This has by far the highest training value. The material includes:

General

- Introduction to Jupyter Notebooks
- Introduction to Python

Part I

- Analytical solutions for the 1–3D acoustic wave equation in homogeneous media
- Analytical solutions for the double-couple point source in 3D
- Analytical solution to Lamb's problem in 3D
- Examples for time reversal and reciprocity
- Numerical Green's functions and convolution

Part II

- Taylor operators for finite-difference calculations
- Finite-difference codes for 1D and 2D (acoustic case)
- Finite-difference code with optimal operators for 1D
- Staggered-grid finite-difference codes for 1D (elastic case)
- Pseudospectral codes (Fourier and Chebyshev) for 1D acoustic and elastic wave propagation
- Finite-element code for static elasticity and 1D elastic wave equation
- Lagrange polynomials, interpolation, derivative
- Legendre polynomials
- Gauss–Lobatto–Legendre collocation points
- Gauss–Lobatto–Legendre integration
- Spectral-element code for elastic wave propagation in 1D
- Finite-volume code for scalar wave propagation and linear systems (elastic wave propagation) in 1D
- Discontinuous Galerkin code for scalar wave propagation and linear systems (elastic wave propagation) in 1D

References

Aki, K. and Richards, P. (1980). *Quantitative Seismology: Theory and Methods.* San Francisco: Freeman.

Aki, K. and Richards, P. (2002). *Quantitative Seismology.* Sausalito, CA: University Science Books.

Al-Attar, D. and Crawford, O. (2016). Particle relabelling transformations in elastodynamics. *Geophys. J. Int.,* **205**, 575–93.

Alford, R. M., Kelly, K. R., and Boore, D. M. (1974). Accuracy of finite-difference modeling of the acoustic wave equation. *Geophysics,* **39**, 834–42.

Alterman, Z., Aboudi, J., and Karal, F. C. (1970). Pulse propagation in a laterally heterogeneous solid elastic sphere. *Geophys. J. R. Astron. Soc.,* **21**, 243–60.

Alterman, Z. and Karal, F. C. (1968). Propagation of elastic waves in layered media by finite-difference methods. *Bull. Seism. Soc. Am.,* **58**, 367–98.

Andrews, D. J. (1973). A numerical study of tectonic stress release by underground explosions. *Bull. Seism. Soc. Am.,* **63**(4), 1375–91.

Andrews, D. J. (1976*a*). Rupture propagation with finite stress in antiplane strain. *J. Geophys. Res.,* **81**(20), 3575–82.

Andrews, D. J. (1976*b*). Rupture velocity of plane strain shear cracks. *J. Geophys. Res.,* **81**(32), 5679–87.

Aochi, H. and Ulrich, T. (2015). A probable earthquake scenario near Istanbul determined from dynamic simulations. *Bull. Seism. Soc. Am.,* **105**(3), 1468–75.

Askan, A. and Bielak, J. (2008). Full anelastic waveform tomography including model uncertainty. *Bull. Seism. Soc. Am.,* **98**(6), 2975–89.

Atkinson, M., Carpene, M., Claus, S., Filgueira, R., Frank, A., Galea, M., Gemuend, A., Igel, H., Klampanos, I., Krause, A., Krischer, L., Leong, S. H., Magnoni, F., Matser, J., Michelini, A., Schwichtenberg, H., Spinuso, A., and Vilotte, J. (2015). VERCE delivers a productive e-Science environment for seismology research. In *IEEE eScience 2015,* <http://www.arxiv.org/pdf/1510.01989.pdf>.

Backus, G. E. (1962). Long-wave elastic anisotropy produced by horizontal layering. *J. Geophys. Res.,* **67**, 4427–40.

Baig, A. M. and Dahlen, F. A. (2004). Travel time biases in random media and the S-wave discrepancy. *Geophys. J. Int.,* **158**, 922–38.

Baig, A. M., Dahlen, F. A., and Hung, S.-H. (2003). Travel times of waves in three-dimensional random media. *Geophys. J. Int.,* **153**(2), 467–82.

Bamberger, A., Chavent, G., Hemons, C., and Lailly, P. (1982). Inversion of normal incidence seismograms. *Geophysics,* **47**, 757–70.

Bao, H., Bielak, J., Ghattas, O., Kallivokas, L. F., O'Hallaron, D. R., Shewchuk, J. R., and Xu, J. (1996). Earthquake ground motion modeling on parallel computers. In *Proceedings of the 1996 ACM/IEEE Conference on Supercomputing,* Supercomputing '96, Washington, DC. IEEE Computer Society.

Ben Hadj Ali, H., Operto, S., Virieux, J., and Sourbier, F. (2009*a*). Efficient 3D frequency-domain full-waveform inversion with phase encodings. In *71st Conference & Technical Exhibition, EAGE, Extended Abstracts,* p. 5812.

Ben Hadj Ali, H., Operto, S., Virieux, J., and Sourbier, F. (2009*b*). Three-dimensional frequency-domain full-waveform inversion with phase encoding. In *79th Annual International Meeting, SEG, Extended Abstracts,* pp. 2288–92.

Ben-Zion, Y. (2003). Appendix 2: Key formulas in earthquake seismology. In *International Handbook of Earthquake and Engineering Seismology, Part B* (ed. W. H. Lee, H. Kanamori, P. C. Jennings, and C. Kisslinger), pp. 1857–75. Amsterdam: Academic Press.

Benjemaa, M., Glinsky-Olivier, N., Cruz-Atienza, V. M., and Virieux, J. (2009). 3D dynamic rupture simulations by a finite volume method. *Geophys. J. Int.*, **178**(1), 541–60.

Benjemaa, M., Glinsky-Olivier, N., Cruz-Atienza, V. M., Virieux, J., and Piperno, S. (2007). Dynamic non-planar crack rupture by a finite volume method. *Geophys. J. Int.*, **171**(1), 271–85.

Bernauer, M., Fichtner, A., and Igel, H. (2014). Reducing non-uniqueness in finite source inversion using rotational ground motions. *J. Geophys. Res.*, **119**(6), doi:10.1002/2014JB011042.

Beyreuther, M., Barsch, R., Krischer, L., and Wassermann, J. (2010). ObsPy: A Python toolbox for seismology. *Seis. Res. Lett.*, **81**, 47–58.

Bielak, J., Ghattas, O., and Bao, H. (1998). Ground motion modeling using 3D finite element methods. In *The Effects of Surface Geology on Seismic Motion* (ed. K. Irikura, K. Kudo, H. Okada, and T. Sasatani), pp. 121–33. Rotterdam: A. A. Balkema.

Bielak, J., Ghattas, O., and Kim, E. (2005). Parallel octree-based finite element method for large-scale earthquake ground motion simulation. *Computer Modeling in Engineering and Sciences*, **10**(2), 99.

Bielak, J., Graves, R. W., Olsen, K. B., Taborda, R., Ramìrez-Guzmàn, L., Day, S. M., Ely, G. P., Roten, D., Jordan, T. H., Maechling, P. J., Urbanic, J., Cui, Y., and Juve, G. (2010). The shakeout earthquake scenario: Verification of three simulation sets. *Geophys. J. Int.*, **180**(1), 375–404.

Bielak, J. and Xu, J. (1999). Earthquake ground motion and structural response in alluvial valleys. *Journal of Geotechnical and Geoenvironmental Engineering*, **125**(5), 413–23.

Bleibinhaus, F., Hole, J. A., and Ryberg, T. (2007). Structure of the California Coast Ranges and San Andreas Fault at SAFOD from seismic waveform inversion and reflection imaging. *J. Geophys. Res.*, **112**, doi:10.1029/2006JB004611.

Bohlen, T. (2002). Parallel 3D viscoelastic finite difference seismic modelling. *Computers & Geosciences*, **28**(8), 887–99.

Boore, D. M. (1970). Love waves in nonuniform waveguides: finite difference calculations. *J. Geophys. Res.*, **1970**, 1512–27.

Booth, D. C. and Crampin, S. (1983). The anisotropic reflectivity technique: theory. *Geophys. J.*, **72**, 755–66.

Bormann, P. (ed.) (2012). *New Manual of Seismological Observatory Practice (NMSOP-2)*. Potsdam, New York: IASPEI, GFZ German Research Centre for Geosciences.

Boxberg, M., Prévost, J. H., and Tromp, J. (2015). Wave propagation in porous media saturated with two fluids. *Transport in Porous Media*, **107**(1), 49–63.

Bracewell, R. (1999). *The Fourier Transform and Its Applications*. 3rd revised edition, Columbus, OH: Mcgraw-Hill.

Braun, J. and Sambridge, M. S. (1995). A numerical method for solving partial differential equations on highly irregular evolving grids. *Nature*, **376**, 655–60.

Brenguier, F., Campillo, M., Hadziioannou, C., Shapiro, N., Nadeau, R., and Larose, E. (2008). Postseismic relaxation along the San Andreas Fault at Parkfield from continuous seismological observations. *Science*, **321**, 1478–81.

Breuer, A., Heinecke, A., Rannabauer, L., and Bader, M. (2015). High-order ADER-DG minimizes energy-and time-to-solution of SeisSol. In *High Performance Computing* (ed. J. M. Kunkel and T. Ludwig), Volume 9137, Lecture Notes in Computer Science, pp. 340–57. Berlin/Heidelberg: Springer International Publishing.

Breuer, A., Heinecke, A., Rettenberger, S., Bader, M., Gabriel, A-A., and Pelties, C. (2014). Sustained petascale performance of seismic simulations with SeisSol on SuperMUC. In *International Supercomputing Conference (ISC) Proceedings*, Volume 8488, Lecture Notes in Computer Science, pp. 1–18. Berlin/Heidelberg: Springer International Publishing.

Brietzke, G. B., Cochard, A., and Igel, H. (2009). Importance of bimaterial interfaces for earthquake dynamics and strong ground motion. *Geophys. J. Int.*, **178**(2), 921–38.

Butzer, S., Kurzmann, A., and Bohlen, T. (2013). 3D elastic full-waveform inversion of small-scale heterogeneities in transmission geometry. *Geophys. Prosp.*, **61**(6), 1238–51.

Capdeville, Y., Guillot, L., and Marigo, J. J. (2010a). 1D non periodic homogenization for the wave equation. *Geophys. J. Int.*, **181**, 897–910.

Capdeville, Y., Guillot, L., and Marigo, J. J. (2010b). 2D nonperiodic homogenization to upscale elastic media for P-SV waves. *Geophys. J. Int.*, **182**, 903–22.

Capdeville, Y., Gung, Y., and Romanowicz, B. (2005). Towards global earth tomography using the spectral element method: a technique based on source stacking. *Geophys. J. Int.*, **162**, 541–54.

Capdeville, Y. and Marigo, J. J. (2007). Second-order homogenization of the elastic wave equation for non-periodic layered media. *Geophys. J. Int.*, **170**, 823–38.

Capdeville, Y. and Marigo, J. J. (2008). Shallow layer correction for spectral element like methods. *Geophys. J. Int.*, **172**, 1135–50.

Capdeville, Y., Romanowicz, B., and To, A. (2003). Coupling spectral elements and modes in a spherical Earth: an extension to the 'sandwich' case. *Geophys. J. Int.*, **154**, 44–57.

Capdeville, Y., Stutzmann, E., Montagner, J.-P., and Wang, N. (2013). Residual homogenization for seismic forward and inverse problems in layered media. *Geophys. J. Int.*, doi:10.1093/gji/ggt102.

Capdeville, Y., Zhao, M., and Cupillard, P. (2015). Fast fourier homogenization for elastic wave propagation in complex media. *Wave Motion*, **54**, 170–86.

Carcione, J. J. M. (2014). *Wave Fields in Real Media: Wave Propagation in Anisotropic, Anelastic, Porous and Electromagnetic Media*. Handbook of Geophysical Exploration. Seismic Exploration. Amsterdam: Elsevier Science & Technology.

Carcione, J. M. and Wang, P. J. (1993). A Chebyshev collocation method for the wave equation in generalized coordinates. *Comp. Fluid. Dyn. J.*, **2**, 269–90.

Casarotti, E., Stupazzini, M., Lee, S., Komatitsch, D., Piersanti, A., and Tromp, J. (2008). CUBIT and seismic wave propagation based upon the spectral-element method: An advanced unstructured mesher for complex 3D geological media. In *Proceedings of the 16th International Meshing Roundtable* (ed. M. Brewer and D. Marcum), pp. 579–97. Berlin/Heidelberg: Springer.

Cerveny, V. (2001). *Seismic Ray Theory*. Cambridge: Cambridge University Press.

Chaljub, E. (2000). *Modèlisation numérique de la propagation d'ondes sismiques à l'échelle du globe*. Doctoral thesis, Université Paris 7.

Chaljub, E., Bard, P. Y., Hollender, F., Theodulidis, N., Moczo, P., Tsuno, S., Kristek, J., Cadet, H., Bielak, J., Moran, S., Beeler, N., Wong, I., Weldon, R., McConnell, V., and Trehu, A. (2010a). Euroseistest numerical simulation project; comparison with local earthquake recordings for validation. *Seismological Research Letters*, **81**(2), 308–09.

Chaljub, E., Capdeville, Y., and Vilotte, J.-P. (2003). Solving elastodynamics in a solid heterogeneous 3-sphere: a spectral element approximation on geometrically non-conforming grids. *J. Comput. Phys.*, **183**, 457–91.

Chaljub, E., Komatitsch, D., Vilotte, J.-P., Capdeville, Y., Valette, B., and Festa, G. (2007). Spectral element analysis in seismology. In *Advances in wave propagation in heterogeneous media* (ed. R.-S. Wu and V. Maupin), Volume 48, Advances in Geophysics, pp. 365–419. London: Elsevier.

Chaljub, E., Maufroy, E., Moczo, P., Kristek, J., Hollender, F., Bard, P.-Y., Priolo, E., Klin, P., de Martin, F., Zhang, Z., Zhang, W., and Chen, X. (2015). 3D numerical simulations of earthquake ground motion in sedimentary basins: testing accuracy through stringent models. *Geophys. J. Int.*, **201**(1), 90–111.

Chaljub, E., Moczo, P., Tsuno, S., Bard, P.-Y., Kristek, J., Käser, M., Stupazzini, M., and Kristekova, M. (2010b). Quantitative comparison of four numerical predictions of 3D ground motion in the Grenoble valley, France. *Bull. Seism. Soc. Am.*, **100**(4), 1427–55.

Chaljub, E. and Tarantola, A. (1997). Sensitivity of SS precursor to topography on the upper-mantle 660-km discontinuity. *Geophys. Res. Lett.*, **24**, 2613–16.

Chapman, C. (2004). *Fundamentals of Seismic Wave Propagation*. Volume 11. Cambridge: Cambridge University Press.

Chen, M., Niu, F., Liu, Q., and Tromp, J. (2015*a*). Mantle-driven uplift of Hangai dome: New seismic constraints from adjoint tomography. *Geophys. Res. Lett.*, **42**(17), 6967–74.

Chen, M., Niu, F., Liu, Q., Tromp, J., and Zheng, X. (2015*b*). Multiparameter adjoint tomography of the crust and upper mantle beneath East Asia: 1. model construction and comparisons. *J. Geophys. Res.*, **120**(3), 1762–86.

Chen, P., Zhao, L., and Jordan, T. H. (2007). Full 3D tomography for the crustal structure of the Los Angeles region. *Bull. Seism. Soc. Am.*, **97**, 1094–120.

Cockburn, B., Karniadakis, G. E., and Shu, C. W. (2000). *Discontinuous Galerkin Methods: Theory, Computation and Applications*. Volume 11. Berlin/Heidelberg: Springer.

Colli, L., Fichtner, A., and Bunge, H.-P. (2013). Full waveform tomography of the upper mantle in the South Atlantic region: Imaging a westward fluxing shallow asthenosphere? *Tectonophysics*, **604**, 26–40.

Cooley, J. W. and Tukey, J. W. (1965). An algorithm for the machine calculation of complex Fourier series. *Math. Comput.*, **19**, 297–301.

Crampin, S. (1984). An introduction to wave propagation in anisotropic media. *Geophys. J. Int.*, **76**(1), 17–28.

Crase, E., Pica, A., Noble, M., McDonald, J., and Tarantola, A. (1990). Robust elastic nonlinear waveform inversion: Application to real data. *Geophysics*, **55**, 527–38.

Cummins, P. R., Geller, R. J., Haori, T., and Takeuchi, N. (1994*a*). DSM complete synthetic seismograms: SH, spherically symmetric case. *Geophys. Res. Lett.*, **21**, 533–36.

Cummins, P. R., Geller, R. J., and Takeuchi, N. (1994*b*). DSM complete synthetic seismograms: P-SV, spherically symmetric case. *Geophys. Res. Lett.*, **21**, 1663–6.

Cummins, P. R., Takeuchi, N., and Geller, R. J. (1997). Computation of complete synthetic seismograms for laterally heterogeneous models using the Direct Solution Method. *Geophys. J. Int.*, **130**, 1–16.

Cupillard, P. and Capdeville, Y. (2010). On the amplitude of surface waves obtained by noise correlation and the capability to recover the attenuation: a numerical approach. *Geophys. J. Int.*, **181**(3), 1687–700.

Cupillard, P., Delavaud, E., Burgos, G., Festa, G., Vilotte, J.-P., Capdeville, Y., and Montagner, J.-P. (2012). RegSEM: a versatile code based on the spectral element method to compute seismic wave propagation at the regional scale. *Geophys. J. Int.*, **188**, 1203–20.

Dablain, M. A. (1986). The application of high-order differencing to the scalar wave equation. *Geophysics*, **51**, 54–66.

Dahlen, F. A. and Tromp, J. (1998). *Theoretical Global Seismology*. Princeton, NJ: Princeton University Press.

Davi, R., O'Brien, G. S., Lokmer, I., Bean, C. J., Lesage, P., and Mora, M. M. (2010). Moment tensor inversion of explosive long period events recorded on Arenal volcano, Costa Rica, constrained by synthetic tests. *Journal of Volcanology and Geothermal Research*, **194**(4), 189–200.

Day, S. M. and Minster, J. B. (1984). Numerical simulation of wavefields using a Pad approximant method. *Geophys. J. R. Astr. Soc.*, **78**, 105–18.

Day, S. M. (1977). *Finite element analysis of seismic scattering problems*. PhD dissertation. University of California, San Diego.

Day, S. M. (1982). Three-dimensional finite difference simulation of fault dynamics; rectangular faults with fixed rupture velocity. *Bull. Seism. Soc. Am.*, **72**(3), 705–27.

Day, S. M. and Boatwright, J. (1982). Three-dimensional simulation of spontaneous rupture; the effect of nonuniform prestress. *Bull. Seism. Soc. Am.*, **72**, **Part A**(6), 1881–902.

de la Puente, J., Ampuero, J.-P., and Käser, M. (2009*a*). Dynamic rupture modeling on unstructured meshes using a discontinuous Galerkin method. *J. Geophys. Res.*, **114**, doi:10.1029/2008JB006271.

de la Puente, J., Dumbser, M., Käser, M., and Igel, H. (2008). Discontinuous Galerkin methods for wave propagation in poroelastic media. *Geophysics*, **73**(5), T77–T97.

de la Puente, J., Käser, M., and Cela, J. M. (2009*b*). SeisSol optimization, scaling and synchronization for local time stepping. In *Science and Supercomputing in Europe* (ed. S. Monfardini), pp. 300–22. Bologna: CINECA.

de la Puente, J., Käser, M., Dumbser, M., and Igel, H. (2007). An arbitrary high order discontinuous Galerkin method for elastic waves on unstructured meshes IV: Anisotropy. *Geophys. J. Int.*, **169**(3), 1210–28.

Delorey, A. A., Frankel, A. D., Liu, P., and Stephenson, W. J. (2014). Modeling the effects of source and path heterogeneity on ground motions of great earthquakes on the Cascadia subduction zone using 3D simulations. *Bull. Seism. Soc. Am.*, **104**(3), 1430–46.

de Martin, F. (2011). Verification of a spectral-element method code for the Southern California earthquake center LOH 3 viscoelastic case. *Bull. Seism. Soc. Am.*, **101**(6), 2855–65.

Dormy, E. and Tarantola, A. (1995). Numerical simulation of elastic wave propagation using a finite volume method. *J. Geophys. Res.*, **100**(B2), 2123–33.

Dougherty, M. E. and Stephen, R. A. (1988). Seismic energy partitioning and scattering in laterally heterogeneous ocean crust. *Pure and Applied Geophysics*, **128**(1), 195–229.

Dumbser, M., Käser, M., and de la Puente, J. (2007*a*). Arbitrary high order finite volume schemes for seismic wave propagation on unstructured meshes in 2D and 3D. *Geophys. J. Int.*, **171**, 665–94.

Dumbser, M., Käser, M., and Toro, E. (2007*b*). An arbitrary high-order discontinuous Galerkin method for elastic waves on unstructured meshes, Part V: Local time stepping and *p*-adaptivity. *Geophys. J. Int.*, **171**, 695–717.

Dumbser, M. and Munz, C. D. (2005*a*). ADER discontinuous Galerkin schemes for aeroacoustics. *Comptes Rendues-Mecanique*, **333**, 683–7.

Dumbser, M. and Munz, C. D. (2005*b*). Arbitrary high-order discontinuous Galerkin schemes. In *Numerical Methods for Hyperbolic and Kinetic Problems* (ed. S. Cordier, T. Goudon, and E. Sonnendrucker), pp. 295–333. Zurich: EMS Publishing House.

Durran, D. R. (1999). *Numerical methods for wave equations in geophysical fluid dynamics*. Berlin/Heidelberg: Springer.

Dziewoński, A. M. and Anderson, D. L. (1981). Preliminary reference Earth model. *Phys. Earth Planet. Inter.*, **25**, 297–356.

Eijkhout, V. (2015). *Introduction to High Performance Scientific Computing*, available at <http://www.lulu.com>.

Emmerich, H. and Korn, M. (1987). Incorporation of attenuation into time-domain computations of seismic wave fields. *Geophysics*, **52**, 1252–64.

Epanomeritakis, I., Akcelik, V., Ghattas, O., and Bielak, J. (2008). A Newton-CG method for large-scale three-dimensional elastic full waveform seismic inversion. *Inverse Problems*, **24**, doi:10.1088/0266–5611/24/3/034015.

Etgen, J. T. and O'Brien, M. J. (2007). Computational methods for large-scale 3D acoustic finite-difference modeling: A tutorial. *Geophysics*, **72**(5), SM223–SM230.

Etienne, V., Chaljub, E., Virieux, J., and Glinsky, N. (2010). An hp-adaptive discontinuous Galerkin finite-element method for 3D elastic wave modelling. *Geophys. J. Int.*, **183**(2), 941–62.

Etienne, V., Hu, G., Operto, S., Virieux, J., Barkved, O. I., and Kommedal, J. H. (2012). Three-dimensional acoustic full waveform inversion; algorithm and application to Valhall. *Conference and Technical Exhibition—European Association of Geoscientists and Engineers*, **74**, Abstract P343.

Ewald, M., Igel, H., Hinzen, K.-G., and Scherbaum, F. (2006). Basin-related effects on ground motion for earthquake scenarios in the Lower Rhine Embayment. *Geophys. J. Int.*, **166**, 197–212.

Eymard, R., Gallouët, T., and Herbin, R. (2000). Finite volume methods. In *Handbook of Numerical Analysis* (ed. P. G. Ciarlet and J. L. Lions), Volume VII, pp. 715–1022. Amsterdam: North Holland.

Faccioli, E., Maggio, F., Quarteroni, A., and Tagliani, A. (1996). Spectral-domain decomposition methods for the solution of acoustic and elastic wave equations. *Geophysics*, **61:4**, 1160–74.

Fehler, M. and Keliher, P. (2011). Seam phase 1: Challenges of subsalt imaging in tertiary basins, with emphasis on deepwater Gulf of Mexico. p. doi:10.1190/1.9781560802945.

Festa, G. and Vilotte, J.-P. (2006). Influence of the rupture initiation on the intersonic transition: Crack-like versus pulse-like modes. *Geophysical Research Letters*, 33(15), doi:10.1029/2006GL026378. L15320.

Fichtner, A. (2009). *Full waveform inversion for structural and source parameters*. Dissertation. Ludwig-Maximilians Universität München.

Fichtner, A. (2010). *Full Seismic Waveform Modelling and Inversion*. Heidelberg: Springer.

Fichtner, A. and Igel, H. (2008). Efficient numerical surface wave propagation through the optimization of discrete crustal models—a technique based on non-linear dispersion curve matching (DCM). *Geophys. J. Int.*, 173, 519–33.

Fichtner, A., Kennett, B. L. N., Igel, H., and Bunge, H.-P. (2008). Theoretical background for continental- and global-scale full-waveform inversion in the time–frequency domain. *Geophys. J. Int.*, 175, 665–85.

Fichtner, A., Kennett, B. L. N., Igel, H., and Bunge, H.-P. (2009a). Full seismic waveform tomography for upper-mantle structure in the Australasian region using adjoint methods. *Geophys. J. Int.*, 179, 1703–25.

Fichtner, A., Kennett, B. L. N., Igel, H., and Bunge, H.-P. (2009b). Spectral-element simulation and inversion of seismic waves in a spherical section of the Earth. *J. Num. An. Ind. Appl. Math.*, 4, 11–22.

Fichtner, A., Kennett, B. L. N., Igel, H., and Bunge, H.-P. (2010). Full waveform tomography for radially anisotropic structure: New insight into present and past states of the Australasian upper mantle. *Earth Planet. Sci. Lett.*, 290, 270–80.

Fichtner, A., Saygin, E., Taymaz, T., Cupillard, P., Capdeville, Y., and Trampert, J. (2013a). The deep structure of the North Anatolian fault zone. *Earth Planet. Sci. Lett.*, 373, 109–17.

Fichtner, A. and Trampert, J. (2011a). Hessian kernels of seismic data functionals based upon adjoint techniques. *Geophys. J. Int.*, 185, 775–98.

Fichtner, A. and Trampert, J. (2011b). Resolution analysis in full waveform inversion. *Geophys. J. Int.*, 187, 1604–24.

Fichtner, A., Trampert, J., Cupillard, P., Saygin, E., Taymaz, T., Capdeville, Y., and Villasenor, A. (2013b). Multiscale full waveform inversion. *Geophys. J. Int.*, 194(1), 534–56.

Fichtner, A. and Villasenor, A. (2015). Crust and upper mantle of the western Mediterranean; constraints from full-waveform inversion. *Earth Planet. Sci. Lett.*, 428, 52–62.

Forbriger, T., Groos, L., and Schäfer, M. (2014). Line-source simulation for shallow-seismic data. part 1: theoretical background. *Geophys. J. Int.*, 198(3), 1387–404.

Fornberg, B. (1996). *A Practical Guide to Pseudospectral Methods*. Cambridge: Cambridge University Press.

Frankel, A. (1989). A review of numerical experiments on seismic wave scattering. *Pure Appl. Geophys.*, 4, 639–85.

Frankel, A. and Vidale, J. (1992). A three-dimensional simulation of seismic waves in the Santa Clara Valley, California, from a Loma Prieta aftershock. *Bull. Seism. Soc. Am.*, 82, 2031–45.

French, S. W. and Romanowicz, B. (2015). Broad plumes rooted at the base of the Earth's mantle beneath major hotspots. *Nature*, 525, 95–99.

Frenje, L. and Juhlin, C. (2000). Scattering attenuation: 2D and 3D finite difference simulations vs. theory. *Journal of Applied Geophysics*, 44(1), 33–46.

Fuchs, K. (1997). *Upper Mantle Heterogeneities from Active and Passive Seismology*. Dordrecht: Springer Netherlands.

Fuchs, K. and Müller, G. (1971). Computation of synthetic seismograms with the reflectivity method and comparison with observations. *Geophys. J. R. Astron. Soc.*, 23, 417–33.

Funaro, D. (1993). *FORTRAN Routines for Spectral Methods*. Modena: Instituto di Analisi Numerica.

Furumura, M., Kennett, B. L. N., and Furumura, T. (1999). Seismic wavefield calculation for laterally heterogeneous Earth models, II: The influence of upper mantle heterogeneity. *Geophys. J. Int.*, 139, 623–44.

Furumura, T. and Kennett, B. L. N. (2005). Subduction zone guided waves and the heterogeneity structure of the subducted plate: intensity anomalies in northern Japan. *J. Geophys. Res.*, 110, doi:10.129/2004JB003486.

Furumura, T., Kennett, B. L. N., and Furumura, M. (1998*a*). Seismic wavefield calculation for laterally heterogeneous whole Earth models using the pseudospectral method. *Geophys. J. Int.*, **135**, 845–60.

Furumura, T., Kennett, B. L. N., and Takenaka, H. (1998*b*). Parallel 3D pseudospectral simulation of seismic wave propagation. *Geophysics*, **63**, 279–88.

Furumura, T., Koketsu, K., and Wen, K.-L. (2002). Parallel PSM/FDM hybrid simulation of ground motions from the 1999 Chi-Chi Taiwan earthquake. *Pure Appl. Geophys.*, **159**, 2133–46.

Galis, M., Moczo, P., and Kristek, J. (2008). A 3D hybrid finite-difference–finite-element viscoelastic modeling of seismic wave motion. *Geophys. J. Int.*, **175**, 153–84.

Gallovic, F., Käser, M., Burjanek, J., and Papaioannou, C. (2010). Three-dimensional modeling of near-fault ground motions with nonplanar rupture models and topography: Case of the 2004 Parkfield earthquake. *J. Geophys. Res.*, **115**(B3), doi:10.1029/2008JB006171.

Galvez, P., Ampuero, J.-P., Dalguer, L. A., Somala, S. N., and Nissen-Meyer, T. (2014). Dynamic earthquake rupture modelled with an unstructured 3D spectral element method applied to the 2011 M9 Tohoku earthquake. *Geophys. J. Int.*, **198**(2), 1222–40.

Gazdag, J. (1981). Modeling of the acoustic wave propagation with transform methods. *Geophysics*, **46**, 854–9.

Geller, R. and Takeuchi, N. (1995). A new method for computing highly accurate DSM synthetic seismograms. *Geophys. J. Int.*, **123**, 449–70.

Geller, R. J. and Takeuchi, N. (1998). Optimally accurate second-order time-domain finite difference scheme for the elastic equation of motion; one-dimensional case. *Geophys. J. Int.*, **135**(1), 48–62.

Gokhberg, A. and Fichtner, A. (2015). Full-waveform inversion on heterogeneous (HPC) systems. *Computers & Geosciences*, dx.doi.org/10.1016/j.cageo.2015.12.013.

Gottschämmer, E. and Olsen, K. B. (2001). Accuracy of the explicit planar free-surface boundary condition implemented in a fourth-order staggered-grid velocity-stress finite-difference scheme. *Bull. Seism. Soc. Am.*, **91**, 617–23.

Graves, R. W. (1993). Modelling three-dimensional site response effects in the Marina District Basin, San Francisco, California. *Bull. Seism. Soc. Am.*, **83**, 1042–63.

Graves, R. W. (1995). Preliminary analysis of long-period basin response in the Los Angeles region from the 1994 Northridge earthquake. *Geophys. Res. Lett.*, **22**(2), 101–4.

Graves, R. W. (1996). Simulating seismic wave propagation in 3D elastic media using staggered finite differences. *Bull. Seism. Soc. Am.*, **86**, 1091–106.

Gualtieri, L., Stutzmann, E., Capdeville, Y., Ardhuin, F., Schimmel, M., Mangeney, A., and Morelli, A. (2013). Modelling secondary microseismic noise by normal mode summation. *Geophysical Journal International*, **193**(3), 1732–45.

Guillier, B., Cornou, C., Kristek, J., Moczo, P., Bonnefoy-Claudet, S., Bard, P. Y., and Fäh, D. (2006). Simulation of seismic ambient vibrations: does the h/v provide quantitative information in 2D–3D structures. In *Third international symposium on the effects of surface geology on seismic motion*, Grenoble, France, Volume 30.

Gupta, H. (ed.) (2011). *Encyclopedia of Solid Earth Geophysics*. New York: Springer Netherlands.

Harder, H. and Hansen, U. (2005). A finite-volume solution method for thermal convection and dynamo problems in spherical shells. *Geophys. J. Int.*, **161**(2), 522–32.

Harris, R. A., Barall, M., Archuleta, R. J., Andrews, D. J., Dunham, E. M., Aagaard, B. T., Ampuero, J. P., Cruz-Atienza, V., Dalguer, L., Day, S. M., Duan, B., Ely, G. P., Gabriel, A., Kaneko, Y., Kase, Y., Lapusta, N., Ma, S., Noda, H., Oglesby, D. D., Olsen, K. B., Roten, D., Song, S., Moran, S., Beeler, N., Wong, I., Weldon, R., McConnell, V., and Trehu, A. (2010). The SCEC–USGS rupture dynamics code comparison exercise. *Seis. Res. Lett.*, **81**(2), 311–12.

Hermann, V., Käser, M., and Castro, C. E. (2011). Non-conforming hybrid meshes for efficient 2D wave propagation using the discontinuous Galerkin method. *Geophys. J. Int.*, **184**(2), 746–58.

Hermeline, F. (1993). Two coupled particle-finite volume methods using Delaunay–Voronoi meshes for the approximation of Vlasov–Poisson and Vlasov–Maxwell equations. *J. Comput. Phys.*, **106**(1), 1–18.

Hesthaven, J. S. and Warburton, T. (2008). *Nodal Discontinuous Galerkin Methods: Algorithms, Analysis, and Applications.* Berlin/Heidelberg: Springer.

Igel, H. (1999). Wave propagation in spherical sections by the Chebyshev spectral method. *Geophys. J. Int.*, **136**(3), 559–66.

Igel, H., Djikpesse, H., and Tarantola, A. (1996). Waveform inversion of marine reflection seismograms for P impedance and Poisson's ratio. *Geophys. J. Int.*, **124**, 363–71.

Igel, H. and Gudmundsson, O. (1997). Frequency-dependent effects on travel times and waveforms of long-period S waves: implications for the scale of mantle heterogeneity. *Phys. Earth Plan. Int.*, **104**(1-3), 229–46.

Igel, H., Käser, M., and Stupazzini, M. (2015). Simulation of seismic wave propagation in media with complex geometries (revised 2nd edition), doi:10.1007/978–3–642–27737–5_468–2. *Encyclopedia of Complexity and System Science* (2nd edition). New York: Springer Verlag.

Igel, H., Mora, P., and Riollet, B. (1995). Anisotropic wave propagation through FD grids. *Geophysics*, **60**, 1203–16.

Igel, H., Nissen-Meyer, T., and Jahnke, G. (2002). Wave propagation in 3D spherical sections. Effects of subduction zones. *Phys. Earth Planet. Inter.*, **132**(1-3), 219–34.

Igel, H., Schreiber, U., Flaws, A., Schuberth, B., Velikoseltsev, A., and Cochard, A. (2005). Rotational motions induced by the M8.1 Tokachi-oki earthquake, September 25, 2003. *Geophys. Res. Lett.*, **32**(8). L08309.

Igel, H. and Weber, M. (1995). SH-wave propagation in the whole mantle using high-order finite differences. *Geophys. Res. Lett.*, **22**(6), 731–4.

Igel, H. and Weber, M. (1996). P-SV wave propagation in the Earth's mantle using finite-differences: application to heterogeneous lowermost mantle structure. *Geophys. Res. Lett.*, **23**, 731–4.

Ikelle, L. T., Yung, S. K., and Daube, F. (1993). 2D random media with ellipsoidal autocorrelation functions. *Geophysics*, **58**(9), 1359–72.

Imperatori, W. and Mai, P. M. (2013). Broadband near-field ground motion simulations in 3D scattering media. *Geophys. J. Int.*, **192**(2), 725–44.

Ismail-Zadeh, A. and Tackley, P. (2010). *Computational Methods for Geodynamics* (1st edn). New York: Cambridge University Press.

Jahnke, G., Igel, H., and Ben-Zion, Y. (2002). Three-dimensional calculations of fault zone guided waves in various irregular structures. *Geophys. J. Int.*, **151**(2), 416–26.

Jahnke, G., Thorne, M. S., Cochard, A., and Igel, H. (2008). Global SH-wave propagation using a parallel ax-isymmetric spherical finite-difference scheme: Application to whole mantle scattering. *Geophys. J. Int.*, **173**(3), 815–26.

Jastram, C. and Tessmer, E. (1994). Elastic modelling on a grid with vertically varying spacing. *Geophys. Prosp.*, **42**(4), 357–70.

Johnson, Lane R. (1974). Green's function for Lamb's problem. *Geophys. J. Int.*, **37**(1), 99–131.

Kaneko, Y., Ampuero, J.-P., and Lapusta, N. (2011). Spectral-element simulations of long-term fault slip: Effect of low-rigidity layers on earthquake-cycle dynamics. *Journal of Geophysical Research: Solid Earth*, **116**(B10), B10313.

Kaneko, Y., Lapusta, N., and Ampuero, J.-P. (2008). Spectral element modeling of spontaneous earthquake rupture on rate and state faults: Effect of velocity-strengthening friction at shallow depths. *Journal of Geophysical Research: Solid Earth*, **113**(B9), doi:10.1029/2007JB005553. B09317.

Käser, M. and Dumbser, M. (2006). An arbitrary high order discontinuous Galerkin method for elastic waves on unstructured meshes, I: the two-dimensional isotropic case with external source terms. *Geophys. J. Int.*, **166**(2), 855–77.

Käser, M., Dumbser, M., de la Puente, J., and Igel, H. (2007a). An arbitrary high order discontinuous Galerkin method for elastic waves on unstructured meshes III: viscoelastic attenuation. *Geophys. J. Int.*, **168**(1), 224–42.

Käser, M. and Gallovic, F. (2008). Effects of complicated 3D rupture geometries on earthquake ground motion and their implications: a numerical study. *Geophys. J. Int.*, **172**(1), 276–92.

Käser, M., Hermann, V., and de la Puente, J. (2008). Quantitative accuracy analysis of the discontinuous Galerkin method for seismic wave propagation. *Geophys. J. Int.*, **173**(3), 990–9.

Käser, M. and Igel, H. (2001). Numerical simulation of 2D wave propagation on unstructured grids using explicit differential operators. *Geophys. Prosp.*, **49**(5), 607–19.

Käser, M., Igel, H., Sambridge, M., and Braun, J. (2001). A comparative study of explicit differential operators on arbitrary grids. *Journal of Computational Acoustics*, **9**(3), 1111–25.

Käser, M., Mai, P. M., and Dumbser, M. (2007b). Accurate calculation of fault-rupture models using the high-order discontinuous Galerkin method on tetrahedral meshes. *Bull. Seism. Soc. Am.*, **97**(5), 1570–86.

Käser, M., Pelties, C., Castro, C. E., Djikpesse, H., and Prange, M. (2010). Wavefield modeling in exploration seismology using the discontinuous Galerkin finite-element method on HPC infrastructure. *The Leading Edge*, **29**(1), 76–85.

Käufl, P., Fichtner, A., and Igel, H. (2013). Probabilistic full waveform inversion based on tectonic regionalization: development and application to the Australian upper mantle. *Geophys. J. Int.*, **193**(1), 437–51.

Kelly, K. R. and Marfurt, J. (ed.) (1990). *Numerical Modeling of Seismic Wave Propagation*. Volume 13, Geophysical Reprint Series. Tulsa, OK: Society of Exploration Geophysicists.

Kelly, K. R., Ward, R. W., Treitel, S., and Alford, R. M. (1976). Synthetic seismograms: a finite difference approach. *Geophysics*, **41**, 2–27.

Kennett, B. L. N. (1983). *Seismic Wave Propagation in Stratified Media*. Cambridge: Cambridge University Press.

Kennett, B. L. N. (2001). *The seismic wavefield, I: Introduction and theoretical development*. Cambridge: Cambridge University Press.

Kim, K., Lees, J. M., Fee, D., Yokoo, A., and Johnson, J. B. (2014, 11). Local volcano infrasound and source localization investigated by 3D simulation. *Seis. Res. Lett.*, **85**(6), 1177–86.

Komatitsch, D. (1997). *Méthodes spectrales et éléments spectraux pour l'équation de l'élastodynamique 2D et 3D en milieu hétérogène*. Doctoral thesis, Université Paris 7.

Komatitsch, D., Barnes, C., and Tromp, J. (2000a). Simulation of anisotropic wave propagation based upon a spectral element method. *Geophysics*, **65**(4), 1251–60.

Komatitsch, D., Barnes, C., and Tromp, J. (2000b). Wave propagation near a fluid–solid interface: A spectral-element approach. *Geophysics*, **65**(2), 623–31.

Komatitsch, D., Coutel, F., and Mora, P. (1996). Tensorial formulation of the wave equation for modelling curved interfaces. *Geophys. J. Int.*, **127**(1), 156–68.

Komatitsch, D. and Tromp, J. (1999). Introduction to the spectral element method for three-dimensional seismic wave propagation. *Geophys. J. Int.*, **139**, 806–22.

Komatitsch, D. and Tromp, J. (2002a). Spectral-element simulations of global seismic wave propagation, part I: validation. *Geophys. J. Int.*, **149**, 390–412.

Komatitsch, D. and Tromp, J. (2002b). Spectral-element simulations of global seismic wave propagation, part II: 3D models, oceans, rotation, and gravity. *Geophys. J. Int.*, **150**, 303–18.

Komatitsch, D. and Vilotte, J. P. (1998). The spectral element method: an effective tool to simulate the seismic response of 2D and 3D geological structures. *Bull. Seism. Soc. Am.*, **88**, 368–92.

Komatitsch, D., Vilotte, J. P., Vai, R., Castillo-Covarrubias, J. M., and Sánchez-Sesma, F. J. (1999). The spectral element method for elastic wave equations: application to 2D and 3D seismic problems. *Int. J. Num. Meth. Engng.*, **45**, 1139–64.

Kosloff, D., Kessler, D., Filho, A. Q., Tessmer, E., Behle, A., and Strahilevitz, R. (1990). Solution of the equations of dynamics elasticity by a Chebyshev spectral method. *Geophysics*, **55**, 748–54.

Kossloff, D. and Bayssal, E. (1982). Forward modelling by Fourier method. *Geophysics*, **47**, 1402–12.

Kossloff, D., Reshef, M., and Loewenthal, D. (1984). Elastic wave calculations by the Fourier method. *Bull. Seism. Soc. Am.*, **74**, 875–91.

Krebs, J., Anderson, J., Hinkley, D., Neelamani, R., Baumstein, A., Lacasse, M. D., and Lee, S. (2009). Fast full-wavefield seismic inversion using encoded sources. *Geophysics*, **74**, WCC177 doi:10.1190/1.3230502.

Kremers, S., Fichtner, A., Brietzke, G. B., Igel, H., Larmat, C., Huang, L., and Käser, M. (2011). Exploring the potential and limitations of the time-reversal imaging of finite seismic sources. *Solid Earth*, **2**, 95–105.

Kremers, S., Wassermann, J., Meier, K., Pelties, C., van Driel, M., Vasseur, J., and Hort, M. (2013). Inverting the source mechanism of Strombolian explosions at Mt. Yasur, Vanuatu, using a multi-parameter dataset. *Journal of Volcanology and Geothermal Research*, **262**, 104–22.

Krischer, L., Fichtner, A., Zukauskaite, S., and Igel, H. (2015*a*). Large-scale seismic inversion framework. *Seis. Res. Lett.*, doi:10.1785/0220140248.

Krischer, L., Megies, T., Barsch, R., Beyreuther, M., Lecocq, T., Caudron, C., and Wassermann, J. (2015*b*). ObsPy: A bridge for seismology into the scientific Python ecosystem. *Computational Science & Discovery*, **8**(1), 014003.

Krischer, L., Smith, J., Lefevre, M., Lei, W., Ruan, Y., Sales de Andrade, E., Podhorszki, N., Bozdag, E., and Tromp, J. (2016). An adaptable seismic data format. *J. Geophys. Res.*. In Press.

Kristek, J. and Moczo, P. (2003). Seismic-wave propagation in viscoelastic media with material discontinuities: A 3D fourth-order staggered-grid finite-difference modeling. *Bull. Seism. Soc. Am.*, **93**(5), 2273–80.

Kristek, J., Moczo, P., and Archuleta, R. J. (2002). Efficient methods to simulate planar free surface in the 3D 4th-order staggered-grid finite-difference schemes. *Stud. Geophys. Geod.*, **46**, 355–81.

Kristekova, M., Kristek, J., Moczo, P., and Day, S. M. (2006). Misfit criteria for quantitative comparison of seismograms. *Bull. Seism. Soc. Am.*, **96**, 1836–50.

Larmat, C., Montagner, J. P., Fink, M., Capdeville, Y., Tourin, A., and Clévédé, E. (2006). Time-reversal imaging of seismic sources and application to the great Sumatra earthquake. *Geophys. Res. Lett.*, **33**(19), doi:10.1029/2006GL026336. L19312.

Lax, P. D. and Wendroff, B. (1960). Systems of conservation laws. *Communications in Pure and Applied Mathematics*, **13**, 217–37.

Lee, S.-J., Chan, Y.-C., Komatitsch, D., Huang, B.-S., and Tromp, J. (2009). Effects of realistic surface topography on seismic ground motion in the Yangmingshan region of Taiwan based upon the spectral-element method and LiDAR DTM. *Bull. Seism. Soc. Am.*, **99**(2), 681–93.

Lee, W. H. K., Kanamori, H., Jennings, P. C., and Kisslinger, C. (eds) (2002). *International Handbook of Earthquake and Engineering Seismology*. Volume 81, Part A, International Geophysics. New York: Academic Press.

Lentas, K., Ferreira, A. M. G., and Valle, M. (2013). Assessment of SCARDEC source parameters of global large (Mw ≥ 7.5) subduction earthquakes. *Geophys. J. Int.*, **195**(3), 1989–2004.

Levander, A. R. (1988). Fourth-order finite difference P-SV seismograms. *Geophysics*, **53**, 1425–36.

Leveque, R. J. (2002). *Finite Volume Methods for Hyperbolic Problems*. Cambridge: Cambridge University Press.

Li, X., Bielak, J., and Ghattas, O. (1994). Seismic response in three-dimensional basin on a CM-2. *Proceedings of the International Conference on Computer Methods and Advances in Geomechanics*, **8**, **Vol. 2**, 929–33.

Lokmer, I., Bean, C. J., Saccorotti, G., and Patanè, D. (2007). Moment-tensor inversion of LP events recorded on Etna in 2004 using constraints obtained from wave simulation tests. *Geophys. Res. Lett.*, **34**(22), doi:10.1029/2007GL031902.

Lozos, J. C., Olsen, K. B., Brune, J. N., Takedatsu, R., Brune, R. J., and Oglesby, D. D. (2015). Broadband ground motions from dynamic models of rupture on the northern San Jacinto fault, and comparison with precariously balanced rocks. *Bull. Seism. Soc. Am.*, **105**(4), 1947–60.

Luo, Y. and Schuster, G. T. (1991). Wave-equation traveltime inversion. *Geophysics*, **56**, 645–53.

Lysmer, J. and Drake, L. (1972). Evolving geometrical and material properties of fault zones in a damage rheology model. In *Methods in Computational Physics II: Seismology* (ed. B. Alder, S. Fernbach, and B. A. Bolt), pp. 181–216. Amsterdam: Academic Press.

Madariaga, R. (1976). Dynamics of an expanding circular fault. *Bull. Seism. Soc. Am.*, **65**, 163–82.

Madariaga, R., Olsen, K., and Archuleta, R. (1998). Modeling dynamic rupture in a 3D earthquake fault model. *Bull. Seism. Soc. Am.*, **88**(5), 1182–97.

Maday, Y. and Patera, A. T. (1989). Spectral element methods for the incompressible Navier-Stokes equations. In *State of the Art Survey in Computational Mechanics* (ed. A. Noor and J. Oden), pp. 71–143. New York: ASME.

Maggi, A., Tape, C., Chen, M., Chao, D., and Tromp, J. (2009). An automated time-window selection algorithm for seismic tomography. *Geophys. J. Int.*, **178**, 257–81.

Magnier, S. A., Mora, P., and Tarantola, A. (1994). Finite differences on minimal grids. *Geophysics*, **59**, 1435–43.

Magnoni, F., Casarotti, E., Michelini, A., Piersanti, A., Komatitsch, D., Peter, D., and Tromp, J. (2013). Spectral-element simulations of seismic waves generated by the 2009 Laquila earthquake. *Bull. Seism. Soc. Am.*, doi:10.1785/0120130106.

Marfurt, K. J. (1984). Accuracy of finite-difference and finite-element modeling of the scalar wave equation. *Geophysics*, **49**, 533–49.

Martini, F., Bean, C. J., Dolan, S., and Marsan, D. (2001). Seismic image quality beneath strongly scattering structures and implications for lower crustal imaging: numerical simulations. *Geophys. J. Int.*, **145**(2), 423–35.

Maufroy, E., Chaljub, E., Hollender, F., Kristek, J., Moczo, P., Klin, P., Priolo, E., Iwaki, A., Iwata, T., Etienne, V., De Martin, F., Theodoulidis, N. P., Manakou, M., Guyonnet-Benaize, C., Pitilakis, K., and Bard, P.-Y. (2015). Earthquake ground motion in the Mygdonian basin, Greece: The E2VP verification and validation of 3D numerical simulation up to 4 Hz. *Bull. Seism. Soc. Am.*, doi:10.1785/0120140228.

May, D. A., Gabriel, A.-A., and Brown, J. (2016). A spectral element discretization on unstructured simplex meshes for elastodynamics. *Geophys. J. Int.*, submitted.

Mazzieri, I., Stupazzini, M., Guidotti, R., and Smerzini, C. (2013). SPEED: SPectral Elements in Elastodynamics with Discontinuous Galerkin: a non-conforming approach for 3D multi-scale problems. *International Journal for Numerical Methods in Engineering*, **95**(12), 991–1010.

Mercerat, E. D., Vilotte, J.-P., and Sanchez-Sesma, F. J. (2006). Triangular spectral element simulation of two-dimensional elastic wave propagation using unstructured triangular grids. *Geophys. J. Int.*, **166**(2), 679–98.

Meschede, M. and Romanowicz, B. (2015). Non-stationary spherical random media and their effect on long-period mantle waves. *Geophys. J. Int.*, **203**(3), 1605–25.

Meschede, M. A., Myhrvold, C. L., and Tromp, J. (2011). Antipodal focusing of seismic waves due to large meteorite impacts on Earth. *Geophys. J. Int.*, **187**(1), 529–37.

Métaxian, J. P., O'Brien, G. S., Bean, C. J., Valette, B., and Mora, M. (2009). Locating volcano-seismic signals in the presence of rough topography: wave simulations on Arenal volcano, Costa Rica. *Geophys. J. Int.*, **179**(3), 1547–57.

Moczo, P., Kristek, J., Franek, P., Chaljub, E., Bard, P.-Y., Tsuno, S., Iwata, T., Iwaki, A., Priolo, E., Klin, P., Aoi, S., Mariotti, C., Bielak, J., Taborda, R., Karaoglu, H., Etienne, V., Virieux, J., Moran, S., Beeler, N., Wong, I., Weldon, R., McConnell, V., and Trehu, A. (2010*a*). Numerical modeling of earthquake ground motion in the Mygdonian basin, Greece; verification of the 3D numerical methods. *Seis. Res. Lett.*, **81**(2), 310. Abstract.

Moczo, P., Kristek, J., and Galis, M. (2004). Simulation of the planar free surface with near-surface lateral discontinuities in the finite-difference modeling of seismic motion. *Bull. Seism. Soc. Am.*, **94**, 760–8.

Moczo, P., Kristek, J., and Galis, M. (2014). *The Finite Difference Modelling of Earthquake Motions*. Cambridge: Cambridge University Press.

Moczo, P., Kristek, J., Galis, M., Chaljub, E., and Etienne, V. (2011). 3D finite-difference, finite-element, discontinuous-Galerkin and spectral-element schemes analysed for their accuracy with respect to P-wave to S-wave speed ratio. *Geophys. J. Int.*, **187**(3), 1645–67.

Moczo, P., Kristek, J., Galis, M., and Pazak, P. (2010*b*). On accuracy of the finite-difference and finite-element schemes with respect to P-wave to S-wave speed ratio. *Geophys. J. Int.*, **182**, 493–510.

Moczo, P., Kristek, J., Vavrycuk, V., Archuleta, R., and Halada, L. (2002). 3D heterogeneous staggered-grid finite-difference modeling of seismic motion with volume harmonic and arithmetic averaging of elastic moduli. *Bull. Seism. Soc. Am.*, **92**, 3042–66.

Mora, P. (1986). Elastic finite differences with convolutional operators. *Stanford Expl. Proj. Rep.*, **48**, 277–89.

Morency, C., Luo, Y., and Tromp, J. (2011). Acoustic, elastic and poroelastic simulations of CO_2 sequestration crosswell monitoring based on spectral-element and adjoint methods. *Geophys. J. Int.*, **185**(2), 955–66.

Muir, F., Dellinger, J., Etgen, J., and Nichols, D. (1992). Modeling elastic fields across irregular boundaries. *Geophysics*, **57**(9), 1189–93.

Nielsen, S. B. and Tarantola, A. (1992). Numerical model of seismic rupture. *Journal of Geophysical Research: Solid Earth*, **97**(B11), 15291–5.

Nissen-Meyer, T., Fournier, A., and Dahlen, F. A. (2007). A two-dimensional spectral-element method for computing spherical-Earth seismograms, I: moment-tensor source. *Geophys. J. Int.*, **168**, 1067–92.

Nissen-Meyer, T., van Driel, M., Stähler, S. C., Hosseini, K., Hempel, S., Auer, L., Colombi, A., and Fournier, A. (2014). Axisem: broadband 3D seismic wavefields in axisymmetric media. *Solid Earth*, **5**(1), 425–45.

Nolet, G. (2008). *A Breviary of Seismic Tomography: Imaging the Interior of the Earth and Sun*. Cambridge: Cambridge University Press.

Obermann, A., Plaès, T., Hadziioannou, C., and Campillo, M. (2016). Volcano topography, structure and intrinsic attenuation; their relative influences on a simulated 3D visco-elastic wavefield. *Geophys. J. Int.*, doi: 10.1093/gji/ggw264.

O'Brien, G. S. and Bean, C. J. (2004). A 3D discrete numerical elastic lattice method for seismic wave propagation in heterogeneous media with topography. *Geophys. Res. Lett.*, **31**(14), doi:10.1029/2004GL020069.

O'Brien, G. S. and Bean, C. J. (2009). Volcano topography, structure and intrinsic attenuation: their relative influences on a simulated 3D visco-elastic wavefield. *Journal of Volcanology and Geothermal Research*, **183**(1-2), 122–36.

Ohminato, T. and Chouet, B. A. (1997). A free-surface boundary condition for including 3D topography in the finite-difference method. *Bull. Seism. Soc. Am.*, **87**, 494–515.

Ollivier-Gooch, C. and Van Altena, M. (2002). A high-order-accurate unstructured mesh finite-volume scheme for the advection–diffusion equation. *J. Comput. Phys.*, **181**(2), 729–52.

Olsen, K. B. and Archuleta, R. J. (1996). Three-dimensional simulation of earthquakes on the Los Angeles fault system. *Bull. Seism. Soc. Am.*, **86**, 575–96.

Olsen, K. B., Archuleta, R. J., and Matarese, J. R. (1995). Three-dimensional simulation of a magnitude 7.75 earthquake on the San Andreas fault. *Science*, **270**, 1628–32.

Olsen, K. B., Day, S. M., Minster, J. B., Cui, Y., Chourasia, A., Okaya, D., Maechling, P., and Jordan, T. (2008). TeraShake2: Spontaneous rupture simulations of Mw 7.7 earthquakes on the southern San Andreas fault. *Bull. Seism. Soc. Am.*, **98**(3), 1162–85.

Olsen, K. B., Madariaga, R., and Archuleta, R. J. (1997). Three-dimensional dynamic simulation of the 1992 Landers earthquake. *Science*, **278**, 834–8.

Pacheco, P. (2011). *An Introduction to Parallel Programming* (1st edn). San Francisco, CA: Morgan Kaufmann Publishers Inc.

Paolucci, R., Mazzieri, I., Smerzini, C., and Stupazzini, M. (2014). Physics-based earthquake ground shaking scenarios in large urban areas. In *Perspectives on European Earthquake Engineering and Seismology* (ed. A. Ansal), Volume 34, Geotechnical, Geological and Earthquake Engineering, pp. 331–59. Berlin/Heidelberg: Springer International Publishing.

Patera, A. T. (1984). A spectral element method for fluid dynamics: laminar flow in a channel expansion. *J. Comput. Phys.*, **54**, 468–88.

Pelties, C., Gabriel, A.-A., and Ampuero, J.-P. (2014). Verification of an ADER-DG method for complex dynamic rupture problems. *Geoscientific Model Development*, **7**(3), 847–66.

Pelties, C., Huang, Y., and Ampuero, J.-P. (2015). Pulse-like rupture induced by three-dimensional fault zone flower structures. *Pure and Applied Geophysics*, **172**(5), 1229–41.

Pelties, C., de la Puente, J., Ampuero, J.-P., Brietzke, G. B., and Käser, M. (2012). Three-dimensional dynamic rupture simulation with a high-order discontinuous Galerkin method on unstructured tetrahedral meshes. *J. Geophys. Res.* (117), doi:10.1029/2011JB008857.

Peter, D., Komatitsch, D., Luo, Y., Martin, R., Le Goff, N., Casarotti, E., Le Loher, P., Magnoni, F., Liu, Q., Blitz, C., Nissen-Meyer, T., Basini, P., and Tromp, J. (2011). Forward and adjoint simulations of seismic wave propagation on fully unstructured hexahedral meshes. *Geophys. J. Int.*, **186**, 721–39.

Pham, D. N., Igel, H., Wassermann, J., Käser, M., de la Puente, J., and Schreiber, U. (2009). Observations and modelling of rotational signals in the P-coda: constraints on crustal scattering. *Bull. Seism. Soc. Am.*, **99**(2B), 1315–32.

Pitarka, A. (1999). 3D elastic finite-difference modeling of seismic motion using staggered grids with nonuniform spacing. *Bull. Seism. Soc. Am.*, **89**(1), 54–68.

Pitarka, A. and Irikura, K. (1996). Modeling 3D surface topography by finite-difference method: Kobe-JMA station site, Japan, case study. *Geophys. Res. Lett.*, **23**, 2729–32.

Pozrikidis, C. (2005). *Finite and spectral element methods using Matlab*. London: Chapman & Hall/CRC.

Pratt, R. G., Shin, C., and Hicks, G. J. (1998). Gauss–Newton and full Newton methods in frequency domain seismic waveform inversion. *Geophys. J. Int.*, **133**, 341–62.

Priolo, E., Carcione, J. M., and Seriani, G. (1994). Numerical simulation of interface waves by high-order spectral modeling techniques. *J. Acoust. Soc. Am.*, **95**:2, 681–93.

Reed, W. H. and Hill, T. R. (1973). Triangular mesh methods for the neutron transport equation. *Technical Report, Los Alamos Scientific Laboratory*, LA-UR-73-479.

Regone, C. J. (2007). Using 3D finite-difference modeling to design wide-azimuth surveys for improved subsalt imaging. *Geophysics*, **72**(5), SM231–SM239.

Reshef, M., Kosloff, D., Edwards, M., and Hsiung, C. (1988). Three-dimensional elastic modeling by the Fourier method. *Geophysics*, **53**, 1184–93.

Rienstra, S. W. and Hirschberg, A. (2016). *An Introduction to Acoustics*. TR-IWDE-92-06, Technische Universiteit Eindhoven, 1992–2004–2015, <http://www.win.tue.nl/ sjoerdr/papers/boek.pdf> (18 January 2016).

Ripperger, J., Igel, H., and Wassermann, J. (2003). Seismic wave simulation in the presence of real volcano topography. *J. Volcanol. Geotherm. Res.*, **128**(1-3), 31–44.

Robertsson, J. O. A., Blanch, J. O., Nihei, K., and Tromp, J. (eds) (2012). *Numerical Modeling of Seismic Wave Propagation: Gridded Two-Way Wave-Equation Methods*. Volume 28, Geophysical Reprint Series. Tulsa, OK: SEG.

Robertsson, J. O. A., Blanch, J. O., and Symes, W. W. (1994). Viscoelastic finite-difference modelling. *Geophysics*, **59**, 1444–56.

Robertsson, J. O. A. and Holliger, K. (1997). A numerical free-surface condition for elastic/viscoelastic finite-difference modeling in the presence of topography. *Phys. Earth Planet. Inter.*, **1041**, 193–211.

Ronchi, C., Ianoco, R., and Paolucci, P. S. (1996). The 'cubed sphere': a new method for the solution of partial differential equations in spherical geometry. *J. Comput. Phys.*, **124**, 93–114.

Rusmanugroho, H. and McMechan, G. A. (2012). 3D, 9C seismic modeling and inversion of Weyburn Field data. *Geophysics*, 77(4), R161–R173.

Sambridge, M. S. (1999*a*). Geophysical inversion with the neighbourhood algorithm, I: Searching a parameter space. *Geophys. J. Int.*, **138**, 479–94.

Sambridge, M. S. (1999*b*). Geophysical inversion with the neighbourhood algorithm, II: Appraising the ensemble. *Geophys. J. Int.*, **138**, 727–46.

Sato, H., Fehler, M. C., and Maeda, T. (2012). *Seismic Wave Propagation and Scattering in the Heterogeneous Earth: Second Edition*. Berlin/Heidelberg: Springer Science.

Schiemenz, A. and Igel, H. (2013). Accelerated 3D full-waveform inversion using simultaneously encoded sources in the time domain: application to Valhall ocean-bottom cable data. *Geophys. J. Int.*, doi:10.1093/gji/ggt362.

Schlue, J. W. (1979). Finite element matrices for seismic surface waves in three-dimensional structures. *Bull. Seism. Soc. Am.*, **69**, 1425–38.

Schubert, G. (ed.) (2015). *Treatise on Geophysics, 2nd Edition*. Amsterdam: Elsevier.

Schuberth, B. S. A., Bunge, H.-P., Steinle-Neumann, G., Moder, C., and Oeser, J. (2009). Thermal versus elastic heterogeneity in high-resolution mantle circulation models with pyrolite composition: High plume excess temperatures in the lowermost mantle. *Geochem. Geophys. Geosyst.*, 10(1), doi:10.1029/2008GC002235.

Schuberth, B. S. A., Zaroli, C., and Nolet, G. (2012). Synthetic seismograms for a synthetic Earth: long-period P- and S-wave traveltime variations can be explained by temperature alone. *Geophys. J. Int.*, 188(3), 1393–412.

Schuberth, B. S. A., Zaroli, C., and Nolet, G. (2015). Traveltime dispersion in an isotropic elastic mantle: strong lower-mantle signal in differential-frequency residuals. *Geophys. J. Int.*, 203(3), 2099–118.

Seriani, G. and Priolo, E. (1994). Spectral element method for acoustic wave simulation in heterogeneous media. *Finite Elements in Analysis and Design*, **16**, 337–48.

Seriani, G. and Su, C. (2012). Wave propagation in highly heterogeneous media by a poly-grid Chebyshev spectral element method. *J. Comp. Ac.*, **20**, doi:10.1142/S0218396X12400048.

Seron, F. J., Sanz, F. J., M., Kindelan, M., and Badal, J. I. (1990). Finite element method for elastic wave propagation. *Comm. Appl. Numerical Methods*, **6**, 359–68.

Shapiro, N. M., Campillo, M., Stehly, L., and Ritzwoller, M. (2005). High resolution surface wave tomography from ambient seismic noise. *Science*, **307**, 1615–18.

Shearer, P. (2009). *Introduction to Seismology*. 2nd edn, Cambridge: Cambridge University Press.

Shen, H. (2014). Interactive notebooks: sharing the code. *Nature*, **515**, 151–2.

Sieminski, A., Trampert, J., and Tromp, J. (2009). Principal component analysis of anisotropic finite-frequency kernels. *Geophys. J. Int.*, **179**, 1186–98.

Sigloch, K., McQuarrie, N., and Nolet, G. (2008). Two-stage subduction history under North America inferred from multiple-frequency tomography. *Nat. Geosc.*, **1**, doi:10.1038/ngeo231.

Sirgue, L., Barkved, O. I., Dellinger, J., Etgen, J., Albertin, U., and Kommedal, J. H. (2010). Full waveform inversion: the next leap forward in imaging at Valhall. *First Break*, 28(4), 65–70.

Sirgue, L., Etgen, J. T., and Albertin, U. (2012). 3D frequency domain waveform inversion using time domain finite difference methods. *Geophysics Reprint Series*, **28**, 75.

Smerzini, C. and Villani, M. (2012). Broadband numerical simulations in complex near-field geological configurations: the case of the 2009 Mw 6.3 l'Aquila earthquake. *Bull. Seism. Soc. Am.*, 102(6), 2436–51.

Snieder, R. (2015). *A Guided Tour of Mathematical Methods for the Physical Sciences*. 3rd edn, Cambridge: Cambridge University Press.

Stehly, L., Cupillard, P., and Romanowicz, B. (2011). Towards improving ambient noise tomography using simultaneously curvelet denoising filters and SEM simulations of seismic ambient noise. *Comptes Rendus Geoscience*, **343**(8–9), 591–9.

Stein, E. M. and Shakarchi, R. (2003). *Fourier Analysis: An Introduction*. Princeton, NJ: Princeton University Press.

Stein, S. and Wysession, M. (2003). *An Introduction to Seismology, Earthquakes, and Earth Structure*. Hoboken, NJ: Blackwell Publishing Ltd.

Strang, G. and Fix, G. J. (1988). *An Analysis of the Finite Element Method*. Cambridge: Wellesley Cambridge Press.

Tadi, M. (2004). Finite volume method for 2D elastic wave propagation. *Bull. Seism. Soc. Am.*, **94**(4), 1500–9.

Tago, J., Cruz-Atienza, V. M., Virieux, J., Etienne, V., and Sanchez-Sesma, F. J. (2012). A 3D hp-adaptive discontinuous Galerkin method for modeling earthquake dynamics. *J. Geophys. Res.*, **117**, doi:10.1029/2012JB009313.

Takemura, S. and Furumura, T. (2013). Scattering of high-frequency P wavefield derived by dense Hi-net array observations in Japan and computer simulations of seismic wave propagations. *Geophys. J. Int.*, doi:10.1093/gji/ggs127.

Tal-Ezer, H., Kosloff, D., and Koren, Z. (1987). An accurate scheme for seismic forward modeling. *Geophys. Prosp.*, **35**, 479–90.

Tape, C., Liu, Q., Maggi, A., and Tromp, J. (2009). Adjoint tomography of the southern California crust. *Science*, **325**, 988–92.

Tape, C., Liu, Q., Maggi, A., and Tromp, J. (2010). Seismic tomography of the southern California crust based upon spectral-element and adjoint methods. *Geophys. J. Int.*, **180**, 433–62.

Tarantola, A. (1986). A strategy for nonlinear elastic inversion of seismic reflection data. *Geophysics*, **51**, 1893–903.

Tarantola, A. (1988). Theoretical background for the inversion of seismic waveforms, including elasticity and attenuation. *Pure Appl. Geophys.*, **128**, 365–99.

Tarantola, A. (2005). *Inverse Problem Theory and Methods for Model Parameter Estimation*, 2nd edition. Philadelphia, PA: Society for Industrial and Applied Mathematics.

Tarantola, A. and Valette, B. (1982). Generalized nonlinear inverse problems solved using the least squares criterion. *Rev. Geophys.*, **20**, 219–32.

Tessmer, E. (1995). 3D seismic modeling of general material anisotopy in the presence of the free surface by a Chebychev spectral method. *Geophys. J. Int.*, **121**, 557–75.

Tessmer, E. (2000). Seismic finite-difference modelling with spatially varying time steps. *Geophysics*, **65**, 1290–93.

Tessmer, E., Kosloff, D., and Behle, A. (1992). Elastic wave propagation simulation in presence of surface topography. *Geophys. J. Int.*, **108**, 621–32.

Thomsen, L. (1986). Weak elastic anisotropy. *Geophysics*, **51**, 1954–66.

Thorne, M. S., Crotwell, H. P., and Jahnke, G. (2013a). An educational resource for visualizing the global seismic wave field. *Seis. Res. Lett.*, **84**, 711–17.

Thorne, M. S., Garnero, E. J., Jahnke, G., Igel, H., and McNamara, A. K. (2013b). Mega ultra low velocity zone and mantle flow. *Earth Planet. Sci. Lett.*, **364**, 59–67.

Tian, Y., Zhou, Y., Sigloch, K., Nolet, G., and Laske, G. (2011). Structure of the North American mantle constrained by simultaneous inversion of multiple-frequency SH, SS, and Love waves. *J. Geophys. Res.*, **116**(B02307), 2156–202.

Titarev, V. A. and Toro, E .F. (2002). ADER: Arbitrary high-order Godunov approach. *J. Sci. Comput.*, **17**, 609–18.

Toomey, A. and Bean, C. J. (2000). Numerical simulation of seismic waves using a discrete particle scheme. *Geophys. J. Int.*, **141**(3), 595–604.

Toyokuni, G. and Takenaka, H. (2006). FDM computation of seismic wavefield for an axisymmetric Earth with a moment tensor point source. *Earth, Planets and Space*, **58**(8), e29–e32.

Trampert, J., Fichtner, A., and Ritsema, J. (2013). Resolution tests revisited: the power of random numbers. *Geophys. J. Int.*, **192**(2), 676–80.

Trefethen, L. N. (2015). *Spectral Methods in MATLAB*. Philadelphia, PA: SIAM.

Tromp, J., Komatitsch, D., Hjörleifsdóttir, V., Liu, Q., Zhu, H., Peter, D., Bozdag, E., McRitchie, D., Friberg, P., Trabant, C., and Hutko, A. (2010). Near real-time simulations of global CMT earthquakes. *Geophys. J. Int.*, **183**(1), 381–9.

Tromp, J., Tape, C., and Liu, Q. (2005). Seismic tomography, adjoint methods, time reversal and banana–doughnut kernels. *Geophys. J. Int.*, **160**, 195–216.

Tsuboi, S., Komatitsch, D., Ji, C., and Tromp, J. (2003). Broadband modeling of the 2002 Denali fault earthquake on the Earth simulator. *Phys. Earth Planet. Inter.*, **139**(3-4), 305–13.

van Driel, M., Krischer, L., Stähler, S., Hosseini, K., and Nissen-Meyer, T. (2015a). *Efficient methods in global seismic wave propagation*. Dissertation, EidgenössischeTechnische Hochschule Zürich, No. 22614.

van Driel, M., Krischer, L., Stähler, S., Hosseini, K., and Nissen-Meyer, T. (2015b). Instaseis: instant global seismograms based on a broadband waveform database. *Solid Earth*, **6**, 701–17.

van Driel, M., Wassermann, J., Nader, M. F., Schuberth, B. S. A., and Igel, H. (2012). Strain rotation coupling and its implications on the measurement of rotational ground motions. *J. Seismol.*, **16**(4), 657–68.

van Driel, M., Wassermann, J., Pelties, C., Schiemenz, A., and Igel, H. (2015c). Tilt effects on moment tensor inversion in the nearfield of active volcanoes. *Geophys. J. Int.* (202), 1711–21.

Versteeg, H. K. and Malalasekera, W. (1995). *An Introduction to Computational Fluid Dynamics: The Finite Volume Method*. Reading, MA: Addison-Wesley.

Virieux, J. (1984). SH wave propagation in heterogeneous media: velocity–stress finite difference method. *Geophysics*, **49**, 1933–42.

Virieux, J. (1986). P–SV wave propagation in heterogeneous media: velocity–stress finite difference method. *Geophysics*, **51**, 889–901.

Virieux, J. and Madariaga, R. (1982). Dynamic faulting studied by a finite difference method. *Bull. Seism. Soc. Am.*, **72**, 345–69.

Virieux, J. and Operto, S. (2009). An overview of full waveform inversion in exploration geophysics. *Geophysics*, **74**, WCC127–WCC152.

Virieux, J., Operto, S., Ben-Hadj-Ali, H., Brossier, R., Etienne, V., Sourbier, F., Giraud, L., and Haidar, Z. (2009). Seismic wave modeling for seismic imaging. *Leading Edge*, **28**(5), 538–44.

Wang, H. (2007). *Source-dependent variations of M7 earthquakes in the Los Angeles Basin*. Dissertation LMU Mńchen: Fakultät für Geowissenschaften.

Wang, H., Igel, H., Gallovič, F., Cochard, A., and Ewald, M. (2008). Source-related variations of ground motions in 3D media: application to the newport–inglewood fault, Los Angeles Basin. *Geophys. J. Int.*, **175**(1), 202–14.

Wang, Y., Takenaka, H., and Furumura, T. (2001). Modelling seismic wave propagation in a two dimensional cylindrical whole Earth model using the pseudospectral method. *Geophys. J. Int.*, **145**(3), 689–708.

Wang, Z. J. (2002). Spectral (finite) volume method for conservation laws on unstructured grids: basic formulation. *J. Comput. Phys.*, **178**(1), 210–51.

Wenk, S., Pelties, C., Igel, H., and Käser, M. (2013). Regional wave propagation using the discontinuous Galerkin method. *Solid Earth* (4), 43–57.

Wilcox, L. C., Stadler, G., Burstedde, C., and Ghattas, O. (2010). A high-order discontinuous Galerkin method for wave propagation through coupled elastic–acoustic media. *J. Comput. Phys.*, **229**(24), 9373–96.

Woodhouse, J. H. and Dziewoński, A. M. (1984). Mapping the upper mantle: Three-dimensional modeling of Earth structure by inversion of seismic waveforms. *J. Geophys. Res.*, **89**, 5953–86.

Yin, J., Kato, N., Miyatake, T., Hirahara, K., Hori, T., and Hyodo, M. (2011). Assessment of the finite element solutions for 3D spontaneous rupture using GeoFEM. *Earth, Planets and Space*, **63**(11), 1119–31.

Zhang, R., Czado, C., and Sigloch, K. (2013). A Bayesian linear model for the high-dimensional inverse problem of seismic tomography. *Annals of Applied Statistics*, 7(2), 1111–38.

Zhu, H., Bozdağ, E., Peter, D., and Tromp, J. (2012). Structure of the European upper mantle revealed by adjoint tomography. *Nat. Geosc.*, 5, 493–8.

Zienkiewicz, O. C. and Taylor, R. L. (1989). *The Finite Element Method, 4th Edn.* Volume 1. New York: McGraw-Hill.

Zienkiewicz, O. C., Taylor, R. L., and Zhu, J. Z. (2013). *The Finite Element Method: Its Basis and Fundamentals, 7th Edn.* Volume 1. Cambridge: Elsevier.

Index